Frontiers in Statistical Quality Control 9

Hans-Joachim Lenz · Peter-Theodor Wilrich ·
Wolfgang Schmid
Editors

Frontiers in Statistical Quality Control 9

Physica-Verlag
A Springer Company

Editors
Prof. Dr. Hans-Joachim Lenz
Freie Universität Berlin
Institut für Statistik und Ökonometrie
Garystraße 21
Germany
ls-lenz@wiwiss.fu-berlin.de

Prof. Dr. Peter-Theodor Wilrich
Freie Universität Berlin
Institut für Statistik und Ökonometrie
Garystraße 21
Germany
wilrich@wiwiss.fu-berlin.de

Prof. Dr. Wolfgang Schmid
Europa-Universität Viadrina
LS für Quantitative Methoden
insbesondere Statistik
Grosse Scharrnstr. 59
15230 Frankfurt
Germany
schmid@euv-frankfurt-o.de

ISBN 978-3-7908-2379-0 e-ISBN 978-3-7908-2380-6
DOI 10.1007/978-3-7908-2380-6
Springer Heidelberg Dordrecht London New York

Library of Congress Control Number: 2010920745

Cover design: WMXDesign GmbH, Heidelberg

Printed on acid-free paper

Physica-Verlag is a brand of Springer-Verlag Berlin Heidelberg
Springer-Verlag is part of Springer Science+Business Media (www.springer.com)

Editorial

The IXth International Workshop on "Intelligent Statistical Quality Control" took place in September 2007 in Beijing, China, and was hosted by Professor Quan-lin Li (chairman) and Professor Wu Su (co-chairman), Department of Industrial Engineering, Tsinghua University, Beijing, P. R. China. The workshop itself was jointly organized by Professors H.-J. Lenz, P.-T. Wilrich and Quan-Lin Li.

The twenty-three papers in this volume are carefully selected, reviewed and revised for this volume, and are divided into three parts: Part I: “On-line Control – Control Charts”, Part II: “On-line Control – Surveillance Sampling and Sampling Plans” and Part III: “Off-line Control”.

In **Part I “On-line Control – Control Charts”**, ***Knoth*** starts with a general view on this topic in his contribution “Control Charting Normal Variance – Reflections, Curiosities, and Recommendations”. The author looks especially at two-sided setups, discusses competing statistics like S and S^2 used for variance monitoring and makes recommendations.

Tsung and ***Wang*** make a plea for adaptive charts in their paper entitled “Adaptive Charting Techniques: Literature Review and Extensions”. They compare the performance of a double-sided directionally variant chart with conventional multivariate charts. The detection power of unpredictable shifts and robustness are the two great advantages of their charts.

Reynolds and ***Stoumbos***† jointly author the paper “Multivariate Monitoring of the Process Mean and Variability Using Combinations of Shewhart and MEWMA Control Charts”. They analyze the problem of simultaneously control charting the mean and the variability of multivariate normal variables. They recommend a combination of MEWMA charts that includes one chart based on the squared deviations from target.

Colosimo, ***Mammarella*** and ***Petrò*** in their paper on “Quality Control of Manufactured Surfaces” study multivariate surface monitoring where the quality characteristic is the response to one or more location variables (in time or space). They present a new method which is based on a combined spatial autoregressive regression model.

In “Statistical Process Control for Semiconductor Manufacturing Processes” ***Higashide***, ***Nishina***, ***Kawamura*** and ***Ishii*** consider SPC for the semiconductor manufacturing chemical industry where automatic process adjustment and process maintenance are widely used. Two case studies are presented related to adjustment and maintenance of auto-correlated processes.

Cheng and ***Thaga*** in their paper "The MAX-CUSUM Chart" propose a single CUSUM control chart capable of detecting changes in both mean and standard deviation. They make a comparative study with other single charts like the Max-EWMA chart and Max chart based on the ARL criterion.

Hryniewicz and ***Szediw*** propose a new control chart based on Kendall's τ in their study entitled "Sequential Signals on a Control Chart Based on Nonparametric Statistical Tests". In the case of a Gaussian auto-regressive production process this chart behaves in a similar way to the well known autocorrelation chart, but it is more robust in non-Gaussian cases.

A slightly different perspective on control charts is taken by ***Golosnoy***, ***Okhrin***, ***Ragulin*** and ***Schmid*** in "On the Application of SPC in Finance". The field of interest is a fast on-line detection of changes of the optimal portfolio of a financial investor. Different types of EWMA and CUSUM control charts are analyzed by an extensive Monte Carlo simulation study using the ARL.

In **Part II "On-line Control – Surveillance Sampling and Sampling Plans"** ***Frisén*** pre-sents a paper on "Principles for Multivariate Surveillance". She reviews general approaches and makes suggestions on the special challenges of evaluating multivariate surveillance methods.

Woodall, ***Grigg*** and ***Burkom*** present an overview paper entitled "Research Issues and Ideas on Health-Related Surveillance" and compare surveillance methods used in health-care with industrial quality control.

In his paper "Surveillance sampling schemes" ***Baillie*** proposes a new type of sampling scheme for the simultaneous acceptance inspection of a number of large lots of similar size. For a beta prior distribution of the process fractions nonconforming, the expected proportion of lots accepted and the expected number of items inspected per lot are derived analytically.

Matsuura and ***Shinozaki*** in "Selective Assembly for Maximizing Profit in the Presence and Absence of Measurement Error" study the selective assembly of two mating components. They assume that the two component dimensions are normally distributed with equal variance, and that measurement error, if any, is Gaussian, too. It is shown numerically that the expected profit based on a density optimal partition decreases with increasing variance of the measurement error.

In his paper "A New Approach to Bayesian Sampling Plans" ***Wilrich*** designs a new (adaptive) Bayesian sampling plan for inspection by attributes based on a beta binomial model. The lot acceptance decision is directly based on the posterior distribution of the fraction nonconforming in the lot. The author illustrates that the adaptive single Bayesian plan dominates the equivalent ISO plans.

In **Part III "Off-line Control"** ***von Collani*** and ***Baur*** in "Stochastic Modelling as a Tool for Quality Assessment and Quality Improvement Illustrated by Means of Nuclear Fuel Assemblies" are concerned with modelling the quality assessment of fuel rods. Their Bernoulli-space model enables an accurate prediction of the performance of fuel rods, and supports a safe increase of the burn-up of nuclear fuel.

Mastrangelo, ***Kumar*** and ***Forrest*** in "Hierarchical Modeling for Monitoring Defects" propose a hierarchical linear model for linking the impact of process variables to defect rates. Process data drawn from the various gates are used to estimate the defect rates. Additionally, the output from the sub-models may be monitored with a control chart that is 'oriented' towards yield.

Göb and ***Müller*** on "Conformance Analysis of Population Means under Restricted Stratified Sampling" analyze risk-based auditing. The authors propose a restricted stratified sampling plan as an alternative for auditing.

In "Data Quality Control based on metric data models" ***Köppen*** and ***Lenz*** consider metric variables linked by the four arithmetic operators due to balance equations. Assuming a multivariate Gaussian distribution and an error in the variables model estimation of the unknown (latent) variables, the authors use MCMC-simulation to determine the "exact" distributions in non-normal cases and under cross-correlation.

Process capability studies are an important part of modern off-line control. ***Spiring*** picks up this topic in his contribution "The Sensitivity of Common Capability Indices to Departures from Normality". The author devises a procedure to analyze the robustness of *Cpw* to departures from normality, and discusses its impact.

In his paper "A note on the estimation of restricted scale parameters of Gamma distributions", ***Chang*** derives an admissible estimator for the scale parameter of a Gamma distribution. Simulation results illustrate the improvement of the new estimator compared with the traditional ones.

Kametani, Nishina and ***Suzuk***i investigate whether or not there exists an empirical relationship between the maturity of quality concerning the environment and the environmental lifestyle. The survey supports such a hypothesis for Japan.

Yasui, ***Ojima*** and ***Suzuki*** reconsider the Box and Meyer statistic in their study "On Identifying Dispersion Effects in Unreplicated Fractional Factorial Experiments". The distribution of the statistic under the null hypothesis is derived for unreplicated fractional factorial experiments. The power of the test for the detection of a single active dispersion effect is evaluated.

Suzuki, ***Yasui*** and ***Ojima*** look at tournament systems in sports. In their paper "Evaluating Adaptive Paired Comparison Experiments" they remind the reader that in the incomplete case forming pairs is crucial. They propose an evaluation method, a new criterion, and give examples.

The study "Approximated Interval Estimation in the Staggered Nested Designs for Precision Experiments" is authored by ***Yamasaki***, ***Okuda***, ***Ojima***, ***Yasui*** and ***Suzuki***. For the interval estimation of reproducibility a staggered nested precision experiment is proposed and evaluated by a Monte-Carlo simulation experiment.

The quality of any workshop is primarily shaped by the quality of papers that are presented at the meeting and their subsequent revision and submission for publication. The editors would like to express their deep gratitude to the members of the scientific programme committee, who did a superb job concerning the recruiting of invited speakers and the refereeing of the papers:

Mr David Baillie, United Kingdom
Prof. Elart von Collani, Germany
Prof. Olgierd Hryniewicz, Poland
Prof. Hans-J. Lenz, Germany
Prof. Quan-lin Li, P. R. China
Prof. Yoshikazu Ojima, Japan
Prof. Peter-Th. Wilrich, Germany
Prof. William H. Woodall, U.S.A.

We would like to thank very much our colleague Quan-lin Li and his students of the Department of Industrial Engineering, Tsinghua University, Beijing, who very efficiently supported the organization of the workshop: Mrs. Qinqin Zhang, Mrs. Rui Liu, Mr. Junjie Wu and Mr. Shi Chen.
Moreover, we again thank Physica-Verlag, Heidelberg, for their continuing efficient collaboration.
Finally, we are very sad to announce that three former participants have passed away: Professors Poul Thyregod, Zachary G. Stoumbos and Edward G. Schilling. Poul Thyregod was a permanent member of the programme committee for many years until he retired. Furthermore, he acted as the host of the third workshop at the Technical University of Denmark, Lyngby, 1986.

Berlin, February 2009

Hans - J. Lenz
Peter - Th. Wilrich
Wolfgang Schmid

Contents

PART III: OFF-LINE CONTROL

Author Index

Baillie, D. H., Chesham, UK
e-mail: david.baillie@which.net

Baur, K., Stochastikon GmbH, Schießhausstr. 15, D-97072 Würzburg, Germany
e-mail: baur@stochastikon.com

Burkom, H. S., National Security Technology Department, The Johns Hopkins University Applied Physics Laboratory, Laurel, MD 20723, U.S.A
e-mail: howard.burkom@jhuapl.edu

Chang, Y.-T., Prof. Dr., Dept. of Social Information, Faculty of Studies on Contemporary Society, Mejiro University, 4-31-1 Shinjuku-ku, Tokyo 161-8539 Japan
e-mail: cho@mejiro.ac.jp

Cheng, S. W., Dr., Department of Statistics, University of Manitoba, Winnipeg, Manitoba R3T 2N2, Canada
e-mail: smiley_cheng@umanitoba.ca

Collani, E. von, Prof. Dr., Universität Würzburg, Volkswirtschaftliches Institut, Sanderring 2, D-97070 Würzburg, Germany
e-mail: collani@mathematik.uni-wuerzburg.de

Colosimo, B. M., Prof. Dr., Dipartimento di Meccanica - Politecnico di Milano, Piazza Leonardo da Vinci 32, 20133 Milano, Italy
e-mail: biancamaria.colosimo@polimi.it

Forrest, D. F., Virginia Institute of Marine Science, Route 1208, Greate Road, P.O.Box 1346, Gloucester Point, VA 23062-1326, U.S.A
e-mail: drf@vims.edu

Frisén, M., Prof., Statistical Research Unit, Department of Economics, University of Gothenburg, P.O.Box 640, SE 40530 Sveden
e-mail: marianne.frisen@statistics.gu.se

Göb, R., Prof. Dr., Universität Würzburg, Institute for Applied Mathematics and Statistics, Sanderring 2, D-97070 Würzburg, Germany
e-mail: goeb@mathematik.uni-wuerzburg.de

Golosnoy, V., Prof. Dr., Institute of Statistics and Econometrics, University of Kiel, Olshausenstr. 40, 24118 Kiel, Germany,
e-mail: vgolosnoy@stat-econ.uni-kiel.de

Grigg, O. A., MRC Biostatistics Unit, Institute of Public Health,
Cambridge, UK
e-mail: olivia.grigg@cantab.net

Higashide, M., NEC Electronics Corp.
Shimonumabe, Nakahara-ku, Kawasaki, Kanagawa 211-8668, Japan
e-mail: masanobu.higashide@necel.com

Hryniewicz, O., Prof. Dr., Systems Research Institute of the Polish Academy of Sciences and Warsaw School of Information Technology , Newelska 6,
01-447 Warsaw, Poland
e-mail: hryniewi@ibspan.waw.pl

Ishii, N., Nagoya Institute of Technology,
Gokiso-cho, Showa-ku, Nagoya 466-8555, Japan
e-mail: naru@nitech.ac.jp

Kametani, T., Nagoya Institute of Technology,
Gokiso-cho, Showa-ku, Nagoya 466-8555, Japan
e-mail: tsuyoshi.kametani@murc.jp

Kawamura, H., Nagoya Institute of Technology,
Gokiso-cho, Showa-ku, Nagoya 466-8555, Japan

Knoth, S., Dr., Advanced Mask Technology Center, Postfach 110161,
D-01330 Dresden, Germany
e-mail: sven.knoth@amtc-dresden.com

Köppen, V., Dr., Institute of Production, Information Systems and Operations Research,
Freie Universität Berlin, Garystr. 21, D-14195 Berlin, Germany
e-mail: koeppen@wiwiss.fu-berlin.de

Kumar, N., Intel Corporation, Hillsboro, OR, U.S.A
e-mail: naveen.kumar@intel.com

Lenz, H.-J., Prof. Dr., Institute of Statistics and Econometrics, Freie Universität Berlin,
Garystr. 21, D-14195 Berlin, Germany
e-mail: hjlenz@wiwiss.fu-berlin.de

Mammarella, F., Dr., Quality Manufacturing Group - Toyota Motor Europe
Avenue du Bourget 60, 1140 Brussels, Belgium
e-mail: federica.mammarella@toyota-europe.com

Mastrangelo, C. M., Prof. Dr., University of Washington,
Box 352650, Seattle, WA 98195-2650, U.S.A
e-mail: mastr@u-washington.edu

Matsuura, S., Research Fellow of the Japan Society for the Promotion of Science,
Graduate School of Science and Technology, Keio University, 3-14-1,
Hiyoshi, Kohoku, Yokohama 223-8522, Japan
e-mail: matsuura@ae.keio.ac.jp

Müller, A., Universität Würzburg, Institute for Applied Mathematics and Statistics,
Sanderring 2, D-97070 Würzburg, Germany

Nishina, K., Prof. Dr., Nagoya Institute of Technology,
Gokiso-cho, Showa-ku, Nagoya 466-8555, Japan
e-mail: nishina@nitech.ac.jp

Ojima, Y., Prof. Dr., Tokyo University of Science, Department of Industrial Administration,
2641 Yamazaki, Noda, Chiba, 278-8510, Japan
e-mail: ojima@ia.noda.tus.ac.jp

Okhrin, I., Department of Statistics, European University Viadrina,
Grosse Scharrnstrasse 59, 15230 Frankfurt (Oder), Germany,
e-mail: iokhrin@euv-frankfurt-o.de

Okuda, M., Ricoh Company, LTD., 810 Shimoimaizumi, Ebina, Kanagawa,
243-0460, Japan
e-mail: michiaki.okuda@nts.rocoh.co.jp

Petrò, S., Dr., Dipartimento di Meccanica - Politecnico di Milano,
Piazza Leonardo da Vinci 32, 20133 Milano, Italy
e-mail: stefano.petro@polimi.it

Ragulin, S., Department of Statistics, European University Viadrina,
Grosse Scharrnstrasse 59, 15230 Frankfurt (Oder), Germany,
e-mail: sragulin@euv-frankfurt-o.de

Reynolds Jr., M. R., Prof. Dr., Department of Statistics, Virginia Tech,
Blacksburg, VA 24061-0439, U.S.A.
e-mail: mrr@vt.edu

Schmid, Wolfgang, Prof. Dr., Department of Statistics, European University Viadrina,
Grosse Scharrnstrasse 59, 15230 Frankfurt (Oder), Germany,
e-mail: schmid@euv-frankfurt-o.de

Shinozaki, N., Prof. Dr., Faculty of Science and Technology, Keio University,
3-14-1, Hiyoshi, Kohoku, Yokohama 223-8522, Japan
e-mail: nshino@ae.keio.ac.jp

Spiring, F., Ph.D., P.Stat., Department of Statistics, The University of Manitoba, Winnipeg, Manitoba R3T 2N2, Canada
e-mail: spiring@cc.umanitoba.ca, fspiring@mts.net

Stoumbos, Z. G., Prof. Dr., Department of Management Science and Information Systems, Rutgers, The State University of New Jersey, Piscataway, NJ 08854-8054, U.S.A.
e-mail: stoumbos@andromeda.rutgers.edu

Suzuki, K., Nagoya Institute of Technology,
Gokiso-cho, Showa-ku, Nagoya 466-8555, Japan

Suzuki, T., Ph.D., Tokyo University of Science, Department of Industrial Administration, 2641 Yamazaki, Noda, Chiba, 278-8510, Japan
e-mail: suzuki@ia.noda.tus.ac.jp

Szediw, A., Systems Research Institute of the Polish Academy of Sciences and Warsaw School of Information Technology , Newelska 6, 01-447 Warsaw, Poland
e-mail: szediw@ibspan.waw.pl

Thaga, K., Dr., Department of Statistics, University of Botswana,
Block 240, Room 216, Gaborone, Botswana
e-mail: thagak@mopipi.ub.bw

Tsung, F., Prof. Dr., Department of Industrial Engineering and Logistics Management, Hong Kong University of Science and Technology, Clear Water Bay, Kowloon, Hong Kong
e-mail: season@ust.hk

Wang, K., Assistant professor, Department of Industrial Engineering,
Tsinghua University, Beijing 100084, P. R. China
e-mail: kbwang@tsinghua.edu.cn

Wilrich, P.-Th., Prof. Dr., Institute of Statistics and Econometrics, Freie Universität Berlin, Garystr. 21, D-14195 Berlin, Germany
e-mail: wilrich@wiwiss.fu-berlin.de

Woodall, W. H., Prof. Dr., Department of Statistics, Virginia Tech,
Blacksburg, VA 24061-0439, U.S.A
e-mail: bwoodall@vt.edu

Yamasaki, M., Tokyo University of Science, Department of Industrial Administration,
2641 Yamazaki, Noda, Chiba, 278-8510, Japan
e-mail: yamasaki.motohiro@md.mt-pharma.co.jp

Yasui, S., Ph.D., Science University of Tokyo, Department of Industrial Administration,
2641 Yamazaki, Noda, Chiba, 278-8510, Japan
e-mail: yasui@ia.noda.tus.ac.jp

Part I

On-line Control

Control Charts

Control Charting Normal Variance – Reflections, Curiosities, and Recommendations

Sven Knoth

Engineering, Advanced Mask Technology Center[†], Dresden, Germany,
Sven.Knoth@amtc-dresden.com, http://www.amtc-dresden.com

Summary. Following an idea of Box, Hunter & Hunter (1978), the consideration of the log of the sample variance S^2 became quite popular in SPC literature concerned with variance monitoring. The sample standard deviation S and the range R are the most common statistics in daily SPC practice. SPC software packages that are used in semiconductor industry offer exclusively R and S control charts. With Castagliola (2005) one new log based transformation started in 2005. Again, the search for symmetry and quasi-normality served as reason to look for a new chart statistic. Symmetry of the chart statistic could help in setting up two-sided control charts. Here, a comparison study is done that looks especially to the two-sided setup, straightens out the view of the available set of competing statistics used for variance monitoring and, eventually, leads to recommendations that could be given in order to choose the right statistic.

1 Introduction

Monitoring the mean of normally distributed random variable is the most popular task within Statistical Process Control (SPC). For setting up the related control charts, one has to identify the underlying variance. Its value influences the chart design and performance heavily. Thus, it seems to be reasonable to check continuously the assumption about a certain variance level. Of course, there are further objectives of applying variance control charts, e. g., in the field of monitoring uniformity.

Before starting with estimating or monitoring variance, one may (or should) take into account that there could be several variance components. This was already considered in, e. g., Yashchin (1994), Woodall & Thomas (1995), and Srivastava (1997). This paper will focus on one variance component.

A further important topic is the amount of data that are collected at a given time point. The usual assumption is that the size of the sample could

[†]AMTC is a joint venture of Qimonda, AMD and Toppan Photomasks.

H.-J. Lenz et al. (eds.), *Frontiers in Statistical Quality Control 9*,
DOI 10.1007/978-3-7908-2380-6_1,

be chosen freely. Three major types could be distinguished here: (i) only one data point per time, (ii) a small (possibly arbitrary) number larger than one, and (iii) a large number. To illustrate all three cases consider two examples from daily practice in a mask shop like AMTC (Advanced Mask Technology Center), a company that is producing photomasks for wafer fabs. On certain monitor masks a larger number (100 and more) of lines are measured in order to monitor both mean level and variation. These lines (the so-called critical dimensions or abbreviated CD) have nominally the same size. Then average and standard deviation are calculated and used in a control chart. For an appropriate SPC setup for the mean one has to regard the mask-to-mask variation, consequently situation (i) is present. For monitoring variation on the mask one has to deal with (iii). These CD values are usually determined with a CD SEM (scanning electron microscope). Now, the CD SEM is monitored by repeating measurements at some fixed locations on a further specific mask. The number of repetitions has to be small, because the measured structures become more and more contaminated so that the CD size will change. Thus, situation (ii) is present. In case (i), typical approaches are relying on the Moving Range or certain measures of deviation from a given mean value μ_0 (see Acosta-Mejía & Pignatiello Jr. (2000) or also Domangue & Patch (1991)). This is not studied here. Nevertheless, it is an interesting and challenging topic. In case (iii) certain approximations will work quite well. In this paper, case (ii) will be studied in deeper detail.

In the first decades of executing SPC, R (range) and MR (moving range) control charts dominated (they survived in a lot of SPC software packages, quality circles etc.). Besides, control charts based on the sample variance S^2 and it's square root S were considered. Later on, EWMA (exponentially weighted moving average) and CUSUM (cumulative sum) control charts were constructed for nearly all considered statistics.

In the late 20th century a number of papers were written that compared several of these charts. It is possible that one of the items on the list *"General Trends and Reseach Ideas"* (in SPC) in Woodall & Montgomery (1999) led to plenty of papers about monitoring variance. Woodall & Montgomery (1999) stated: *"There has been a trend toward more research on monitoring process variability, but more work is needed on this topic."* Thereby, a couple of transformations such as the popular $\log S^2$ and the recently introduced $a+b\log(S^2+c)$ (a,b,c are chosen to get nearly normality; Castagliola (2005) was the beginning of a whole family of publications) should improve shape and, hopefully, the performance of the considered schemes.

Eventually, by monitoring variance one could be interested in detecting increases and/or decreases. While an increased variance level indicates mostly some trouble that should be detected and removed, a decreased variance could weaken the performance of a simultaneously operated mean chart.

Now, this paper will focus on

- EWMA (exponentially weighted moving average) charts
- in two-sided fashion.

The latter is chosen because all these normalizing transformations are motivated by the search for symmetry (and normality). Essentially, this seems to be reasonable only for two-sided schemes. Given the two-sided case, EWMA control charts are more practicable by design. For applying CUSUM charts, one has to combine two single charts. And this is, of course, a disadvantage for application despite the (slightly) better performance. However, some remarks about CUSUM charting for monitoring variance are given in the remaining part of this section.

There is a large number of papers about CUSUM charts for monitoring normal variance. Refer to Amin & Wilde (2000) for a Crosier-type CUSUM chart based on $\log S^2$. In Acosta-Mejía, Pignatiello Jr. & Rao (1999) and already Tuprah & Ncube (1987) several CUSUM charts based on S, S^2, and R are compared. See also Box & Ramírez (1991), Srivastava & Chow (1992), and recently Poetrodjojo, Abdollahian & Debnath (2002) for more comparison studies. Already in Page (1963) CUSUM charts deploying R were analyzed. Hawkins (1981) considered a CUSUM chart for a normalized version of $|(X-\mu_0)/\sigma_0|^{1/2}$. Acosta-Mejía et al. (1999) reviewed a couple of more normal approximations. Lowry, Champ & Woodall (1995) compared CUSUM-type variance charts (using S^2, S, and R) with classical S and R Shewhart charts extended with runs rules (different from the Western Electric recommended ones). Finally, not only in Chang & Gan (1995) and in Poetrodjojo et al. (2002) it was pointed out that S^2 CUSUM is the optimal scheme in the one-sided setup if looking at the famous worst case criterion due to Lorden (1971). See Table 1 for a small example.

Table 1. Slightly modified and shortened update of Table 5 in Chang & Gan (1995) – the EWMA schemes are also one-sided and equipped with a lower reflecting barrier, see Knoth (2005) for more details.

	CUSUM-S^2	CUSUM-ln S^2	EWMA-S^2	EWMA-ln S^2
	$k_h = 1.285$	$k_h^{\log} = 0.309$	$\lambda = 0.15$	$\lambda^{\log} = 0.28$
σ	$h_h = 2.922$	$h_h^{\log} = 1.210$	$c = 2.4831$	$c^{\log} = 1.4085$
1	100.0	100.0	100.0	100.0
1.1	27.9	30.2	27.9	30.0
1.2	12.8	13.8	12.9	13.8
1.3	**7.75**	**8.15**	**7.86**	**8.26**
1.4	5.47	5.63	5.57	5.76
1.5	4.22	4.29	4.30	4.43
2	2.08	2.11	2.11	2.22

The boldly written rows in both tables (1 and 2) are indicating the case the chart was optimized for. For further examples see, e. g., Lowry et al. (1995) who (numerically) demonstrated that CUSUM charts based on S^2 dominate those based on S or R. This is not really surprising because of the support by theoretical results for the S^2 version. To illustrate the same for detecting decrease see Table 2.

Table 2. Update of Table 2 of Acosta-Mejía et al. (1999).

σ	CUSUM-S^2 $k_l = 0.793$ $h_l = 2.332$	CUSUM-ln S^2 $k_l^{\log} = 0.375$ $h_l^{\log} = 4.389$
1	199.67	200.02
.9	38.40	41.51
.8	**14.16**	**15.78**
.7	8.25	8.86
.6	5.97	5.88

To make it clear, it is evident that for the one-sided charts the original sample variance, S^2, should be used.

For fixed mean μ_0 one should use instead of $S^2 = 1/(n-1)\sum(X_i - \bar{X})^2$ the statistic $\tilde{S}^2 = 1/n\sum(X_i - \mu_0)^2$. The latter was called CP CUSUM (change point CUSUM) in Acosta-Mejía et al. (1999).

Finally, Srivastava & Chow (1992) gave also some results for a Shiryaev-Roberts procedure for monitoring normal variance.

2 Two-sided EWMA charts for monitoring variance

The first papers dealing with EWMA control charts for monitoring the variance came from Wortham & Ringer (1971) and, a decade later, Sweet (1986), where only rough recommendations for the control chart design parameters were given. Thereafter, Domangue & Patch (1991), Crowder & Hamilton (1992), MacGregor & Harris (1993), and Mittag, Stemann & Tewes (1998) investigated EWMA control charts based on S^2, S and the natural log of the S^2. Ng & Case (1989) evaluated EWMA charts based on R and MR. Additionally, in papers by Srivastava (1994), Gan (1995), Reynolds Jr. & Stoumbos (2001), and Knoth (2007) the joint monitoring of mean and variance with EWMA schemes was considered.

It is interesting that EWMA smoothing of $\log S^2$ (the natural log) has reached great popularity. There are different reasons for this phenomenon. Box et al. (1978) and others had recommended this transformation and so started Crowder & Hamilton (1992) with an EWMA chart smoothing $\log S^2$ instead of S^2 itself or S and R, respectively. Mainly, it is the transition from a scale-change model to a level-change model that motivates the log-transformation. Now, changes in the scale do not affect the variance of the new chart statistic. Furthermore, $\log S^2$ is nearly normally distributed and so one can transfer the known results from the EWMA mean control chart. Finally, the distribution of $\log S^2$ is more symmetric than that of S^2 so that two-sided control charts are simpler to design. This attained a certain climax with papers following Castagliola (2005) who tuned $\log S^2$ by choosing suitable constants a, b, and c to get with $a + b \log(S^2 + c)$ a nearly normally distributed random variable. By the way, Castagliola (2005) did not simply choose a, b, and c to create a statistic that has in the in-control case mean 0, variance 1, and – as one might expect – skewness 0. He considered a three-parameter log-normal distribution, that is, a random variable Y where $a+b \log(Y+c)$ is exactly standard normally distributed. Then, a, b, and c are chosen to match the first three moments of Y and S^2. The resulting variable $a + b \log(S^2 + c)$, however, is skewed and has a different curtosis than a normal distribution. See Table 3 for some numerical results.

Here, all these different statistics will be compared and studied in more detail. To begin with, some basic notation will be introduced now.

Let $\{X_{ij}\}$ be a sequence of subgroups of independent and normally distributed data. Each subgroup i consists of n observations $X_{i1}, \ldots, X_{in}$. The subgroup size n is larger than 1 (called case (ii) in the Introduction). The following change point model is considered for the variance σ^2:

$$\sigma^2 = \begin{cases} \sigma_0^2 & , \ i < m \\ \sigma_1^2 \neq \sigma_0^2 & , \ i \geq m \end{cases} .$$

The parameters σ_0^2 and σ_1^2 denote the in- and out-of-control values of the monitored variance. Then, based on the sequence of sequentially observed data one wants to detect the unknown integer value m, which is called the *change point.* In the sequel, the subscript m will refer to the actual change point position. Thus, e. g., $E_m(\cdot)$ denotes the expectation under a change point m ($m = \infty$ stands for no change at all). Without loss of generality the variance σ_0^2 is set to 1.

The basic idea of EWMA control charts is to smooth an appropriate statistic such as the average for mean monitoring. For variance monitoring, the following statistics will be studied.

$$R_i = \max_j X_{ij} - \min_j X_{ij} \, ,$$

$$S_i^2 = \frac{1}{n-1} \sum_{j=1}^{n} \left(X_{ij} - \bar{X}_i\right)^2 \qquad \left(\bar{X}_i = \frac{1}{n} \sum_{j=1}^{n} X_{ij}\right) ,$$

$$S_i = \sqrt{S_i^2} \quad , \ lS_i = \log S_i^2 \quad , \ abcS_i = a + b \log(S_i^2 + c) \, .$$

To get a first impression about the differences, the density functions for all 5 competitors are plotted in Figure 1 for $n = 5$. In order to allow better judgment, the densities are standardized under the in-control model. While it is quite simple to get numerical results for all statistics derived from S^2 (here the function `dchisq()` in the statistics software R was used), it is quite demanding for the (normal) range R. In standard text books one can find tables (and formulas) that give the mean and the standard deviation of R for given sample size n – the famous constants d_2 and d_3. It is not easy to find results for the cumulative distribution (cdf) or density function (pdf) of R. Here, a result of Bland, Gilbert, Kapadia & Owen (1966) was taken:

$$P(R/\sigma \leq r) = \int_{\infty}^{\infty} n\,\phi(x)\big(\Phi(x+r) - \Phi(x)\big)^{n-1}\,dx\,.$$

To get the density, the term above was differentiated by r and – as already done for the cdf – the integral was evaluated with numerical quadrature.

In Figure 1 the pdf for all five statistics is drawn for 3 process states, in-control ($\sigma = 1$) and two out-of-control states ($\sigma = 0.8$ and $\sigma = 1.25$). In the in-control case, the third moment γ_1 and the fourth moment γ_2 (here the excess curtosis is calculated) of all considered chart statistics, except of S^2, are collected in Table 3. Recall that the first and second moment do not affect the control chart performance. For S^2, $\gamma_1 = \sqrt{8/(n-1)}$ and $\gamma_2 = 12/(n-1)$. The

Table 3. Third and fourth moments of the distribution of $\log S^2$, Castagliola's *abcS*, S, and R for different sample sizes n. Thereby, γ_2 denotes the so-called excess curtosis. A normally distributed variable would have $\gamma_2 = 0$.

n	$\log S^2$		*abcS*		S		R	
	γ_1	γ_2	γ_1	γ_2	γ_1	γ_2	γ_1	γ_2
3	-1.1395	2.4000	0.5572	-0.3206	0.6311	0.2451	0.6461	0.2866
4	-0.9170	1.6125	0.3752	-0.3947	0.4857	0.1082	0.5230	0.1884
5	**-0.7802**	**1.1875**	**0.2746**	**-0.3803**	**0.4057**	**0.0593**	**0.4656**	**0.1694**
6	-0.6879	0.9312	0.2119	-0.3478	0.3542	0.0370	0.4350	0.1699
7	-0.6209	0.7626	0.1697	-0.3142	0.3179	0.0251	0.4176	0.1761
8	-0.5699	0.6442	0.1398	-0.2837	0.2906	0.0181	0.4073	0.1840
9	-0.5293	0.5569	0.1176	-0.2572	0.2692	0.0136	0.4012	0.1921
10	-0.4962	0.4901	0.1007	-0.2344	0.2519	0.0106	0.3976	0.2002
11	-0.4686	0.4374	0.0874	-0.2147	0.2374	0.0085	0.3957	0.2076
12	-0.4450	0.3949	0.0768	-0.1978	0.2252	0.0070	0.3950	0.2144
13	-0.4247	0.3597	0.0681	-0.1834	0.2147	0.0058	0.3949	0.2233
14	-0.4069	0.3303	0.0610	-0.1703	0.2055	0.0049	0.3953	0.2258
15	-0.3911	0.3053	0.0550	-0.1591	0.1974	0.0042	0.3962	0.2330

sample variance S^2 (see also Figure 1) is obviously the "least" normal one. The case $n = 5$ (the n chosen in Figure 1 and written in bold in Table 3) leads to $\gamma_1 = \sqrt{2}$ and $\gamma_2 = 3$ which are considerably larger than the corresponding values in Table 3.

The densities of S, R, and *abcS* look quite similar. Only the numbers in Table 3 demonstrate that *abcS* is less skewed and S's curtosis is closer to a normal one. The original log transformation, the $\log S^2$ is "less" normal than the previous candidates. There is no reason to prefer $\log S^2$ in terms of symmetry or normality. Its charm comes from the scale to a level change

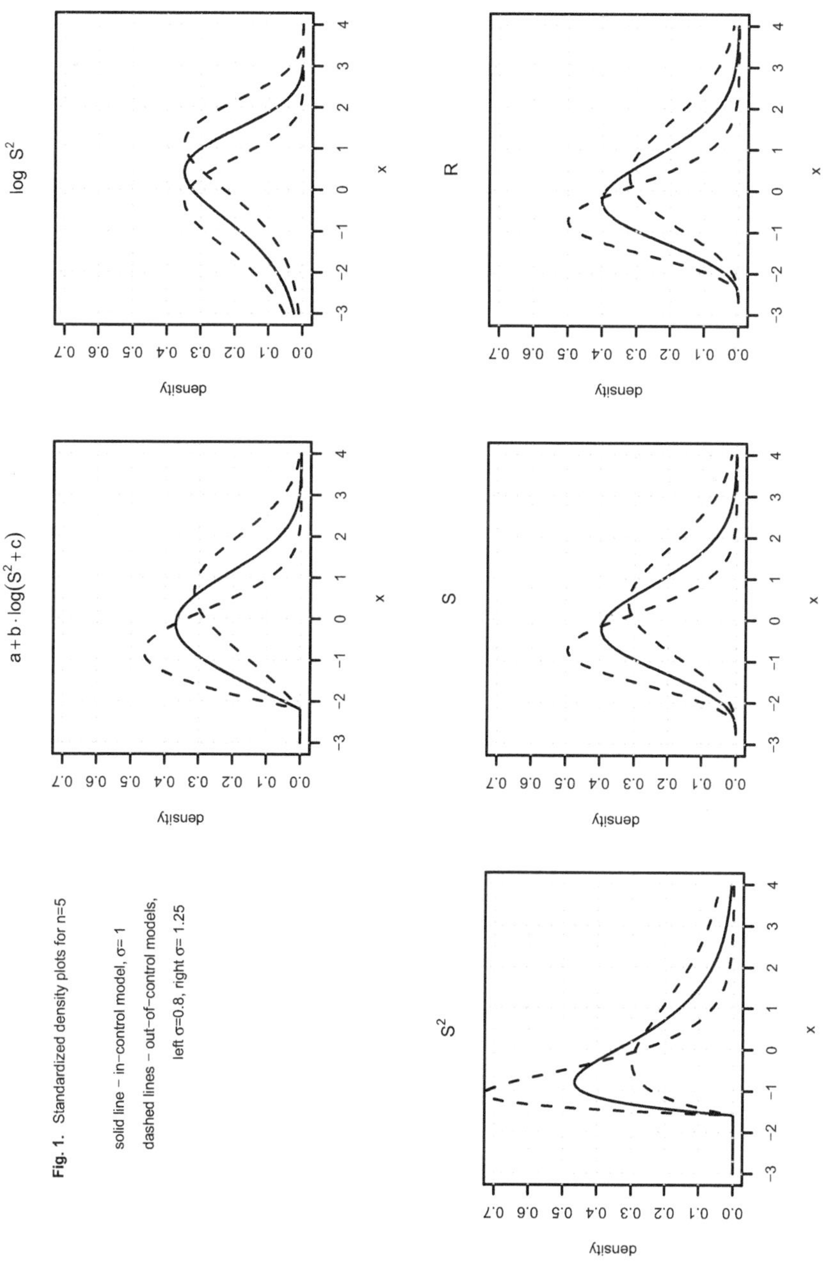

Fig. 1. Standardized density plots for n=5

solid line – in-control model, σ= 1

dashed lines – out-of-control models, left σ=0.8, right σ= 1.25

model transformation. Thus, the competing process states are only shifts of the chart statistic density. Eventually, S^2 distinguishes itself by the widest variety among the three given states. Moreover, it is the "least" normal statistic. Apparently, it is too early to predict the most powerful basis for an EWMA control chart.

By writing V_i as a dummy for the related variance estimator, the EWMA sequence with smoothing constant $\lambda \in (0,1]$ is given by:

$$Z_0 = z_0 = E_\infty(V_i)\,, \tag{1}$$

$$Z_i = (1-\lambda)\,Z_{i-1} + \lambda\,V_i \quad , \; i \geq 1\,. \tag{2}$$

Note that both Crowder & Hamilton (1992) and Castagliola (2005) started their EWMA schemes not at $E_\infty(V_i)$, but at $(\sigma_0^2 = 1)$

$$z_0 = \begin{cases} \log\sigma_0^2 = 0 & (\log S^2) \\ a + b\,\log(\sigma_0^2 + c) = 0.211 & (abcS) \end{cases},$$

which are larger than

$$E_\infty(V_i) = \begin{cases} E_\infty(\log S^2) = -0.2704 & (\log S^2) \\ E_\infty\big(a + b\,\log(S^2 + c)\big) = 0.0075 & (abcS) \end{cases}.$$

Thus, they gave their schemes a head-start for detecting increases. Here, all schemes are started from their in-control mean level $E_\infty(Z_i) = E_\infty(V_i)$.

As already announced, two-sided EWMA charts will be considered. Thus, an alarm is given as soon as

$$Z_i < c_l \vee Z_i > c_u\,.$$

Or, written as stopping time,

$$L = \min\big\{i \in \mathbb{N} : Z_i \notin [c_l, c_u]\big\}\,.$$

Recall that (using the independence)

$$Z_i = (1-\lambda)\,\sigma_0^2 + \lambda\sum_{j=1}^{i}(1-\lambda)^{i-j}\,V_j\,,$$

$$Var(Z_i) = \frac{\lambda}{2-\lambda}\big(1-(1-\lambda)^{2i}\big)\,Var(V_i)\,.$$

The thresholds given in the above alarm rule are not normalized.

All EWMA control charts will be evaluated in terms of their Average Run Length (ARL). This is nothing else than $E_1(L)$ and $E_\infty(L)$ for the (special) change point $m = 1$ and no-change point situation, respectively.

Many papers were written about calculating the ARL. Most of the numerical approaches could be embedded into the class of procedures for solving

Fredholm integral equations of the second kind. Thereby, the following integral equation that characterizes the ARL $\mathcal{L}$ as function of the starting value $z_0 = z$ (see (1)) should be solved.

$$\mathcal{L}(z) = 1 + \int_{c_l}^{c_u} \mathcal{L}(x) \frac{1}{\lambda} f\left(\frac{x-(1-\lambda)\,z}{\lambda}\right) dx \quad , \; z \in [c_l, c_u] \,. \tag{3}$$

For more details about this integral equation see Crowder (1987), Champ & Rigdon (1991), or Knoth (2005). Popular approaches for solving (3) are

- applying mid point rule to the integral (equivalent to the famous Markov chain approach due to Brook & Evans (1972) and already proposed by Page (1963)) or
- evaluating the integral with Gauss-Legendre quadrature (see Crowder (1987) for the first application to EWMA ARL) or Simpson rule.

As pointed out in Knoth (2005), these (and other) approaches are not accurate for EWMA charts built on characteristics with bounded support such as for S^2, S, R, and also $abcS = a + b\,\log(S^2 + c)$. Collocation is the approach that is used here. For more computational details see Knoth (2005). Castagliola (2005) did not mention how he actually had solved the corresponding integral equation for $abcS$.

In the following section, an ARL based comparison is done for all five two-sided EWMA charts in order to find out an appropriate choice of the statistic to be deployed in the EWMA smoothing.

3 Comparison study

The variance EWMA control charts under consideration are tuned for two situations. In the first case, the chart should detect small changes as fast as possible. The related out-of-control σ values are $4/5 = 0.8$ and $5/4 = 1.25$ (recall that $\sigma_0 = 1$). In the second case these two values are $2/3 \approx 0.667$ and $3/2 = 1.5$. Moreover, the so-called in-control ARL, $E_\infty(L)$, should be 500 for all schemes. To ensure a certain symmetry of the ARL function $\mathcal{L}_\sigma = \mathcal{L}_\sigma(z = z_0)$ in σ, the idea of unbiased ARL functions proposed by Acosta-Mejía et al. (1999) is applied. The idea is simple: The maximum of the ARL function should be attained for $\sigma = \sigma_0$. This was realized like in Knoth (2005). It has to be mentioned that none of the five schemes provides symmetry by choosing symmetrical control limits (c_l, c_u) – see Table 5 for the final control limits. The smoothing parameter λ was chosen to give the smallest value for

$$\mathcal{L}_{0.8} + \mathcal{L}_{1.25} \quad \text{and} \quad \mathcal{L}_{0.667} + \mathcal{L}_{1.5} \quad \text{, respectively.}$$

This led to the following values for λ searched on $\{0.02, 0.03, \ldots, 0.99\}$.

It is comfortable that these values do not differ considerably. In Figure 2 an illustration is given for the optimal λ search. Here, the ARL functions for

Table 4. Optimal values for smoothing λ.

case	statistic				
	R	S^2	S	lS	$abcS$
$\mathcal{L}_{0.8} + \mathcal{L}_{1.25}$	0.06	0.07	0.06	0.05	0.07
$\mathcal{L}_{0.667} + \mathcal{L}_{1.5}$	0.15	0.17	0.15	0.12	0.16

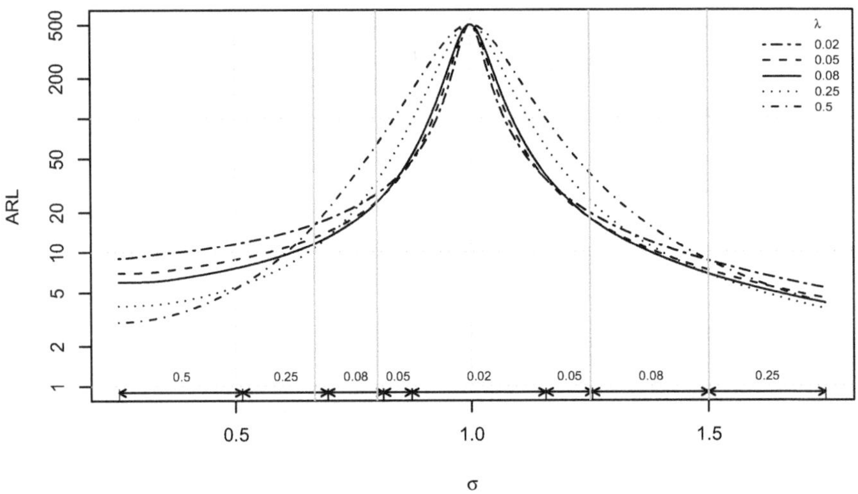

Fig. 2. Comparison of ARL profiles for S^2 EWMA control charts with various λ values. The arrows in the bottom line indicate the corresponding area of optimality. The gray vertical lines mark the σ values used for the tuning.

various λ of only the S^2 based EWMA chart are drawn. As usual, the smaller the change that should be detected, the smaller λ has to be chosen.

Now, the ARL $\mathcal{L}_\sigma$ function is calculated for $\sigma \in [0.25, 1.75]$. In Figure 3 all 5 schemes were compared on a log scale for the ARL values (the log scale was already used in Figure 2). Their control limits are listed in Table 5.

Table 5. Control limits and "center" lines for $\mathcal{L}_{0.8} + \mathcal{L}_{1.25}$ optimal charts.

chart parameter	statistic				
	R	S^2	S	lS	$abcS$
c_u	2.757	1.425	1.109	0.054	0.529
$E_\infty(V_i)$	2.326	1	0.940	-0.270	0.008
c_l	1.944	0.688	0.788	-0.618	-0.468

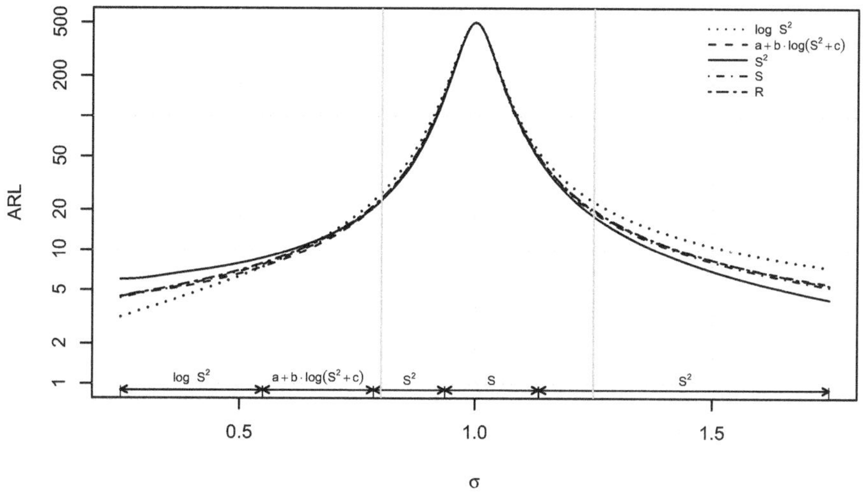

Fig. 3. ARL profiles of all 5 EWMA charts – the $\mathcal{L}_{0.8} + \mathcal{L}_{1.25}$ case. The arrows in the bottom line indicate the corresponding area of optimality. The gray vertical lines mark the σ values used for the tuning.

Despite or because of the log scale, no big differences between the five competitors could be seen (except the tails). Therefore, in Figure 4 the difference between the S^2 scheme and each of the competitors is plotted (on original scale). Additionally, in Table 6 some ARL numbers are collected.

From Figure 4 one can conclude that the S^2 EWMA procedure does a good job for a large range of possible values for σ. The two log based variants provide the best behavior for small σ values. However, the $\log S^2$ EWMA has the worst performance for detecting increases and small up to moderate decreases. Even the R EWMA is better. The numbers in Table 6 support these conclusions. Based on the results given, one would take S^2 or S. To put it in other words, there is no reason to apply the more artificially looking log based charts in practice. Additionally, for small sample sizes like $n = 5$, the range R provides similar power like the other statistics.

How does the picture change by optimizing for larger changes? See now Figure 5. Again, on the log scale the profiles look quite similar. Therefore, Figure 6 displays the difference between S^2 EWMA ARL and the remaining schemes.

Now, the S^2 variant beats all competitors for a long range within $\sigma < \sigma_0$. S gives the best performance for small increases. For $\sigma > 1.3$ all schemes, except $\log S^2$, are very close to each other. Again, one may conclude that there is no need to choose one of the log schemes.

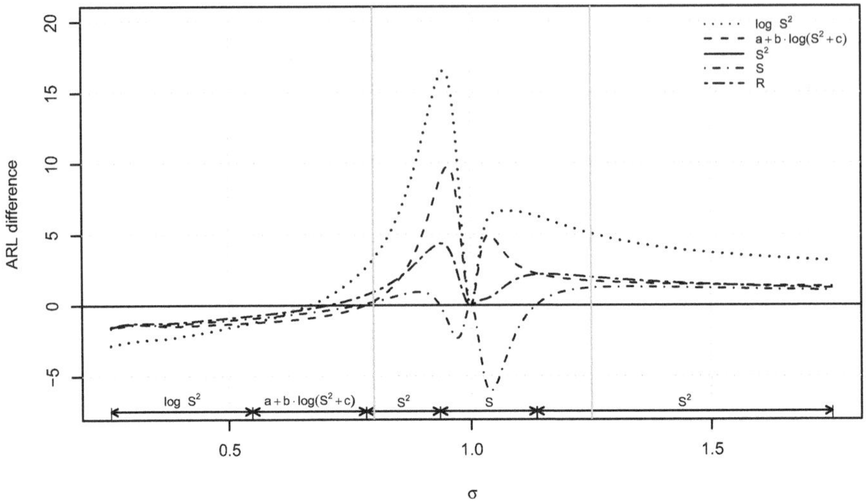

Fig. 4. Difference between the ARL function of all 5 EWMA charts and S^2 EWMA – the $\mathcal{L}_{0.8} + \mathcal{L}_{1.25}$ case. The arrows in the bottom line indicate the corresponding area of optimality. The gray vertical lines mark the σ values used for the tuning.

Table 6. ARL table for the $\mathcal{L}_{0.8} + \mathcal{L}_{1.25}$ competition.

σ	statistic				
	lS^2	$abcS^2$	S^2	S	R
0.5	**6.348**	6.617	7.943	6.881	7.033
0.6	8.977	**8.651**	9.779	9.004	9.215
0.7	13.97	**12.72**	13.48	13.09	13.43
0.75	18.59	**16.63**	17.03	16.90	17.37
0.8	26.65	23.66	**23.40**	23.62	24.32
0.9	85.00	78.31	**73.43**	74.34	76.78
1.0			500.000		
1.1	84.25	80.39	77.66	**76.12**	79.54
1.2	30.76	27.02	**25.19**	26.19	27.30
1.25	23.02	19.65	**17.95**	19.16	19.92
1.3	18.46	15.40	**13.79**	15.06	15.63
1.4	13.39	10.80	**9.308**	10.55	10.92
1.5	10.67	8.396	**6.976**	8.155	8.424
1.6	8.978	6.928	**5.564**	6.677	6.887

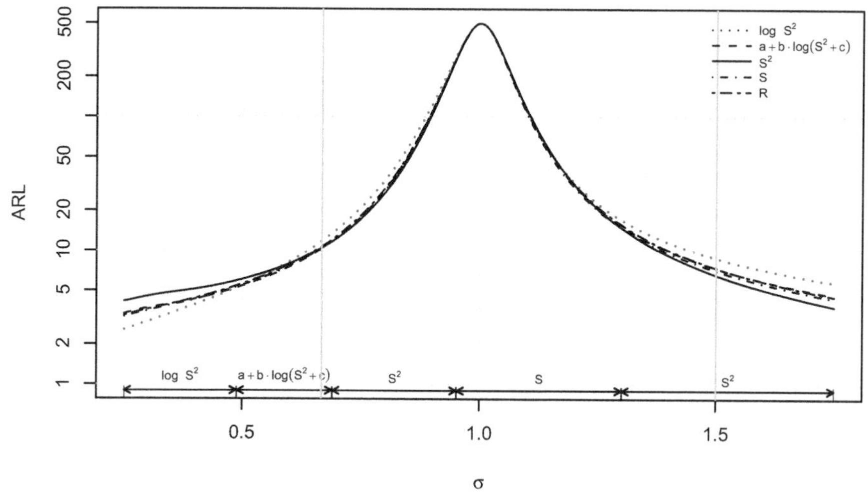

Fig. 5. ARL profiles of all 5 EWMA charts – the $\mathcal{L}_{0.667} + \mathcal{L}_{1.5}$ case. The arrows in the bottom line indicate the corresponding area of optimality. The gray vertical lines mark the σ values used for the tuning.

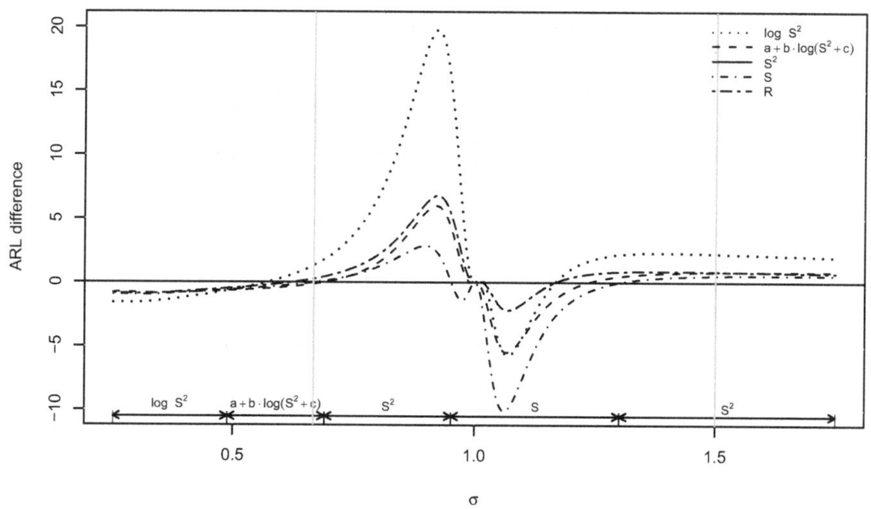

Fig. 6. Difference between the ARL function of all 5 EWMA charts and S^2 EWMA – the $\mathcal{L}_{0.667} + \mathcal{L}_{1.5}$ case. The arrows in the bottom line indicate the corresponding area of optimality. The gray vertical lines mark the σ values used for the tuning.

The ARL profiles in Figure 3 and 5 are nearly symmetric for all five competitors with ARL values that are slightly smaller for $\sigma < \sigma_0$ than for $\sigma > \sigma_0$. Thus, by utilizing the concept of unbiased ARL function the potential control chart user would get a nearly symmetric scheme for all considered charts. The question remains open whether it is reasonable to ask for a control chart design that provides ARL performance symmetric in $\sigma = \sigma_0$. The results obtained here and their practical meaning will be summarized in the next section.

4 Conclusions

It was and is quite popular to deploy the log transformation in order to get an appropriate symmetric control chart for monitoring normal variance. It turned out that the log schemes are not more symmetric than the older schemes based on S^2, S, and R. Moreover, the best performance in terms of the ARL profile is given by the S^2 and S EWMA charts. The first log example, $\log S^2$ is even beaten by R. Generally speaking, each of the charts, except the $\log S^2$, is usable. In practice, variation is mostly evaluated in terms of S or R. Then nothing could be said against their usage in an EWMA chart. If the sample size n increases, then S is the favorite in practice. Eventually, take the S^2 scheme and get the best.

References

Acosta-Mejía, C. A. & Pignatiello Jr., J. J. (2000). Monitoring process dispersion without subgrouping, *Journal of Quality Technology* **32**(2): 89–102.

Acosta-Mejía, C. A., Pignatiello Jr., J. J. & Rao, B. V. (1999). A comparison of control charting procedures for monitoring process dispersion, *IIE Transactions* **31**: 569–579.

Amin, R. W. & Wilde, M. (2000). Two-sided CUSUM control charts for variability, *Allgemeines Statistisches Archiv* **84**(3): 295–313.

Bland, R., Gilbert, R., Kapadia, C. & Owen, D. (1966). On the distributions of the range and mean range for samples from a normal distribution, *Biometrika* **53**: 245–248.

Box, G. E. P., Hunter, W. G. & Hunter, J. S. (1978). *Statistics for Experimenters: An Introduction to Design, Data Analysis, and Model Building*, John Wiley & Sons.

Box, G. & Ramírez, J. (1991). Sequential methods om statistical process monitoring, *Technical Report 65*, Center for Quality and Productivity Improvement, University of Wisconsin, 610 Walnut Street, Madison, Wisconsin 53705.

Brook, D. & Evans, D. A. (1972). An approach to the probability distribution of CUSUM run length, *Biometrika* **59**(3): 539–549.

Castagliola, P. (2005). A new S^2-EWMA control chart for monitoring process variance, *Quality and Reliability Engineering International* **21**: 781–794.

Champ, C. W. & Rigdon, S. E. (1991). A comparison of the Markov chain and the integral equation approaches for evaluating the run length distribution of quality control charts, *Commun. Stat. Simula. Comput.* **20**(1): 191–204.

Chang, T. C. & Gan, F. F. (1995). A cumulative sum control chart for monitoring process variance, *Journal of Quality Technology* **27**(2): 109–119.

Crowder, S. V. (1987). A simple method for studying run-length distributions of exponentially weighted moving average charts, *Technometrics* **29**: 401–407.

Crowder, S. V. & Hamilton, M. D. (1992). An EWMA for monitoring a process standard deviation, *Journal of Quality Technology* **24**(1): 12–21.

Domangue, R. & Patch, S. (1991). Some omnibus exponentially weighted moving average statistical process monitoring schemes, *Technometrics* **33**(3): 299–313.

Gan, F. F. (1995). Joint monitoring of process mean and variance using exponentially weighted moving average control charts, *Technometrics* **37**: 446–453.

Hawkins, D. M. (1981). A cusum for a scale parameter, *Journal of Quality Technology* **13**(4): 228–231.

Knoth, S. (2005). Accurate ARL computation for EWMA-S^2 control charts, *Statistics and Computing* **15**(4): 341–352.

Knoth, S. (2007). Accurate ARL calculation for EWMA control charts monitoring simultaneously normal mean and variance, *Sequential Analysis* **26**(3): 251–264.

Lorden, G. (1971). Procedures for reacting to a change in distribution, *Ann. Math. Stat.* **42**(6): 1897–1908.

Lowry, C. A., Champ, C. W. & Woodall, W. H. (1995). The performance of control charts for monitoring process variation, *Commun. Stat. Simula. Comput.* **24**(2): 409–437.

MacGregor, J. F. & Harris, T. J. (1993). The exponentially weighted moving variance, *Journal of Quality Technology* **25**: 106–118.

Mittag, H.-J., Stemann, D. & Tewes, B. (1998). EWMA-Karten zur Überwachung der Streuung von Qualitätsmerkmalen, *Allgemeines Statistisches Archiv* **82**: 327–338.

Ng, C. H. & Case, K. E. (1989). Development and evaluation of control charts using exponentially weighted moving averages, *Journal of Quality Technology* **21**: 242–250.

Page, E. S. (1963). Controlling the standard deviation by Cusums and warning lines, *Technometrics* **5**: 307–315.

Poetrodjojo, S., Abdollahian, M. A. & Debnath, N. C. (2002). Optimal cusum schemes for monitoring variability, *International Journal of Mathematics and Mathematical Sciences* **32**: 1–15.

Reynolds Jr., M. R. & Stoumbos, Z. G. (2001). Monitoring the process mean and variance using individual observations and variable sampling intervals, *Journal of Quality Technology* **33**(2): 181–205.

Srivastava, M. S. (1994). Comparison of CUSUM and EWMA procedures for detecting a shift in the mean or an increase in the variance, *J. Appl. Stat. Sci.* **1**(4): 445–468.

Srivastava, M. S. (1997). CUSUM procedure for monitoring variability, *Commun. Stat., Theory Methods* **26**(12): 2905–2926.

Srivastava, M. S. & Chow, W. (1992). Comparison of the CUSUM procedure with other procedures that detect an increase in the variance and a fast accurate approximation for the ARL of the cusum procedure, *Technical Report 9210*, University of Toronto, Department of Statistics.

Sweet, A. L. (1986). Control charts using coupled exponentially weighted moving averages, *IEEE Transactions on Information Theory* **18**: 26–33.

Tuprah, K. & Ncube, M. (1987). A comparison of dispersion quality control charts, *Sequential Analysis* **6**(2): 155–163.

Woodall, W. H. & Montgomery, D. C. (1999). Research issues and ideas in statistical process control, *Journal of Quality Technology* **31**(4): 376–386.

Woodall, W. H. & Thomas, E. V. (1995). Statistical process control with several components of common cause variability, *IIE Transactions* **27**: 757–764.

Wortham, A. W. & Ringer, L. J. (1971). Control via exponential smoothing, *The Logistics Review* **7**(32): 33–40.

Yashchin, E. (1994). Monitoring variance components, *Technometrics* **36**(4): 379–393.

Adaptive Charting Techniques: Literature Review and Extensions

Fugee Tsung[1] and Kaibo Wang[2]

[1]Department of Industrial Engineering and Logistics Management, Hong Kong University of Science and Technology, Clear Water Bay, Kowloon, Hong Kong

[2]Department of Industrial Engineering, Tsinghua University, Beijing 100084, P. R. China

Summary. The continuous development of SPC is driven by challenges arising from practical applications across diverse industries. Among others, adaptive charts are becoming more and more popular due to their capability in tackling these challenges by learning unknown shifts and tracking time-varying patterns. This chapter reviews recent development of adaptive charts and classifies them into two categories: those with variable sampling parameters and those with variable design parameters. This review focuses on the latter group and compares their charting performance. As an extension to conventional multivariate charts, this work proposes a double-sided directionally variant chart. The proposed chart is capable of detecting shifts having the same or opposite directions as the reference vector and is more robust to processes with unpredictable shift directions.

1 Introduction

As an efficient tool for monitoring process status and helping identify assignable causes, Statistical Process Control (SPC) has been successfully applied in applications across diverse industries, including manufacturing, financial service (Tsung *et al.* (2007)), healthcare (Woodall (2006)), medical research and others. With more and more challenges arising from these real world scenarios, SPC techniques have been evolving continuously from the days of Shewhart (1931), when the concept of control charts was first introduced.

The challenges facing SPC are multi-faceted. Under the seemingly contrary goal of pursuing lower false alarm rates and higher sensitivity, SPC algorithms have to handle complex while practical situations like unknown shift magnitudes, high dimensionalities and process dynamics. Some control charts can be tuned to be more sensitive to specific shifts magnitudes. However, when true shift magnitudes are unknown or not constant, either assumptions have to be made regarding shift sizes or capable algorithms have to be designed to estimate such information. The curse of dimensionality is a well-known obstacle to achieving better charting performance. Process dynamics, especially when confounded with unknown parameters and high dimensionalities, have given serious challenges to the SPC

H.-J. Lenz et al. (eds.), *Frontiers in Statistical Quality Control 9*,
DOI 10.1007/978-3-7908-2380-6_2, © Springer-Verlag Berlin Heidelberg 2010

community. In Section 2, features and current solutions to these issues will be summarized.

So far, one of the best ways to tackle the above challenging problems is to use control charts equipped with adaptive capability. Literally, adaptive control charts are smarter than ordinary charts in the sense that they can update themselves to fit for different situations. Therefore, even if initial settings of these charts are not accurate, they are expected to achieve optimality via self-adjustment. In Section 3, adaptive control charts with variable design parameters are extensively reviewed and some general guidelines for implementing these charts are provided.

In Section 4, a new multivariate adaptive control chart is proposed, which fills one of the gaps we identified in our review. Finally, Section 5 concludes this chapter with recommendations on future research topics.

2 The Challenges facing SPC

Due to the fast development of manufacturing and sensing technology, as well as the continuously increasing demand for high service levels in banking, healthcare and other industries, SPC are encountering challenges driven by versatile practical applications. In this section, we summarize several challenges along which we see active SPC research. In Section 3, we review different types of adaptive control charts, which have been proposed to tackle these challenges.

2.1 Detecting unknown or mixed-range shifts

The primary criterion for evaluating any SPC schemes is their capability in detecting process changes quickly and correctly. Given a fixed false alarm rate, the one with the shortest delay in detecting a specific shift is usually favored. While it is relatively easier to tune a control scheme to be sensitive to a specific shift, as the CUSUM chart presented in Section 2, true shifts may have unpredictable sizes and directions. Most control charts are good at detecting only shifts with particular magnitudes. For example, Shewhart-type charts, including univariate I-chart, $\overline{x}$-chart and multivariate Hotelling's T^2 chart, are more sensitive to large shifts than to small shifts; while some other schemes, such as the exponentially weighted moving average (EWMA) chart, cumulative sum (CUSUM) chart, and multivariate versions of these charts, multivariate EWMA (MEWMA) and multivariate CUSUM (MCUSUM), are more sensitive to shifts with small sizes.

The way that historical information is incorporated influences the performance of a control chart in detecting shifts of different sizes. Usually, if both latest and past observations are put together for monitoring, shifts with small sizes are more likely to be identified. The foregoing EWMA, CUSUM, MEWMA and MCUSUM are all this type of charts that accumulate historical information in certain ways to improve sensitivity. On the contrary, if merely the newest observations are studied, large shifts can be detected with a shorter delay since the information conveyed by these observations is not averaged out across a time span. The aforementioned Shewhart-type control charts possess this feature and are therefore more suitable for detecting large shifts.

More fundamentally, the settings of parameters dominate the performance of control charts over specific shift ranges (Montgomery (2005)). For example, the EWMA chart with a small smoothing parameter is more sensitive to small shifts than the same chart with a large smoothing parameter; the CUSUM chart with a certain reference value is most sensitive to that specific shift size.

Obviously, if the shift magnitude of a process is unknown or keeps varying over time, none of the above schemes, which are designed for either small or large shifts, can perform uniformly well. One possible treatment is to combine multiple charts to take care shifts in all ranges. For example, applying Shewhart control limits to EWMA or CUSUM charts to detect both large and small shifts (Lucas (1982), Albin *et al.* (1997), Chih-Min *et al.* (2000), Reynolds and Stoumbos (2005)).

Recently, adaptive EWMA charts (Capizzi and Masarotto (2003)) and adaptive CUSUM charts (Sparks (2000), Shu and Jiang (2006)) have been proposed and studied. By varying design parameters of the conventional charts, adaptive charts have been seen to be a promising tool for detecting unknown or mixed-range shifts. A more detailed review of adaptive control charts will be presented in next section.

2.2 Dealing with high dimensionalities

As processes in both manufacturing and service industries are getting more and more complex, dozens or even hundreds of correlated quality characteristics or process variables may need to be monitored simultaneously, which becomes another challenge to SPC. The false alarm rate of a chart that monitors multiple variables is much higher than when it monitors a single variable. If the false alarm rate is controlled by loosening control limits, the sensitivity of the chart will in turn be harmed. Due to the curse of dimensionality, the effect of shifts in one variable becomes less prominent when the variable is banded together with others.

Besides using the aforementioned Hotelling's T^2 chart for large shifts or the MEWMA and MCUSUM charts for small shifts, dimension reduction is a major technique to deal with high dimensional processes. Among others, principal component analysis (PCA), partial least square (PLS) and independent component analysis (ICA) are the most popular ways to extract a smaller set of variables that are representative enough for process monitoring.

PCA-based control charts try to find effective principal components (PCs), which are linear combinations of original process variables to represent the original high dimensional process. The number of PCs is usually much smaller than the number of original variables. Therefore, control charts can avoid high dimensionalities by monitoring these PCs only (Jackson (1991), Nomikos and Macgregor (1995), Mastrangelo *et al.* (1996), Runger and Alt (1996), Jolliffe (2002)). Recently, Tsung (2000) proposed to monitor dynamic processes with dynamic PCA. Instead of using fixed principal components, Tsung (2000) proposed to conduct PCA online, which is more suitable for processes with time-varying shifts. The application of dynamic PCA is also found in Yoo *et al.* (2002) and Choi *et al.* (2006).

If one is interested in monitoring a group of correlated variables, PCA might be utilized first to find representative components. While if the influence of these variables on quality characteristics is known, PLS can be performed first to identify a set of latent variables that are capable of better predicting quality variables (Macgregor *et al.* (1994)). The latent variables are linear combination of original process variables and can be monitored against process changes (Palm *et al.* (1997)). Nomikos and Macgregor (1995) introduced PLS-based methods that integrate initial setup condition, process variables and product variables together for process monitoring. A discussion on choosing latent variables in PLS was given by Li *et al.* (2002).

Ding *et al.* (2006) presented an example in which ICA outperforms PCA during data reduction. Different from PCA that tries to maximize explainable variation, ICA searches for components that can cluster data into distinct groups. Lee *et al.* (2003) integrated ICA with MEWMA for process monitoring; Albazzaz and Wang (2004) applied ICA-based SPC techniques to batch operations. Satisfactory results have been demonstrated by the authors.

It is becoming gradually common for profiles to be collected to characterize product quality or process status in many processes. Profiles reflect functional relationships between a response variable and one or more explanatory variables. Compared with a multivariate dataset, profiles contain even more data point and have to be modeled as an extremely high dimensional problem. Both model-based and model-free methods have been proposed for profile monitoring (see Woodall *et al.* (2004) for an extensive survey).

As a mathematical tool in signal processing, wavelet transformation has been applied to filter and decompose profile-type variables or quality characteristics (Ganesan *et al.* (2004), Jeong *et al.* (2006)). Wavelet transformation has the potential to reconstruct signals on multiple scales. Therefore, SPC schemes can be developed to monitor wavelet coefficients of different resolutions to detect slow or fast changes in underlying processes (Bakshi (1998), Ding *et al.* (2006)). Even though different wavelet basis functions can be chosen, such functions do not take engineering knowledge of original signals into consideration. Instead, model-based methods have been developed to model profile data for SPC. For example, Williams *et al.* (2007) advocated fitting nonlinear models to profiles first and then monitoring fitted parameters.

Recently, Wang and Tsung (2005) introduced an application where the quality of a surface is to be monitored. The authors borrowed profile-monitoring techniques to monitor product surface images. However, much information has been lost when modeling a surface as a profile. New technologies are to be developed for similar applications.

In Section 3, we review the most recent research on adaptive multivariate charts, which is more effective in monitoring high-dimensional processes.

2.3 Monitoring processes with dynamic means

A usual assumption behind most SPC schemes is that the normal status of the target process is stable. Means of manufacturing processes are usually assumed to be constant at a specific level when no assignable causes exist. However, there are situations under which a "normal" operational status changes over time. For example, the via etch process studied by Spitzlsperger *et al.* (2005) has drifting mean due to aging effects and its mean resets every time chamber maintenance is conducted. Such natural drifting trends must be decoupled from sampled signals so that true process faults can be identified.

To monitoring processes with dynamic means, adaptive control charts have been successfully applied in the literature. Section 3 gives examples of adaptive charts for monitoring such processes.

2.4 Detecting dynamic shifts

Dynamic shifts could be generated due to process autocorrelation, inertia properties or the existence of feedback controller (Tucker *et al.* (1993), Wang and Tsung (2007b)). Compared to sustained shifts, shifts that change over time are more difficult to capture. One way to detect time-varying shifts is to calculate smoothed trends while ignore local details. Atienza *et al.* (2002) proposed to use a backward CUSUM chart to monitoring autocorrelated observations. However, if the dynamic shifts happen to oscillate around the in-control process means, the smoothed signals lose useful information for shift detection.

Some researchers appeal to model-based methods (Lin and Adams (1996), Montgomery and Mastrangelo (1991), Pandit and Wu (1993), Apley and Tsung (2002)). By fitting time series models first, process shifts can be estimated by one-step-ahead predictions. Control charts based on residuals have been thoroughly investigated. For example, Lu and Reynolds (1999) applied EWMA charts to monitor residuals obtained from a model-based forecasting. However, the performance of model-based methods depends heavily on the accuracy of the time series models. Poor estimates of model parameters inevitably lead to poor forecasting accuracy.

Most of the above research focuses on detecting dynamic shifts in univariate processes. Apley and Tsung (2002), Tsung and Apley (2002), Wang and Tsung (2007a) and Wang and Tsung (2007b) studied the detection of dynamic shifts in multivariate processes. Since the performance of some multivariate charts is dominated by not only shift magnitudes but also shift directions, we will investigate this issue extensively in following sections and propose a new double-sided directionally variant chart for multivariate process monitoring.

Table 1 summarizes the above discussion and classifies the literature based on their primary purposes.

Table 1. Recent development of SPC along different directions

Objective	New Techniques and Literature
Detecting unknown or mixed-range shifts	Univariate charts: adaptive EWMA (Capizzi and Masarotto (2003)), adaptive CUSUM (Sparks (2000)) Multivariate charts: adaptive T^2 chart (Wang and Tsung (2007a), Wang and Tsung (2007b))
Dealing with high dimensionalities	Dynamic PCA (Tsung (2000), Yoo et al. (2002), Choi et al. (2006)) ICA (Ding et al. (2006)) Wavelet-based methods (Ganesan et al. (2004), Jeong et al. (2006)); Profile-targeted methods Linear profile models (Kang and Albin (2000), Kim et al. (2003)) Nonlinear profile models (Williams et al. (2007)) Surface (Wang and Tsung (2005))
Monitoring processes with dynamic means	Batch process monitoring (Spitzlsperger et al. (2005))
Detecting dynamic shifts from a stable process	Cuscore (Shu et al. (2002)); RF-Cuscore (Han and Tsung (2006)) Dynamic T^2 (Tsung and Apley (2002)); Adaptive T^2 (Wang and Tsung (2007a), Wang and Tsung (2007b))

3 Adaptive Control Charts for Process Monitoring

In control chart design and implementation, there are two sets of parameters need to be determined. The first set includes the sample size and the sampling interval, which can be classified as sampling parameters since settings of these parameters directly influence the way a control chart is operated; the second set contains design parameters of charting statistics, for example, the reference parameter that a CUSUM control chart is designed for, the intended shift magnitude that an EWMA control is designed for.

Adaptive control charts, in the same fashion, can be classified into two categories: those with adaptive sampling parameters and those with adaptive design parameters. Several review works on adaptive control charts is seen in the literature. Tagaras (1998) reviewed the development of adaptive charts until 1998. However, the author surveyed mainly univariate charts. In addition, only sampling parameters, including the sample size, the sampling interval, are allowed to be variable (the control limits are allowed to change with these parameters as well). Another broad review presented by Woodall and Montgomery (1999) covered some methods with variable sampling schemes. Zimmer *et al.* (2000) compared the performance of Shewhart-type control charts with adaptive sample sizes and/or sample intervals.

In practice, using control charts with adaptive sampling parameters needs the cooperation of operators, since their ordinary working pace or procedures might be interrupted or altered. In contrast, adaptive design parameters can be easily realized as long as SPC schemes are implemented with the aid of computers. Automated software can be designed to adapt to continuously updated parameter without affecting practical operational procedures. In this review, we focus on the second category of adaptive charts, that is, control charts with adaptive design parameters.

Adapting design parameters of control charts is driven by practical situations in which true process parameters are critical to charting performance while impossible to obtain. In the following, we review adaptive control charts and group them based on the way that critical design parameters are estimated and updated.

3.1 Recursive estimation of in-control mean/variance/covariance parameters

Some researchers have classified the control chart implementation procedure into two distinct stages: Phase I and Phase II (Woodall (2000)). In Phase I, historical data are collected and parameters are estimated. Control charts are then set up based on estimated parameters. Phase II involves online monitoring and signaling (Woodall (2000), Wang and Tsung (2005)).

However, as noted by Spitzlsperger *et al.* (2005), obtaining data in Phase I to reach acceptable level is too costly for some applications. While if insufficient number of samples is used in Phase I, large uncertainties may be seen in parameter estimation. Jones (2002) studied the design of EWMA charts with estimated parameters; Jones *et al.* (2004) evaluated the run length performance of CUSUM charts when parameters are estimated. Jensen *et al.* (2006) reviewed the impact of parameter estimation on control chart properties. Aside from all the work that tries to quantify the impact of parameter estimation uncertainties on ARL performance, a seemingly better alternative is to improve estimation accuracy by using samples collected in Phase II. This idea has been adopted by some adaptive control charts.

Spitzlsperger *et al.* (2005) applied Hotelling's T^2 chart to monitor a via etch process on a dual-frequency capacitive coupled parallel plate machine. Due to the aging effects and periodic chamber maintenance, some variables show slow drifts and quick jumps periodically.

Let $\mathbf{z}_i$ be a vector of the i th sample, $i = 1,...,n$. Each element of $\mathbf{z}_i$ is standardized with respect to sample mean μ_j and standard deviation σ_j. However, as μ_j and σ_j are unknown in practice, the authors estimated and updated μ_j via

$$\hat{\mu}_j = \lambda x_{ij} + (1-\lambda)\hat{\mu}_{j-1},$$

and updated σ_j via

$$\hat{\sigma}_j^2 = \frac{p-2}{p-1}\hat{\sigma}_{j-1}^2 + \frac{1}{p}(x_{ij} - \hat{\mu}_j)^2 ,$$

where λ is a smoothing parameter. By recursive estimation of such parameters, the resulting chart is capable of compensating aging effects and detecting true process faults.

3.2 Recursive estimation of out-of-control mean shifts

Unlike process parameters that can be estimated in Phase I, true process shifts are only available when the process is already running and assignable causes have occurred. As some control charts can be designed to be most sensitive to a specific shift, online estimation of process shifts becomes crucial to those charts (Capizzi and Masarotto (2003), Sparks (2000)).

As noted by Sparks (2000), the CUSUM statistic can only be optimized if accurate information on a specific sustained process shift is known. The derivations in Section 2 also demonstrate the way that CUSUM relies on a specific reference parameter, δ. Therefore, under situations where such information is not available, algorithms must be designed to predict the future value of δ and tune the chart for predicted values.

Sparks (2000) proposed an adaptive CUSUM chart, which monitors the following statistics:

$$z_t = \max\left[0, z_{t-1} + \left(x_t - \delta_t / 2\right)/h(\delta_t)\right].$$

Compared with the conventional CUSUM chart, the adaptive CUSUM procedure replaces the constant term, δ , by time-varying statistics, δ_t. A new function $h(\delta_t)$ is added to maintain a constant control limit. The shift magnitude, δ_t, is online updated via an EWMA-type equation,

$$\delta_t = \max(wx_{t-1} + (1-w)\delta_{t-1}, \delta_{\min}) .$$

The author suggested taking $\delta_{\min} = 0.5$ for detecting smaller shifts and $\delta_{\min} = 1.0$ for detecting shifts larger than 1.0. The ARL performance of the adaptive CUSUM chart is studied by Shu and Jiang (2006).

As the smoothing parameter of a traditional EWMA control chart is usually chosen according to intended shift magnitudes, Capizzi and Masarotto (2003) proposed an adaptive EWMA procedure and suggested using a variable smoothing parameter, which is determined based on estimated shift magnitudes. In specific, the adaptive EWMA takes the following form,

$$z_t = (1 - w(e_t))z_{t-1} + w(e_t)x_t ,$$

where $e_t = x_t - z_t$, and $w(e_t)$ is a function of e_t. For small values of e_t, $w(e_t)$ takes a relative small value; while when e_t is large, the value of $w(e_t)$ becomes large accordingly. Therefore, the adaptive EWMA can adjust its smoothing parameters according to estimated shift sizes.

3.3 Recursive estimation of shift directions of multivariate processes

Shifts in multivariate processes are characterized by not only shift magnitudes but also shift directions. Some multivariate control charts can therefore be designed to be sensitive to shifts along a specific direction. If the ARL performance of a chart depends on the mean vector and variance-covariance matrix only through the non-centrality parameter, the chart is directionally invariant. In contrast, if the performance is also influenced by shift direction, the chart is directionally variant (Lowry and Montgomery (1995), Wang and Tsung (2007a)). MEWMA, MCUSUM and Hotelling's T^2 charts all belong to the latter category.

Hawkins (1993), Jiang (2004b) and Zhou *et al.* (2005) considered the following format for a directionally variant chart:

$$T_d^2 = \mathbf{d}^T \mathbf{\Sigma}^{-1} \mathbf{x}_t > h, \tag{1}$$

where $\mathbf{d}$ is a constant vector that indicates the direction along which the control chart is optimized. Jiang (2004a) proposed a U_0 chart and a U_∞ chart for monitoring feedback-controlled processes. The U_0 and U_∞ charts can be proved to be directionally variant charts designed for shifts identified for representing transient and steady-state status. However, in order to implement the above charts, specific and constant shift directions must be identified.

Zhou *et al.* (2005) proposed to combine both directionally variant and invariant charts together for process monitoring. The directionally variant charts are designed for most likely directions along which shifts may occur; the directional invariant chart is in place to take care of general shifts:

$$\begin{cases} T_1^2 = \mathbf{d}_1^T \mathbf{\Sigma}^{-1} \mathbf{x}_t > h_1 \\ T_2^2 = \mathbf{d}_2^T \mathbf{\Sigma}^{-1} \mathbf{x}_t > h_2 \\ T^2 = \mathbf{x}_t^T \mathbf{\Sigma}^{-1} \mathbf{x}_t > h_3 \end{cases}$$

If any one of the charts signals, the process is diagnosed as out-of-control. Similar to the chart in Equation (1), Zhou *et al.* (2005)'s method requires some specific shifts to be known in advance. Furthermore, the simultaneously functioning multiple charts have overlapped regions. For example, a large shift along direction $\mathbf{d}_1$ is expected to trigger both T_1^2 and T^2 charts to signal. Such an overlapped design can obviously harm the overall sensitivity of the scheme.

Recently, Wang and Tsung (2007b) and Wang and Tsung (2007a) proposed an adaptive T^2 chart:

$$T^2 = \mathbf{d}_t^T \mathbf{\Sigma}^{-1} \mathbf{x}_t - \frac{1}{2} \mathbf{d}_t^T \mathbf{\Sigma}^{-1} \mathbf{d}_t > h, \tag{2}$$

where $\mathbf{d}_t$ is a time-varying vector that indicates the direction along which the control chart is optimized for.

As $\mathbf{d}_t$ is unknown in practice, Wang and Tsung (2007a) proposed two ways of forecasting $\mathbf{d}_t$. The first one is model-based forecasting. An ARMA(1,1) time-series model was fit to the data sequence and the one-step-ahead prediction is obtained, which is used to estimate $\mathbf{d}_t$. The second one is model-free EWMA smoothing. The authors update $\mathbf{d}_t$ recursively by

$$\mathbf{d}_t = \lambda \mathbf{x}_t + (1-\lambda)\mathbf{d}_{t-1}.$$

Based on simulation results, the authors found that EWMA-based forecasting is more robust to model misspecification.

To tackle strong oscillations found in feedback-controlled processes, Wang and Tsung (2007b) proposed an oscillated EWMA for shift forecasting. Similar to the EWMA forecasting with uses exponentially decaying weights, the oscillated EWMA uses decaying weights with alternative signs. That is,

$$\mathbf{d}_t = \lambda \mathbf{x}_t - (1-\lambda)\mathbf{d}_{t-1}.$$

When applied to a data sequence with signals go up and down alternatively, the oscillated EWMA can pickup such trend and enhance useful signals for fault detection.

Table 2 summarizes the adaptive charts that have variable design parameters.

Table 2 Adaptive charts with variable design parameters

Adaptive parameters	Control charts and representative literature
Shift magnitude	Adaptive EWMA (Capizzi and Masarotto (2003)) Adaptive CUSUM (Sparks (2000)) GLRT (Apley and Shi (1999))
Process mean/covariance matrix	Adaptive PCA (Tsung (2000)) Hotelling's T^2 (Spitzlsperger et al. (2005))
Shift pattern and shift direction	Dynamic T^2 (Tsung and Apley (2002)) Adaptive T^2 (Wang and Tsung (2007a), Wang and Tsung (2007b))

One fundamental assumption behind the above study is that the process is undergoing a sustained mean shift. Therefore, directionally variant charts show good performance in designated areas and the adaptive T^2 chart with EWMA forecasting exhibit satisfactory results in detecting small shifts. However, if process shifts are difficult to be estimated accurately, the performance of the above charts will be deteriorated. In the following section, we propose a double-sided directionally variant chart that takes the advantageous of both directionally variant charts and adaptive charts. The charting performance of the new chart is expected to be superior to other charts when double-sided shifts are to be detected.

4 A double-sided directionally variant chart for multivariate processes

It has been noted in previous sections that the multiple-chart scheme proposed by Zhou *et al.* (2005) has overlapped detection regions and failed to fully utilize the capability of all control charts. The adaptive T^2 chart due to Wang and Tsung (2007a) is good at detecting shifts that can be forecasted by EWMA equations. However, at any step, the adaptive chart is still a one-sided chart. If true process mean shifts are difficult to be estimated via EWMA, such as the feedback-controlled process discussed by Wang and Tsung (2007b), the resulting performance may be seriously deteriorated.

To improve the performance of the adaptive T^2 chart, there are two possible solutions: one way is to find intelligent forecasting algorithms so that the shifts can be estimated more accurately; the other way is to design robust charts to reduce the adverse effect of estimation uncertainties. Accurate forecasting algorithms may be case-dependent. For example, the oscillated EWMA proposed by Wang and Tsung (2007b) is suitable for strongly oscillated signals. Designing a robust control chart, on the other hand, may be a generally fitted solution.

Aiming to compensate the low efficiency of one-sided directionally variant chart, we propose a double-sided directionally variant chart. This chart takes care of both positive and negative shifts. This chart preserves the flexibility of the adaptive T^2 chart in estimating process shifts while uses doubled-sided statistics to avoid missing unexpected shifts in opposite directions.

We start from the directionally variant chart in Equation (1). Let $\mathbf{d}_{\sim}$ be an estimated shift direction and is standardized, $\mathbf{d}_{\sim}^{T}\boldsymbol{\Sigma}^{-1}\mathbf{d}_{\sim}=1$, let $\mathbf{d}_{\sim}^{-}$ be the opposite of $\mathbf{d}_{\sim}$, $\mathbf{d}_{\sim}=-\mathbf{d}_{\sim}^{-}$. We define two charting statistics,

$$T_1^2=\mathbf{d}_{\sim}^{T}\boldsymbol{\Sigma}^{-1}\mathbf{x}_t$$

and

$$T_2^2=(\mathbf{d}_{\sim}^{-})^{T}\boldsymbol{\Sigma}^{-1}\mathbf{x}_t$$

We know that the chart T_1^2 is capable of detecting all shifts that satisfy

$$\boldsymbol{\mu}^{T}\boldsymbol{\Sigma}^{-1}\mathbf{d}_{\sim}>0 \tag{3}$$

while T_2^2 is capable of detecting all shifts that satisfy

$$\boldsymbol{\mu}^{T}\boldsymbol{\Sigma}^{-1}\mathbf{d}_{\sim}^{-}>0 \tag{4}$$

Since $\mathbf{d}_{\sim}=-\mathbf{d}_{\sim}^{-}$ holds, the region characterized by Equations (3) and (4) covers all possible shifts.

We now define the double-sided directionally variant chart as

$$T_D^2 = \max(\mathbf{d}_{\sim}^T \mathbf{\Sigma}^{-1}\mathbf{x}_t, (\mathbf{d}_{\sim}^-)^T \mathbf{\Sigma}^{-1}\mathbf{x}_t) > h$$

As T_1^2 and T_2^2 follow the same statistical properties and $\mathbf{d}_{\sim} = -\mathbf{d}_{\sim}^-$, the above equation can be further simplified as

$$T_D^2 = |\, \mathbf{d}_{\sim}^T \mathbf{\Sigma}^{-1}\mathbf{x}_t \,| > h$$

Analogous to the adaptive T^2 chart, we estimate the shift via EWMA updating

$$\mathbf{d}_t = (1-w)\mathbf{d}_{t-1} + w\mathbf{x}_t \,.$$

Once $\mathbf{d}_t$ is obtained, $\mathbf{d}_{\sim}$ will be derived as

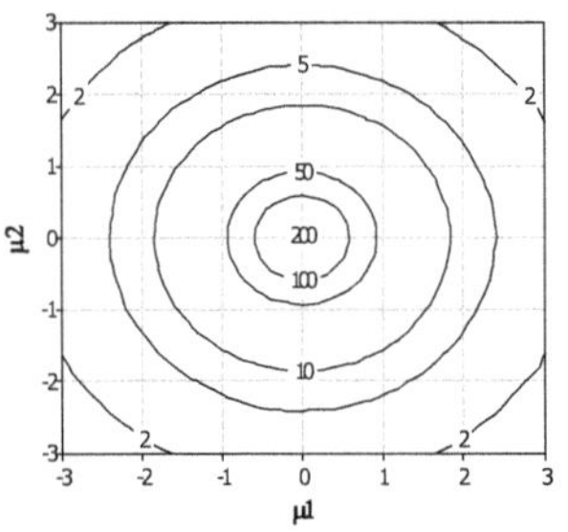

(a) Adaptive T^2

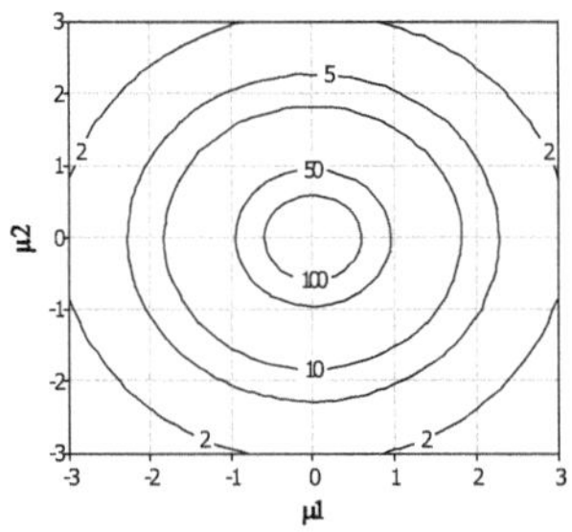

(b) Double-sided directionally variant chart

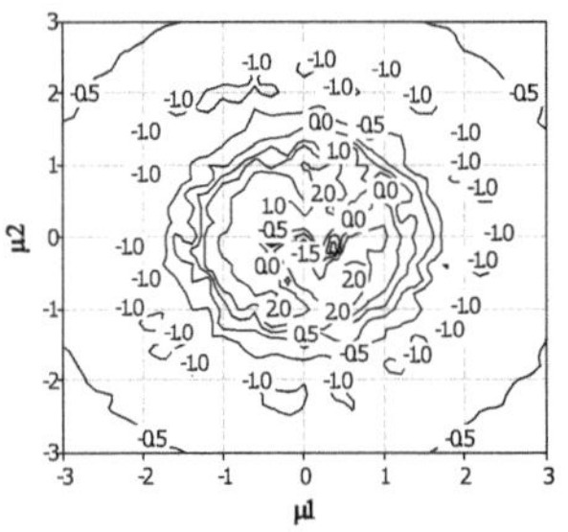

(c) ARL difference between the double-sided chart and the adaptive T^2 chart

Figure 1 Performance comparison between the double-sided directionally variant chart and the adaptive T^2 chart

$$\mathbf{d}_{\sim} = \frac{\mathbf{d}_t}{\sqrt{\mathbf{d}_t^T \boldsymbol{\Sigma}^{-1} \mathbf{d}_t}},$$

which is standardized to satisfy $\mathbf{d}_{\sim}^T \boldsymbol{\Sigma}^{-1} \mathbf{d}_{\sim} = 1$ at each step.

We now conduct simulations to investigate the performance of the newly proposed double-sided chart and compare it with the adaptive T^2 charts. A bivariate process with an identity covariance matrix is studied. Different from the above studies, dynamic shifts rather than sustained and constant shifts are applied to the process. At each step, shift vectors, with a sustained shift magnitude, are flipped with a probability of 50% to simulate shifts that EWMA cannot forecast accurately. The ARL contour plots of the adaptive T^2 chart are shown in Figure 1 (a), which is seen to expand away from the center and implies that the sensitivity of this chart has decreased. Figure 1 (b) shows the ARL performance of the double-sided directionally variant chart under the dynamic shifts described above. The ARL contour plots show that even though the proposed chart employs directionally variant statistics, the chart itself is invariant to shift directions. Compared with the adaptive T^2 chart, as Figure 1 (c) shows, the proposed chart is advantageous in detecting moderate and large shifts.

4 Conclusions

This chapter has summarized the recent challenges that are facing SPC, including unknown or mixed-range shifts, high dimensionalities, and process dynamics. To better tackle these challenges, adaptive control charts with variable sampling and design parameters have been reviewed. In specific, we have emphasized those charts with adaptive design parameters.

Adaptive control charts are ideal for meeting the above challenges due to its fundamental capabilities: learning and tracking. The learning capability of a control chart makes it possible for unknown shifts to be estimated. As a result, control charts can be optimized to detect such shifts. The tracking capability makes it possible for dynamic shifts to be captured.

This chapter has also proposed a double-sided directionally variant T^2 chart. The newly proposed chart possesses the flexibility of an adaptive T^2 chart and is more robust to shift oscillated seriously and cannot be accurately forecasted via EWMA.

It is learned from our review that research efforts are still needed to design better learning and tracking algorithms. Current EWMA-based or time-series-model-based methods are not sufficient for general purposes. Such issues are important topics for future research.

References

Albazzaz, H., and Wang, X. Z. (2004). "Statistical process control charts for batch operations based on independent component analysis." *Industrial & Engineering Chemistry Research*, 43(21), 6731-6741.

Albin, S. L., Kang, L., and Shea, G. (1997). "An X and EWMA chart for individual observations." *Journal of Quality Technology*, 29(1), 41-48.

Apley, D. W., and Shi, J. (1999). "The GLRT for statistical process control of autocorrelated processes." *IIE Transactions*, 31(12), 1123-1134.

Apley, D. W., and Tsung, F. (2002). "The autoregressive T-2 chart for monitoring univariate autocorrelated processes." *Journal of Quality Technology*, 34(1), 80-96.

Atienza, O. O., Tang, L. C., and Ang, B. W. (2002). "A CUSUM scheme for autocorrelated observations." *Journal of Quality Technology*, 34(2), 187-199.

Bakshi, B. R. (1998). "Multiscale PCA with application to multivariate statistical process monitoring." *AIChE Journal*, 44(7), 1596-1610.

Capizzi, G., and Masarotto, G. (2003). "An adaptive exponentially weighted moving average control chart." *Technometrics*, 45(3), 199-207.

Chih-Min, F., Ruey-Shan, G., Shi-Chung, C., and Chih-Shih, W. (2000). "SHEWMA: an end-of-line SPC scheme using wafer acceptance test data." *Semiconductor Manufacturing, IEEE Transactions on*, 13(3), 344-358.

Choi, S. W., Martin, E. B., Morris, A. J., and Lee, I. B. (2006). "Adaptive multivariate statistical process control for monitoring time-varying processes." *Industrial & Engineering Chemistry Research*, 45(9), 3108-3118.

Ding, Y., Zeng, L., and Zhou, S. Y. (2006). "Phase I analysis for monitoring nonlinear profiles in manufacturing processes." *Journal of Quality Technology*, 38(3), 199-216.

Ganesan, R., Das, T. K., and Venkataraman, V. (2004). "Wavelet-based multiscale statistical process monitoring: A literature review." *IIE Transactions*, 36(9), 787-806.

Han, D., and Tsung, F. (2006). "A reference-free Cuscore chart for dynamic mean change detection and a unified framework for charting performance comparison." *Journal of the American Statistical Association*, 101(473), 368-386.

Hawkins, D. M. (1993). "Regression adjustment for variables in multivariate quality control." *Journal of Quality Technology*, 25(3), 170-182.

Jackson, J. E. (1991). *A User's Guide to Principal Components*, Wiley, New York.

Jensen, W. A., Jones-Farmer, L. A., Champ, C. W., and Woodall, W. H. (2006). "Effects of parameter estimation on control chart properties: A literature review." *Journal of Quality Technology*, 38(4), 349-364.

Jeong, M. K., Lu, J. C., Huo, X. M., Vidakovic, B., and Chen, D. (2006). "Wavelet-based data reduction techniques for process fault detection." *Technometrics*, 48(1), 26-40.

Jiang, W. (2004a). "A joint monitoring scheme for automatically controlled processes." *IIE Transactions*, 36(12), 1201-1210.

Jiang, W. (2004b). "Multivariate control charts for monitoring autocorrelated processes." *Journal of Quality Technology*, 36(4), 367-379.

Jolliffe, I. T. (2002). *Principal component analysis*, Springer, New York.

Jones, L. A. (2002). "The statistical design of EWMA control charts with estimated parameters." *Journal of Quality Technology*, 34(3), 277-288.

Jones, L. A., Champ, C. W., and Rigdon, S. E. (2004). "The run length distribution of the CUSUM with estimated parameters." *Journal of Quality Technology*, 36(1), 95-108.

Kang, L., and Albin, S. L. (2000). "On-line monitoring when the process yields a linear profile." *Journal of Quality Technology*, 32(4), 418-426.

Kim, K., Mahmoud, M. A., and Woodall, W. H. (2003). "On the monitoring of linear profiles." *Journal of Quality Technology*, 35(3), 317-328.

Lee, J. M., Yoo, C., and Lee, I. B. (2003). "Statistical process monitoring with multivariate exponentially weighted moving average and independent component analysis." *Journal of Chemical Engineering of Japan*, 36(5), 563-577.

Li, B. B., Morris, J., and Martin, E. B. (2002). "Model selection for partial least squares regression." *Chemometrics and Intelligent Laboratory Systems*, 64(1), 79-89.

Lin, W. S. W., and Adams, B. M. (1996). "Combined control charts for forecast-based monitoring schemes." *Journal of Quality Technology*, 28(3), 289-301.

Lowry, C. A., and Montgomery, D. C. (1995). "A review of multivariate control charts." *IIE Transactions*, 27(6), 800-810.

Lu, C. W., and Reynolds, M. R., Jr. (1999). "EWMA control charts for monitoring the mean of autocorrelated processes." *Journal of Quality Technology*, 31(2), 166-188.

Lucas, J. M. (1982). "Combined Shewhart-CUSUM quality control schemes." *Journal of Quality Technology*, 14, 51-59.

Macgregor, J. F., Jaeckle, C., Kiparissides, C., and Koutoudi, M. (1994). "Process Monitoring and Diagnosis by Multiblock Pls Methods." *Aiche Journal*, 40(5), 826-838.

Mastrangelo, C. M., Runger, G. C., and Montgomery, D. C. (1996). "Statistical process monitoring with principal components." *Quality and Reliability Engineering International*, 12, 203-210.

Montgomery, D. C. (2005). *Introduction to statistical quality control*, John Wiley, Hoboken, N.J.

Montgomery, D. C., and Mastrangelo, C. M. (1991). "Some statistical process control methods for autocorrelated data." *Journal of Quality Technology*, 23(3), 179-193.

Nomikos, P., and Macgregor, J. F. (1995). "Multivariate SPC Charts for Monitoring Batch Processes." *Technometrics*, 37(1), 41-59.

Palm, A. C., Rodriguez, R. N., Spiring, F. A., and Wheeler, D. J. (1997). "Some perspectives and challenges for control chart methods." *Journal of Quality Technology*, 29(2), 122-127.

Pandit, S. M., and Wu, S. M. (1993). *Time series and system analysis with applications*, Krieger Pub. Co., Malabar, Florida.

Reynolds, M. R., Jr., and Stoumbos, Z. G. (2005). "Should exponentially weighted moving average and cumulative sum charts be used with Shewhart limits?" *Technometrics*, 47(4), 409-424.

Runger, G. C., and Alt, F. B. (1996). "Choosing principal components for multivariate statistical process control." *Communications in Statistics-Theory and Methods*, 25(5), 909-922.

Shewhart, W. A. (1931). *Economic control of quality of manufactured product*, D. Van Nostrand, New York.

Shu, L. J., Apley, D. W., and Tsung, F. (2002). "Autocorrelated process monitoring using triggered Cuscore charts." *Quality and Reliability Engineering International*, 18, 411-421.

Shu, L. J., and Jiang, W. (2006). "A Markov chain model for the adaptive CUSUM control chart." *Journal of Quality Technology*, 38(2), 135-147.

Sparks, R. S. (2000). "CUSUM charts for signaling varying location shifts." *Journal of Quality Technology*, 32(2), 157-171.

Spitzlsperger, G., Schmidt, C., Ernst, G., Strasser, H., and Speil, M. (2005). "Fault detection for a via etch process using adaptive multivariate methods." *IEEE Transactions on Semiconductor Manufacturing*, 18(4), 528-533.

Tagaras, G. (1998). "A survey of recent developments in the design of adaptive control charts." *Journal of Quality Technology*, 30(3), 212-231.

Tsung, F. (2000). "Statistical monitoring and diagnosis of automatic controlled processes using dynamic PCA." *International Journal of Production Research*, 38(3), 625-637.

Tsung, F., and Apley, D. W. (2002). "The dynamic T-2 chart for monitoring feedback-controlled processes." *IIE Transactions*, 34(12), 1043-1053.

Tsung, F., Zhou, Z. H., and Jiang, W. (2007). "Applying Manufacturing Batch Techniques to Fraud Detection with Incomplete Customer Information." *IIE Transactions*, 39, 671-680.

Tucker, W. T., Faltin, F. W., and VanderWiel, S. A. (1993). "Algorithmic statistical process-control - an elaboration." *Technometrics*, 35(4), 363-375.

Wang, K., and Tsung, F. (2005). "Using Profile Monitoring Techniques for a Data-rich Environment with Huge Sample Size." *Quality and Reliability Engineering International*, 21, 677-688.

Wang, K., and Tsung, F. (2007a). "An Adaptive T2 Chart for Monitoring Dynamic Systems." *Journal of Quality Technology*, Accepted.

Wang, K., and Tsung, F. (2007b). "Monitoring feedback-controlled processes using adaptive T2 schemes." *International Journal of Production Research*, 45(23), 5601-5619.

Williams, J. D., Woodall, W. H., and Birch, J. B. (2007). "Statistical Monitoring of Nonlinear Product and Process Quality Profiles." *Quality and Reliability Engineering International*, 23(8), 925-941.

Woodall, W. H. (2000). "Controversies and contradictions in statistical process control." *Journal of Quality Technology*, 32(4), 341-350.

Woodall, W. H. (2006). "The use of control charts in health-care and public-health surveillance." *Journal of Quality Technology*, 38(2), 89-104.

Woodall, W. H., and Montgomery, D. C. (1999). "Research issues and ideas in statistical process control." *Journal of Quality Technology*, 31(4), 376-386.
Woodall, W. H., Spitzner, D. J., Montgomery, D. C., and Gupta, S. (2004). "Using control charts to monitor process and product quality profiles." *Journal of Quality Technology*, 36(3), 309-320.
Yoo, C. K., Choi, S. W., and Lee, I. B. (2002). "Dynamic monitoring method for multiscale fault detection and diagnosis in MSPC." *Industrial & Engineering Chemistry Research*, 41(17), 4303-4317.
Zhou, S., Jin, N., and Jin, J. (2005). "Cycle-based signal monitoring using a directionally variant multivariate control chart system." *IIE Transactions*, 37(11), 971-982.
Zimmer, L. S., Montgomery, D. C., and Runger, G. C. (2000). "Guidelines for the application of adaptive control charting schemes." *International Journal of Production Research*, 38(9), 1977-1992.

Multivariate Monitoring of the Process Mean and Variability Using Combinations of Shewhart and MEWMA Control Charts

Marion R. Reynolds, Jr.[1], and Zachary G. Stoumbos[2]

[1]Virginia Polytechnic Institute and State University, Department of Statistics, Blacksburg, VA 24061-0439, USA
mrr@vt.edu

[2]Rutgers, The State University of New Jersey, Piscataway, NJ 08854-8054, USA

Summary. Control charts are considered for the problem of simultaneously monitoring the mean and variability of a multivariate process when the joint distribution of the process variables is multivariate normal. We investigate sets of univariate EWMA charts used in combination with sets of Shewhart charts or in combination with sets of EWMA charts based on the squared deviations of the observations from target. We also investigate the MEWMA control chart used in combination with a multivariate Shewhart control chart or with a form of MEWMA-type chart based on the squared deviations from target. We conclude that a combination of multivariate charts gives somewhat better average performance than a combination of sets of univariate charts, and that a combination of MEWMA charts that includes one based on the squared deviations from target gives the best overall performance.

1 Introduction

When the quality of a process is characterized by a normal random variable, it is important to detect special causes that may change the mean μ and/or the variance σ^2. The traditional Shewhart control charts are effective for detecting large shifts in process parameters, but EWMA control charts are much more effective for detecting small shifts. When the objective is to detect both small and large shifts in process parameters, it is frequently recommended that EWMA charts be used in combination with a Shewhart chart (see, e.g., Lucas and Saccucci (1990)). Reynolds and Stoumbos (2001, 2004, 2005) have recently shown that a very effec-

H.-J. Lenz et al. (eds.), *Frontiers in Statistical Quality Control 9*,
DOI 10.1007/978-3-7908-2380-6_3, © Springer-Verlag Berlin Heidelberg 2010

tive combination for detecting both small and large shifts in μ and σ^2 is a combination of the standard EWMA chart for monitoring μ and an EWMA chart of squared deviations from target for monitoring σ^2.

In many applications process quality may be characterized by p variables that are assumed to have a multivariate normal distribution with mean vector $\boldsymbol{\mu}$ and covariance matrix $\boldsymbol{\Sigma}$. In this setting a simple and still widely used approach to process monitoring is to use sets of univariate charts applied to each of the p variables. However, this approach may not perform well because the correlation between the variables is not used effectively.

The first control chart based on a multivariate statistic was a Shewhart-type chart proposed by Hotelling (1947). This chart suffers from the same disadvantage as other Shewhart charts in that it is not effective for detecting small shifts in process parameters. A multivariate analog of the EWMA chart is the multivariate EWMA (MEWMA) chart developed by Lowry, Woodall, Champ, and Rigdon (1991).

For the problem of monitoring $\boldsymbol{\mu}$ and $\boldsymbol{\Sigma}$, Reynolds and Cho (2006), Reynolds and Kim (2007), and Reynolds and Stoumbos (2008) have recently investigated combinations of the standard MEWMA chart for monitoring $\boldsymbol{\mu}$ and MEWMA-type charts based on the squared deviations from target for monitoring $\boldsymbol{\Sigma}$.

As an extension from the univariate setting, it has been recommended that in the multivariate setting the Shewhart and MEWMA charts be used together in combination. Reynolds and Stoumbos (2008) investigated the performance of this type of combination relative to combinations involving the squared deviations from target.

The objective of this paper is to extend the investigation of Reynolds and Stoumbos (2008) involving combinations of Shewhart and MEWMA charts and combinations of the standard MEWMA chart and MEWMA-type charts based on the squared deviations from target. This extension includes an investigation of the performance of combinations of sets of univariate Shewhart and EWMA charts applied to each variable. This extension also includes an investigation of the performance of control charts with and without the use of regression adjustment of the variables (Hawkins (1991, 1993)). It is assumed that the objective of monitoring is to detect small or large changes in the process mean and small or large increases in process variability. Reynolds and Stoumbos (2008) considered control chart performance averaged over all shift directions, and here we also consider control chart performance for specific shift directions for the mean and variability.

2 Definitions of the Control Charts

Suppose that a process with p variables of interest will be monitored by taking samples of size $n = 1$ at each sampling point, using a sampling interval of one time unit. Let $\boldsymbol{\sigma}$ represent the vector of standard deviations of the p variables, and let $\boldsymbol{\mu}_0$, $\boldsymbol{\Sigma}_0$, and $\boldsymbol{\sigma}_0$ represent the in-control values for $\boldsymbol{\mu}$, $\boldsymbol{\Sigma}$, and $\boldsymbol{\sigma}$, respectively. We assume that these in-control parameter values are known (or estimated with negligible error). We will usually refer to $\boldsymbol{\mu}_0$ as the target, even though, in practice, some of the components of $\boldsymbol{\mu}_0$ may correspond to estimated in-control means, rather than to specified target values.

At sampling point $k = 1, 2, \ldots$, let X_{ki} represent the observation for variable i $(i = 1, 2, \ldots, p)$, and let the corresponding standardized observation be

$$Z_{ki} = (X_{ki} - \mu_{0i}) / \sigma_{0i} ,$$

where μ_{0i} is the ith component of $\boldsymbol{\mu}_0$, and σ_{0i} is the ith component of $\boldsymbol{\sigma}_0$. Let $\boldsymbol{\Sigma}_{Z0}$ be the in-control covariance matrix of $(Z_{k1}, Z_{k2}, \ldots, Z_{kp})$.

We investigate a number of different combinations of Shewhart and MEWMA charts, and use some simple shorthand notation to keep track of the different statistics and charts. In particular, we use the notation from Reynolds and Cho (2006), Reynolds and Kim (2007), and Reynolds and Stoumbos (2008), where "S" is used for Shewhart charts, "E" for EWMA charts, and "M" for MEWMA-type charts. We also investigate sets of univariate control charts used in the multivariate setting, so we use "U" to indicate sets of univariate charts. The acronyms and descriptions of the control charts being considered are listed in Table 1 for convenient reference.

Consider first the set of univariate Shewhart charts for the p individual variables. In particular, a signal is given at sample k if $\max_i |Z_{ki}|$ exceeds an upper control limit (UCL). We will refer to this set of univariate charts as the USZ chart, where the "U" indicates univariate, "S" indicates Shewhart, and "Z" indicates that the chart is based on the Z_{ki} statistics. At sampling point k let the EWMA statistic of standardized observations for variable i be

$$E_{ki}^{Z} = (1-\lambda) E_{k-1,i}^{Z} + \lambda Z_{ki}, \quad i = 1, 2, \ldots, p ,$$

where $E_{0i} = 0$ and λ is a weighting or tuning parameter satisfying $0 < \lambda \le 1$ (the superscript "Z" is used to indicate that this EWMA statistic is based on Z_{ki}). If an EWMA chart is used for each of the p variables, then a signal would be given at sampling point k if $\max_i c_\infty^{-1/2} |E_{ki}^{Z}| >$ UCL, where $c_\infty = \lambda/(2-\lambda)$ is the asymptotic variance of E_{ki}^{Z} as $k \to \infty$. This set of univariate EWMA charts will be called the UEZ chart.

At sampling point k let the EWMA statistic of squared standardized deviations from target for variable i be

$$E_{ki}^{RZ^2} = (1-\lambda)\max\{E_{k-1,i}^{RZ^2}, 1\} + \lambda Z_{ki}^2, \quad i = 1, 2, \ldots, p,$$

where $E_{0i}^{Z^2} = 1$ and $0 < \lambda \leq 1$. The superscript "Z^2" is used to indicate that this is an EWMA statistic based on the squares of the Z statistics. The "R" is used to indicate that a reset is used to set the statistic back to 1 (the in-control expected value of Z_{ki}^2) whenever it drops below 1. A signal is given a sample k if $\max_i (2c_\infty)^{-1/2} E_{ki}^{RZ^2} > \text{UCL}$. We call this chart the UERZ^2 chart.

Table 1. Symbols and Descriptions of the Control Charts Being Considered

Chart	Description
USZ	A set of univariate Shewhart charts of the standardized observations (Z)
USA	A set of univariate Shewhart charts of the regression adjusted observations (A)
UEZ	A set of univariate EWMA charts of the standardized observations
UEA	A set of univariate EWMA charts of the regression adjusted observations
UERZ^2	A set of univariate EWMA charts of the squared standardized deviations from target (Z^2)
UERA^2	A set of univariate EWMA charts of the squared regression adjusted deviations from target (A^2)
SZ	A multivariate Shewhart chart of the standardized observations (Hotelling's T^2 chart)
MZ	An MEWMA chart of the standardized observations
M$_2$RZ^2	An MEWMA-type chart of the squared deviations from target
M$_2$RA^2	An MEWMA-type chart of the squared regression adjusted deviations from target

Reynolds and Cho (2006) found that using regression adjusted variables (Hawkins (1991, 1993)) instead of the original standardized variables tends to improve the performance of some control charts. Regression adjustment of the variables involves using linear regression to adjust the value of each variable in an observation vector based on the values of the other variables in this vector. When the "Z" variables in the USZ, UEZ, and UERZ^2 charts are replaced with the corresponding regression adjusted versions of these variables (represented by "A"), we will refer to the charts as the USA, UEA, and UERA^2 charts, respectively.

When $\mathbf{\Sigma}_0$ is known the multivariate Shewhart-type control chart proposed by Hotelling (1947) for monitoring $\boldsymbol{\mu}$ is based on a statistic that is a quadratic form of $(Z_{k1}, Z_{k2}, \ldots, Z_{kp})$ (see Eq. (1) in Reynolds and Cho (2006)). We will refer to this control chart as the SZ chart (the absence of a "U" before "S" indicates that this chart is not a set of univariate Shewhart charts).

The MEWMA control chart for monitoring $\boldsymbol{\mu}$ is based on a statistic that is a quadratic form of $(E_{k1}^{Z}, E_{k2}^{Z}, \ldots, E_{kp}^{Z})$ (see Eq. (9) in Reynolds and Cho (2006)). Call this chart the MZ chart.

Reynolds and Cho (2006) proposed control charts for monitoring $\mathbf{\Sigma}$ using two forms of MEWMA-type statistics based on squared standardized deviations from target. One of these forms has the in-control expectation subtracted from the EWMA statistic, and the other does not. Here we consider the second form of MEWMA-type statistic because it seems to have slightly better properties. This statistic is based on a quadratic form of $(E_{k1}^{RZ^2}, E_{k2}^{RZ^2}, \ldots, E_{kp}^{RZ^2})$ (see Reynolds and Kim (2007)). Call this chart the M_2RZ^2 chart, where the subscript "2" indicates the second form of MEWMA statistic.

The MZ and SZ charts are not affected if regression adjusted variables are used instead of the original variables (with the appropriate change in the covariance matrix), so there is no need to use regression adjustment with these charts. However, the M_2RZ^2 chart is affected if regression adjusted variables are used instead of the original variables (assuming that the variables are correlated). Replacing the original standardized variables in the M_2RZ^2 chart with the regression adjusted variables gives a chart that will be called the M_2RA^2 chart.

The SZ and MZ charts are usually referred to as charts for monitoring $\boldsymbol{\mu}$, although they are sensitive to shifts in $\mathbf{\Sigma}$. Similarly, the charts based on squared deviations from target are referred to as charts for $\mathbf{\Sigma}$ (or for $\boldsymbol{\sigma}$), although they are sensitive to shifts in $\boldsymbol{\mu}$. The control charts based on squared deviations from target are very effective for detecting changes in $\mathbf{\Sigma}$, but they also have the important advantage that they are very sensitive to large shifts in $\boldsymbol{\mu}$.

3 The Setup for Evaluation and Comparison of Charts

The evaluation and comparison of different control charts will be done using the same general setup used by Reynolds and Cho (2006), Reynolds and Kim (2007), and Reynolds and Stoumbos (2008) (this allows for comparisons with results in these papers).

The *average time to signal* (ATS) is defined to be the expected length of time from the start of process monitoring until a signal is generated. We set up all charts so that the in-control ATS is 800 time units. When two or more charts are used together in combination, the control limits of each chart were adjusted to give the same individual in-control ATS for each chart, with the in-control ATS of the combination at 800.

The ability of control charts to detect sustained shifts in process parameters is measured by the steady-state ATS (SSATS), which is defined as the expected length of time from the random point in time that the shift occurs until the time that a signal is generated. The SSATS is computed assuming that the control statistics have reached their steady-state distribution by the random time point that the shift occurs. The SSATS also allows for the possibility that the shift can occur randomly in a time interval between samples.

Most ATS and SSATS results in this paper are based on simulation with 1,000,000 simulation runs. Steady state properties were obtained by simulating the operation of the control charts for 400 in-control observation vectors, and then introducing a shift.

We present numerical results here for the case of p = 4 variables, when all pairs of variables have the same correlation ρ, where $\rho = 0$ or 0.9. These two values of ρ are intended to represent two extreme cases of independence and high positive correlation. The values of λ used here are 0.02600 and 0.11989, corresponding to values used in previous papers.

Shifts in $\boldsymbol{\mu}$ are expressed in terms of the standardized mean shift vector, defined as $\mathbf{v} = (\nu_1, \nu_2, \ldots, \nu_p)'$, where $\nu_i = (\mu_i - \mu_{0i}) / \sigma_{0i}$, $i = 1, 2, \ldots, p$, and the size of the shift is measured in terms of the non-centrality parameter

$$\delta = \sqrt{\mathbf{v}'\boldsymbol{\Sigma}_{Z0}^{-1}\mathbf{v}} \ .$$

The out-of-control properties of the SZ and MZ charts depend on $\mathbf{v}$ only through δ, but this is not true for the other charts being considered. We initially consider (in Tables 2-7) an average SSATS averaged over all shift directions (obtained by simulating random shift directions), and later give (in Tables 8 and 9) the SSATS for some specific shift directions.

The numerical results presented here for shifts in $\boldsymbol{\Sigma}$ are for increases in $\boldsymbol{\sigma}$, with the assumption that the correlations between the variables do not change. Shifts in $\boldsymbol{\sigma}$ are expressed in terms of $\boldsymbol{\gamma} = (\gamma_1, \gamma_2, \ldots, \gamma_p)'$, where $\gamma_i = \sigma_i / \sigma_{0i}$, $i = 1, 2, \ldots, p$, and the size of the shift in $\boldsymbol{\sigma}$ is measured in terms of

$$\psi = 1 + \sqrt{(\boldsymbol{\gamma} - \mathbf{1})'(\boldsymbol{\gamma} - \mathbf{1})} = 1 + (\Sigma_{i=1}^{p} (\gamma_i - 1)^2)^{1/2}.$$

The in-control case of $\boldsymbol{\sigma} = \boldsymbol{\sigma}_0$ corresponds to $\boldsymbol{\gamma} = \mathbf{1}$ and $\psi = 1$. The SSATS of the charts considered here depends on the direction of the shift $\boldsymbol{\gamma}$, so an average SSATS is used initially, where the average is taken over all shift directions corresponding to increases in one or more components of $\boldsymbol{\sigma}$. The SSATS is also given later for some specific shift directions.

4 Numerical Results

Table 2 gives average SSATS values for some charts and chart combinations when $p = 4$, $n = 1$, $\rho = 0$, and $\lambda = 0.02600$. The first row of the table gives the in-control ATS (approximately 800), and the following rows give SSATS values for shifts in $\boldsymbol{\mu}$ indexed by δ, and shifts in $\boldsymbol{\sigma}$ indexed by ψ.

The column labeled [1] is for the USZ chart (the set of univariate Shewhart charts for individuals). We see that, as expected, this chart is not effective for detecting small shifts in $\boldsymbol{\mu}$. We also see that it effectively detects large increases in $\boldsymbol{\sigma}$. From column [2] for the UEZ chart we see that this chart is much more effective that the USZ chart for small shifts in $\boldsymbol{\mu}$, but is not as effective for large shifts in $\boldsymbol{\mu}$ or for increases in $\boldsymbol{\sigma}$. From column [3] for the UERZ^2 chart we see that this chart is effective for detecting increases in $\boldsymbol{\sigma}$, and is also very effective for detecting large shifts in $\boldsymbol{\mu}$. In fact, the UERZ^2 chart in column [3] is much more effective than the UEZ chart in column [2] for large shifts in $\boldsymbol{\mu}$.

Column [4] in Table 2 is for the UEZ and USZ combination. We see that using the Shewhart and EWMA charts together gives good performance over a wide range of shifts in $\boldsymbol{\mu}$. Column [5] is for the UEZ and UERZ^2 combination. This combination also gives good performance over a wide range of shifts in $\boldsymbol{\mu}$, and gives much better performance than the UEZ and USZ combination for small increases in $\boldsymbol{\sigma}$. Thus, to improve the performance of the EWMA chart for $\boldsymbol{\mu}$ in detecting large shifts, this EWMA chart can be used either with the Shewhart chart or with the EWMA chart of squared deviations. The EWMA chart of squared deviations seems to be a better choice because it has much better performance for detecting small increases in $\boldsymbol{\sigma}$.

These conclusions obtained from columns [1]-[5] in Table 2 are similar to the conclusions obtained by Reynolds and Stoumbos (2005) in the univariate case. Thus, we see that these conclusions carry over to the multivariate case when using sets of univariate charts for each variable.

Columns [6]-[8] of Table 2 are for the SZ, MZ, and M_2RZ^2 charts, respectively. The conclusions from columns [6]-[8] are similar to the conclusions from columns [1]-[3]. In particular, the Shewhart chart is good for large shifts in $\boldsymbol{\mu}$ or large increases in $\boldsymbol{\sigma}$, the EWMA chart is good for small shifts in $\boldsymbol{\mu}$, and the squared deviation chart is good for large shifts in $\boldsymbol{\mu}$ and any increase in $\boldsymbol{\sigma}$.

Table 2. Average SSATS for Some Control Chart Combinations for Shifts in $\boldsymbol{\mu}$ and $\boldsymbol{\sigma}$ when $p = 4$, $n = 1$, $\rho = 0$, and $\lambda = 0.02600$.

		USZ	UEZ	UERZ^2	UEZ USZ	UEZ UERZ^2	SZ	MZ	M_2RZ^2	MZ SZ	MZ M_2RZ^2
δ	ψ	[1]	[2]	[3]	[4]	[5]	[6]	[7]	[8]	[9]	[10]
0	1.0	799.9	800.1	800.1	800.2	800.1	800.0	800.0	800.0	800.0	800.0
0.2	1.0	743.7	206.8	669.3	263.3	259.6	736.1	187.0	634.2	235.0	230.1
0.4	1.0	614.1	69.4	419.1	82.9	82.0	590.5	61.5	366.8	71.5	70.6
0.6	1.0	463.2	38.6	202.1	43.9	43.4	427.4	34.1	171.1	38.1	37.7
0.8	1.0	329.5	26.4	94.4	29.4	29.0	290.8	23.4	80.1	25.6	25.2
1.0	1.0	226.7	20.0	50.0	22.0	21.5	190.8	17.7	42.9	19.2	18.8
1.2	1.0	154.4	16.1	30.4	17.5	16.9	123.9	14.3	26.3	15.3	14.8
1.4	1.0	105.2	13.4	20.5	14.4	13.8	80.2	11.9	17.9	12.6	12.0
1.6	1.0	72.2	11.5	14.9	12.2	11.4	52.5	10.2	13.0	10.6	10.0
2.0	1.0	35.1	9.0	8.9	9.2	8.1	23.4	8.0	7.8	7.8	7.1
2.5	1.0	15.5	7.0	5.5	6.6	5.6	9.5	6.2	4.8	5.5	4.9
3.0	1.0	7.5	5.8	3.7	4.8	4.0	4.3	5.1	3.2	3.7	3.4
4.0	1.0	2.3	4.2	2.0	2.3	2.2	1.3	3.8	1.8	1.5	1.9
5.0	1.0	1.0	3.4	1.3	1.1	1.4	0.7	3.0	1.1	0.7	1.2
8.0	1.0	0.5	2.1	0.5	0.5	0.6	0.5	1.8	0.5	0.5	0.5
0	1.2	247.6	378.7	92.8	268.5	113.5	224.3	363.0	72.2	246.7	87.8
0	1.4	91.6	213.9	32.5	105.5	37.9	80.7	203.1	26.0	94.4	30.0
0	1.6	42.2	136.3	18.2	50.0	20.8	36.5	129.3	14.8	43.8	16.7
0	1.8	23.1	95.0	12.3	27.6	13.9	19.6	90.0	10.1	23.8	11.3
0	2.0	14.3	70.6	9.1	17.1	10.2	12.0	66.8	7.5	14.5	8.4
0	2.5	6.1	40.1	5.3	7.2	5.8	5.1	37.9	4.4	6.0	4.9
0	3.0	3.5	26.6	3.6	4.1	4.0	3.0	25.1	3.1	3.4	3.4
0	4.0	1.8	14.8	2.1	2.0	2.3	1.5	14.0	1.9	1.7	2.0
0	5.0	1.2	9.8	1.5	1.3	1.6	1.1	9.2	1.4	1.1	1.5
0	10.0	0.6	3.0	0.7	0.6	0.7	0.6	2.8	0.7	0.6	0.7

Column [9] in Table 2 is for the MZ and SZ chart combination, and column [10] is for the MZ and M_2RZ^2 chart combination. As in the comparison of columns [4] and [5], comparing columns [9] and [10] shows that the MZ and M_2RZ^2 chart combination and the MZ and SZ chart combination have roughly similar performance, except that the MZ and SZ chart combination is a little better for very large shifts in $\boldsymbol{\mu}$, and the MZ and M_2RZ^2 chart combination is much better for small increases in $\boldsymbol{\sigma}$.

Comparing a column from [1]-[5] for the sets of univariate charts with the corresponding column from [6]-[10] for the multivariate charts shows that the multivariate charts are uniformly better in terms of average performance. Thus, even though the variables are independent, using the multivariate charts gives better performance than using sets of univariate charts. However, the differences are not extremely large. For example, the UEZ and UEZ^2 chart combination requires an average of 21.5 time units to detect a shift in $\boldsymbol{\mu}$ of size $\delta = 1.0$, while the MZ and M_2RZ^2 chart combination requires an average of 18.8 time units. Some practitioners prefer using sets of univariate charts because these charts are familiar and easier to interpret. The results given here show that using sets of univariate charts entails a moderate loss of efficiency compared to using multivariate charts.

We next consider the case of $\rho = 0.9$ in Table 3, where SSATS values are given for four pairs of charts, where each pair has charts with and without regression adjustment of the variables. In particular, column [1] is for the USZ chart (based on the original standardized variables) and column [2] is for the USA chart (based on the regression adjusted variables). We see that using regression adjusted variables gives much better average performance. The same conclusion is obtained for the other three pairs of columns in Table 3. Thus, when $\rho = 0.9$, we use regression adjustment of the variables in future comparisons of charts and chart combinations.

Table 4 has the same structure as Table 2, except that $\rho = 0.9$ and regression adjusted variables are used instead of the original variables (except for the SZ and MZ charts, which are not affected by regression adjustment).

The conclusions obtained from Table 4 are similar to the conclusions from Table 2. In particular, the performance of the set of univariate EWMA charts for $\boldsymbol{\mu}$ in detecting large shifts can be improved by using it with either with the set of Shewhart charts or with the set of EWMA charts of squared deviations. The set of EWMA charts of squared deviations have much better performance for detecting small increases in $\boldsymbol{\sigma}$, so it seems to be a better choice. The same conclusion is obtained when the MZ and M_2RA^2 chart combination is compared to the MZ and SZ chart combination: the M_2RA^2 chart based on squared deviations gives much better performance for detecting small increases in $\boldsymbol{\sigma}$.

In Table 4 the conclusion about the average performance of the sets of univariate charts relative to the multivariate charts is similar to the conclusion from Table 2 where $\rho = 0$. In particular, using sets of univariate charts (based on regression adjusted variables) entails a moderate loss of efficiency compared to using multivariate charts. However, plotting regression adjusted variables in the sets of univariate charts may reduce the advantage of easy interpretation.

Table 3. Average SSATS for Some Control Charts with and without Regression Adjustment for Shifts in $\boldsymbol{\mu}$ and $\boldsymbol{\sigma}$ when $p = 4$, $n = 1$, $\rho = 0.9$, and $\lambda = 0.02600$.

δ	ψ	USZ [1]	USA [2]	UEZ [3]	UEA [4]	$UERZ^2$ [5]	$UERA^2$ [6]	M_2RZ^2 [7]	M_2RA^2 [8]
0	1.0	800.0	800.0	800.1	800.0	800.0	800.0	800.0	800.0
0.2	1.0	789.1	731.8	504.0	182.7	762.1	644.6	763.7	617.5
0.4	1.0	756.8	577.2	254.4	61.7	709.8	363.7	715.5	332.1
0.6	1.0	710.1	410.3	143.0	34.7	630.9	158.4	642.3	145.6
0.8	1.0	653.1	276.2	91.6	23.9	535.5	71.9	552.2	67.1
1.0	1.0	590.9	181.6	65.0	18.1	436.2	38.6	456.1	36.3
1.2	1.0	528.4	118.6	49.8	14.6	344.6	24.0	362.9	22.6
1.4	1.0	468.0	78.1	40.2	12.2	266.6	16.5	281.5	15.5
1.6	1.0	411.1	52.1	33.6	10.5	204.0	12.1	214.2	11.3
2.0	1.0	313.0	24.4	25.3	8.2	119.6	7.3	123.4	6.8
2.5	1.0	219.9	10.5	19.3	6.4	65.0	4.5	65.2	4.2
3.0	1.0	154.5	5.0	15.6	5.3	38.8	3.1	38.3	2.9
4.0	1.0	78.0	1.6	11.3	3.9	18.0	1.7	17.4	1.6
5.0	1.0	41.2	0.8	8.8	3.1	10.4	1.1	10.1	1.0
8.0	1.0	8.0	0.5	5.3	1.9	3.7	0.5	3.5	0.5
0	1.2	261.9	210.0	418.3	351.8	108.6	78.2	111.9	67.9
0	1.4	99.3	62.2	244.5	173.9	38.2	24.8	38.2	21.5
0	1.6	47.3	24.9	158.9	100.0	21.3	13.1	21.2	11.4
0	1.8	26.8	12.7	112.3	64.9	14.4	8.4	14.1	7.3
0	2.0	17.1	7.6	84.2	45.8	10.7	6.0	10.5	5.3
0	2.5	7.8	3.3	48.4	24.1	6.3	3.3	6.2	3.0
0	3.0	4.7	2.0	32.2	15.3	4.4	2.3	4.3	2.0
0	4.0	2.6	1.1	18.0	8.1	2.7	1.4	2.7	1.3
0	5.0	1.8	0.9	11.9	5.2	2.0	1.0	2.0	1.0
0	10.0	0.8	0.6	3.7	1.7	1.0	0.6	0.9	0.6

Tables 5 and 6 show the same charts as Tables 2 and 4, respectively, except that $\lambda = 0.11989$ instead of $\lambda = 0.02600$. We see that when the charts for monitoring $\boldsymbol{\mu}$ are based on a larger value of λ, they have better performance for large shifts, so the improvement in performance from adding a Shewhart chart or a chart based on squared deviations from target is not quite as dramatic as in the case of $\lambda = 0.02600$. However, the improvement in performance is still significant.

Table 4. Average SSATS for Some Control Chart Combinations with Regression Adjustment for Shifts in $\boldsymbol{\mu}$ and $\boldsymbol{\sigma}$ when $p=4$, $n=1$, $d=1$, $\rho=0.9$, and $\lambda=0.02600$.

		USA	UEA	$UERA^2$	UEA USA	UEA $UERA^2$	SZ	MZ	M_2RA^2	MZ SZ	MZ M_2RA^2
δ	ψ	[1]	[2]	[3]	[4]	[5]	[6]	[7]	[8]	[9]	[10]
0	1.0	800.0	800.0	800.0	800.0	800.0	800.0	800.0	800.0	800.0	800.0
0.2	1.0	731.8	182.7	644.6	230.5	226.9	736.1	187.0	617.5	235.0	230.4
0.4	1.0	577.2	61.7	363.7	72.6	71.9	590.5	61.5	332.1	71.5	70.5
0.6	1.0	410.3	34.7	158.4	39.1	38.7	427.4	34.1	145.6	38.1	37.5
0.8	1.0	276.2	23.9	71.9	26.4	26.0	290.8	23.4	67.1	25.6	25.1
1.0	1.0	181.6	18.1	38.6	19.8	19.3	190.8	17.7	36.3	19.2	18.6
1.2	1.0	118.6	14.6	24.0	15.7	15.1	123.9	14.3	22.6	15.3	14.5
1.4	1.0	78.1	12.2	16.5	13.0	12.2	80.2	11.9	15.5	12.6	11.7
1.6	1.0	52.1	10.5	12.1	10.9	10.1	52.5	10.2	11.3	10.6	9.6
2.0	1.0	24.4	8.2	7.3	8.1	7.0	23.4	8.0	6.8	7.8	6.7
2.5	1.0	10.5	6.4	4.5	5.7	4.7	9.5	6.2	4.2	5.5	4.4
3.0	1.0	5.0	5.3	3.1	3.9	3.3	4.3	5.1	2.9	3.7	3.1
4.0	1.0	1.6	3.9	1.7	1.7	1.9	1.3	3.8	1.6	1.5	1.7
5.0	1.0	0.8	3.1	1.1	0.9	1.2	0.7	3.0	1.0	0.7	1.1
8.0	1.0	0.5	1.9	0.5	0.5	0.5	0.5	1.8	0.5	0.5	0.5
0	1.2	210.0	351.8	78.2	230.9	94.9	165.5	317.3	67.9	185.8	81.2
0	1.4	62.2	173.9	24.8	72.6	28.6	45.1	152.7	21.5	53.3	24.6
0	1.6	24.9	100.0	13.1	29.7	14.8	18.1	88.4	11.4	21.5	12.7
0	1.8	12.7	64.9	8.4	15.1	9.4	9.5	57.9	7.3	11.1	8.2
0	2.0	7.6	45.8	6.0	8.9	6.7	5.9	41.3	5.3	6.8	5.8
0	2.5	3.3	24.1	3.3	3.7	3.7	2.7	22.1	3.0	3.0	3.3
0	3.0	2.0	15.3	2.3	2.2	2.5	1.7	14.1	2.0	1.9	2.2
0	4.0	1.1	8.1	1.4	1.2	1.5	1.0	7.6	1.3	1.1	1.4
0	5.0	0.9	5.2	1.0	0.9	1.1	0.8	4.9	1.0	0.8	1.0
0	10.0	0.6	1.7	0.6	0.6	0.6	0.6	1.6	0.6	0.6	0.6

If a Shewhart chart or a squared deviation chart is going to be relied upon to provide fast detection of large shifts in $\boldsymbol{\mu}$, then this suggests that λ in the EWMA or MEWMA chart for $\boldsymbol{\mu}$ can be taken to be relatively small to also provide fast detection of small shifts in $\boldsymbol{\mu}$. Table 7 contains SSATS values for some chart combinations when the charts for $\boldsymbol{\mu}$ have $\lambda=0.02600$, and the charts for $\boldsymbol{\sigma}$ have $\lambda=0.11989$. Results for both $\rho=0$ and $\rho=0.9$ are given, and for each value of ρ the lowest SSATS for each shift is shown in bold. From Table 7 we see that using $\lambda=0.11989$ in the charts based on squared deviations improves the ability to detect large shifts to the point where there is little difference between the combinations with the squared deviation charts and the combinations with the Shewhart charts. This increase in λ does, of course, decrease the ability to detect small increases in $\boldsymbol{\sigma}$, and also slightly decrease the ability to detect intermediate shifts in $\boldsymbol{\mu}$. However, the combinations with the squared deviation charts are still a bit better than the combinations with the Shewhart charts for detecting small shifts in $\boldsymbol{\mu}$, and much better for detecting small increases in $\boldsymbol{\sigma}$.

Table 5. Average SSATS for Some Control Chart Combinations for Shifts in $\boldsymbol{\mu}$ and $\boldsymbol{\sigma}$ when $p = 4$, $n = 1$, $\rho = 0$, and the MZ chart has $\lambda = 0.11989$.

δ	ψ	USZ [1]	UEZ [2]	UERZ^2 [3]	UEZ USZ [4]	UEZ UERZ^2 [5]	SZ [6]	MZ [7]	M_2RZ^2 [8]	MZ SZ [9]	MZ M_2RZ^2 [10]
0	1.0	799.9	800.1	800.0	800.0	800.0	800.0	799.9	800.0	799.9	800.0
0.2	1.0	743.7	411.8	721.3	502.4	497.5	736.1	378.2	699.2	468.5	462.1
0.4	1.0	614.1	136.5	535.2	186.1	183.5	590.5	114.3	485.8	153.5	151.1
0.6	1.0	463.2	55.2	328.7	72.1	71.3	427.4	44.6	278.1	56.1	55.4
0.8	1.0	329.5	28.8	177.2	35.3	35.0	290.8	23.4	141.9	27.6	27.3
1.0	1.0	226.7	18.2	92.2	21.3	21.1	190.8	15.0	71.0	16.9	16.8
1.2	1.0	154.4	13.0	49.6	14.7	14.6	123.9	10.8	37.7	11.9	11.8
1.4	1.0	105.2	10.0	28.7	11.1	11.0	80.2	8.4	22.0	9.1	9.0
1.6	1.0	72.2	8.1	18.2	8.8	8.7	52.5	6.9	14.1	7.3	7.2
2.0	1.0	35.1	5.9	9.0	6.2	6.0	23.4	5.0	7.1	5.2	5.0
2.5	1.0	15.5	4.3	4.9	4.4	4.2	9.5	3.7	3.9	3.6	3.5
3.0	1.0	7.5	3.4	3.1	3.3	3.0	4.3	3.0	2.5	2.6	2.5
4.0	1.0	2.3	2.4	1.6	1.8	1.7	1.3	2.1	1.3	1.3	1.4
5.0	1.0	1.0	1.9	1.0	1.0	1.0	0.7	1.6	0.8	0.7	0.9
8.0	1.0	0.5	1.1	0.5	0.5	0.5	0.5	1.0	0.5	0.5	0.5
0	1.2	247.6	303.6	154.5	245.9	179.5	224.3	281.8	118.6	224.7	143.0
0	1.4	91.6	138.3	45.9	91.9	54.1	80.7	127.0	34.2	82.1	40.7
0	1.6	42.2	75.7	21.0	43.0	24.2	36.5	69.0	16.0	37.9	18.4
0	1.8	23.1	47.3	12.4	23.9	14.0	19.6	43.0	9.7	20.8	10.9
0	2.0	14.3	32.5	8.4	15.1	9.4	12.0	29.6	6.7	12.9	7.5
0	2.5	6.1	16.5	4.5	6.6	4.9	5.1	15.0	3.7	5.6	4.0
0	3.0	3.5	10.4	2.9	3.8	3.2	3.0	9.4	2.5	3.2	2.7
0	4.0	1.8	5.5	1.7	2.0	1.9	1.5	5.0	1.5	1.7	1.6
0	5.0	1.2	3.6	1.2	1.3	1.3	1.1	3.2	1.1	1.1	1.2
0	10.0	0.6	1.2	0.6	0.6	0.7	0.6	1.1	0.6	0.6	0.6

Comparing columns [4] and [8] of Table 7 with the corresponding columns of Tables 5 and 6 shows that using $\lambda = 0.02600$ in the chart for $\boldsymbol{\mu}$ and $\lambda = 0.11989$ in the chart for $\boldsymbol{\sigma}$, as opposed to using $\lambda = 0.11989$ in both charts, significantly improves the ability to detect small increases in $\boldsymbol{\mu}$ ($\delta \leq 0.8$), but hurts the ability to detect other shifts in $\boldsymbol{\mu}$ or $\boldsymbol{\sigma}$. Thus, if small shifts in $\boldsymbol{\mu}$ are of particular concern, then it is better to use the small value of λ in the chart for $\boldsymbol{\mu}$.

Table 6. Average SSATS for Some Control Chart Combinations with Regression Adjustment for Shifts in $\boldsymbol{\mu}$ and $\boldsymbol{\sigma}$ when $p = 4$, $n = 1$, $\rho = 0.9$, and $\lambda = 0.11989$.

		USA	UEA	UERA^2	UEA USA	UEA & UERA^2	SZ	MZ	M$_2$RA^2	MZ SZ	MZ M$_2$RA^2
δ	ψ	[1]	[2]	[3]	[4]	[5]	[6]	[7]	[8]	[9]	[10]
0	1.0	800.0	800.0	800.1	799.9	800.1	800.0	800.0	800.0	800.0	800.2
0.2	1.0	731.8	364.2	702.2	454.2	450.1	736.1	378.2	685.1	468.5	461.1
0.4	1.0	577.2	111.7	483.6	150.3	148.7	590.5	114.3	452.0	153.5	150.7
0.6	1.0	410.3	45.0	269.5	57.4	56.8	427.4	44.6	242.3	56.1	55.3
0.8	1.0	276.2	24.1	133.3	28.9	28.6	290.8	23.4	117.4	27.6	27.2
1.0	1.0	181.6	15.6	65.9	17.9	17.7	190.8	15.0	57.6	16.9	16.7
1.2	1.0	118.6	11.3	35.1	12.6	12.5	123.9	10.8	30.6	11.9	11.7
1.4	1.0	78.1	8.8	20.7	9.6	9.5	80.2	8.4	18.2	9.1	8.9
1.6	1.0	52.1	7.2	13.5	7.7	7.6	52.5	6.9	11.9	7.3	7.1
2.0	1.0	24.4	5.2	7.0	5.5	5.2	23.4	5.0	6.2	5.2	4.9
2.5	1.0	10.5	3.9	3.9	3.8	3.6	9.5	3.7	3.5	3.6	3.3
3.0	1.0	5.0	3.1	2.5	2.8	2.5	4.3	3.0	2.3	2.6	2.3
4.0	1.0	1.6	2.2	1.3	1.5	1.4	1.3	2.1	1.2	1.3	1.3
5.0	1.0	0.8	1.7	0.8	0.8	0.9	0.7	1.6	0.7	0.7	0.8
8.0	1.0	0.5	1.0	0.5	0.5	0.5	0.5	1.0	0.5	0.5	0.5
0	1.2	210.0	269.4	126.7	209.1	148.1	165.5	227.3	109.8	165.7	125.0
0	1.4	62.2	103.1	31.7	63.0	37.0	45.1	83.3	26.8	46.2	30.3
0	1.6	24.9	50.4	13.6	25.9	15.4	18.1	41.0	11.6	19.0	12.8
0	1.8	12.7	29.6	7.9	13.4	8.7	9.5	24.5	6.8	10.1	7.4
0	2.0	7.6	19.5	5.3	8.1	5.8	5.9	16.5	4.6	6.3	5.0
0	2.5	3.3	9.5	2.8	3.5	3.0	2.7	8.3	2.5	2.9	2.7
0	3.0	2.0	5.8	1.9	2.1	2.0	1.7	5.2	1.7	1.8	1.8
0	4.0	1.1	3.1	1.2	1.2	1.2	1.0	2.8	1.1	1.1	1.1
0	5.0	0.9	2.0	0.9	0.9	0.9	0.8	1.9	0.8	0.8	0.9
0	10.0	0.6	0.8	0.6	0.6	0.6	0.6	0.8	0.6	0.6	0.6

The performance of the MZ and SZ charts does not depend on the direction of the shift in $\boldsymbol{\mu}$, but it does depend on the direction of the shift in $\boldsymbol{\sigma}$. The performance of the other charts considered here depends on the direction of the shift in $\boldsymbol{\mu}$ and the shift in $\boldsymbol{\sigma}$.

Table 7. Average SSATS for Some Control Chart Combinations with Regression Adjustment for Shifts in **μ** and **σ** when $p = 4$, $n = 1$, $\rho = 0$ or 0.9, the Charts for **μ** have $\lambda = 0.02600$, and the Charts for **σ** have $\lambda = 0.11989$.

		ρ = 0				ρ = 0.9			
		UEZ USZ	UEZ $UERZ^2$	MZ SZ	MZ M_2RZ^2	UEA USA	UEA $UERA^2$	MZ SZ	MZ M_2RA^2
δ	ψ	[1]	[2]	[3]	[4]	[5]	[6]	[7]	[8]
0	1.0	800.2	800.0	800.0	800.0	800.0	800.1	800.0	800.0
0.2	1.0	263.3	262.6	235.0	**233.2**	230.5	**229.5**	235.0	233.1
0.4	1.0	82.9	82.5	71.5	**71.2**	72.6	72.4	71.5	**71.1**
0.6	1.0	43.9	43.7	38.1	**37.9**	39.1	38.9	38.1	**37.8**
0.8	1.0	29.4	29.1	25.6	**25.3**	26.4	26.1	25.6	**25.2**
1.0	1.0	22.0	21.6	19.2	**18.8**	19.8	19.5	19.2	**18.6**
1.2	1.0	17.5	17.0	15.3	**14.7**	15.7	15.1	15.3	**14.4**
1.4	1.0	14.4	13.7	12.6	**11.9**	13.0	12.1	12.6	**11.5**
1.6	1.0	12.2	11.3	10.6	**9.7**	10.9	9.8	10.6	**9.2**
2.0	1.0	9.2	7.8	7.8	**6.6**	8.1	6.6	7.8	**6.1**
2.5	1.0	6.6	5.0	5.5	**4.1**	5.7	4.1	5.5	**3.7**
3.0	1.0	4.8	3.4	3.7	**2.7**	3.9	2.7	3.7	**2.5**
4.0	1.0	2.3	1.8	1.5	**1.4**	1.7	1.4	1.5	**1.3**
5.0	1.0	1.1	1.1	**0.7**	0.9	0.9	0.9	**0.7**	0.8
8.0	1.0	**0.5**	**0.5**	**0.5**	**0.5**	**0.5**	**0.5**	**0.5**	**0.5**
0	1.2	268.5	188.3	246.7	**148.8**	230.9	155.7	185.8	**134.1**
0	1.4	105.5	56.9	94.4	**42.1**	72.6	38.7	53.3	**32.2**
0	1.6	50.0	25.2	43.8	**18.9**	29.7	16.0	21.5	**13.4**
0	1.8	27.6	14.5	23.8	**11.1**	15.1	9.0	11.1	**7.6**
0	2.0	17.1	9.7	14.5	**7.6**	8.9	6.0	6.8	**5.1**
0	2.5	7.2	5.0	6.0	**4.1**	3.7	3.1	3.0	**2.7**
0	3.0	4.1	3.3	3.4	**2.7**	2.2	2.0	1.9	**1.8**
0	4.0	2.0	1.9	1.7	**1.6**	1.2	1.2	**1.1**	**1.1**
0	5.0	1.3	1.3	**1.1**	1.2	0.9	0.9	**0.8**	0.9
0	10.0	**0.6**	0.7	**0.6**	**0.6**	**0.6**	**0.6**	**0.6**	**0.6**

Table 8 gives SSATS values for some charts with and without regression adjustment for some specific shift directions when $\rho = 0.9$. In Table 8 the shift direction is determined by using "+", "0", and "-" to indicate that a parameter has increased, stayed constant, or decreased, respectively. For example, "+ + + +" for

Table 8. SSATS for Some Control Charts with and without Regression Adjustment for Specific Shift Directions for $\boldsymbol{\mu}$ and $\boldsymbol{\sigma}$ when $p = 4$, $n = 1$, $\rho = 0.9$, and $\lambda = 0.02600$.

δ	ψ	means	st dev	USZ [1]	USA [2]	UEZ [3]	UEA [4]	$UERZ^2$ [5]	$UERA^2$ [6]	M_2RZ^2 [7]	M_2RA^2 [8]
0	1.0	0 0 0 0	0 0 0 0	800.0	800.0	800.1	800.0	800.0	800.0	800.0	800.0
0.4	1.0	+ + + +	0 0 0 0	408.4	791.5	44.2	535.2	212.9	760.1	210.7	757.3
1.0	1.0	+ + + +	0 0 0 0	80.0	751.0	13.8	206.1	21.8	674.6	21.4	684.0
2.0	1.0	+ + + +	0 0 0 0	9.1	634.4	6.3	71.2	4.6	483.2	4.5	482.4
3.0	1.0	+ + + +	0 0 0 0	2.0	498.3	4.1	40.6	2.0	289.8	1.9	283.1
5.0	1.0	+ + + +	0 0 0 0	0.5	278.8	2.4	21.8	0.7	89.5	0.7	79.4
0.4	1.0	+ + + 0	0 0 0 0	723.9	584.3	193.5	61.4	650.3	361.2	651.8	340.6
1.0	1.0	+ + + 0	0 0 0 0	472.7	181.3	50.7	17.7	276.7	34.1	261.2	36.8
2.0	1.0	+ + + 0	0 0 0 0	181.2	21.6	20.4	7.9	51.8	6.5	43.4	7.0
3.0	1.0	+ + + 0	0 0 0 0	70.4	4.1	12.6	5.1	18.4	2.8	15.3	1.9
5.0	1.0	+ + + 0	0 0 0 0	14.0	0.7	7.1	3.0	5.8	1.0	4.9	1.0
0.4	1.0	+ + 0 0	0 0 0 0	752.7	582.6	241.7	69.5	700.0	405.1	704.7	348.0
1.0	1.0	+ + 0 0	0 0 0 0	564.4	208.8	63.0	20.8	380.8	58.7	380.6	42.3
2.0	1.0	+ + 0 0	0 0 0 0	264.8	38.0	24.5	9.4	79.7	10.6	69.0	7.5
3.0	1.0	+ + 0 0	0 0 0 0	114.8	9.3	15.0	6.1	26.5	4.4	22.3	3.1
5.0	1.0	+ + 0 0	0 0 0 0	25.0	1.2	8.4	3.5	8.0	1.5	6.7	1.1
0.4	1.0	+ 0 0 0	0 0 0 0	762.0	570.3	233.1	56.1	715.9	328.2	724.2	318.2
1.0	1.0	+ 0 0 0	0 0 0 0	596.6	160.5	56.4	16.3	377.0	28.5	419.4	32.3
2.0	1.0	+ 0 0 0	0 0 0 0	277.7	16.7	22.0	7.4	63.0	5.6	66.4	6.3
3.0	1.0	+ 0 0 0	0 0 0 0	109.7	3.1	13.5	4.8	20.8	2.4	21.0	2.7
5.0	1.0	+ 0 0 0	0 0 0 0	19.8	0.6	7.6	2.8	6.4	0.8	6.4	0.9
0.4	1.0	+ + - -	0 0 0 0	770.1	578.3	314.2	68.3	733.7	399.0	738.6	341.6
1.0	1.0	+ + - -	0 0 0 0	640.8	204.0	93.6	20.6	551.6	56.9	558.6	40.8
2.0	1.0	+ + - -	0 0 0 0	399.5	36.6	35.8	9.3	249.3	10.3	231.3	7.3
3.0	1.0	+ + - -	0 0 0 0	240.1	8.8	22.0	6.0	101.0	4.3	86.5	3.0
5.0	1.0	+ + - -	0 0 0 0	91.2	1.2	12.3	3.5	24.9	1.5	21.8	1.0
0.4	1.0	+ - 0 0	0 0 0 0	769.5	572.5	274.1	59.0	732.2	350.7	737.9	323.6
1.0	1.0	+ - 0 0	0 0 0 0	634.5	171.8	67.4	17.4	512.3	34.7	542.1	33.6
2.0	1.0	+ - 0 0	0 0 0 0	369.3	21.1	26.5	7.9	155.1	6.7	173.6	6.5
3.0	1.0	+ - 0 0	0 0 0 0	193.5	4.2	16.4	5.1	47.8	2.8	49.2	2.7
5.0	1.0	+ - 0 0	0 0 0 0	53.8	0.7	9.3	2.9	12.3	1.0	11.6	0.9
0	1.4	0 0 0 0	+ + + +	118.7	95.6	257.6	211.4	51.9	36.3	51.9	27.2
0	2.0	0 0 0 0	+ + + +	23.9	16.2	98.3	72.9	14.3	10.1	14.3	7.8
0	3.0	0 0 0 0	+ + + +	6.3	3.8	39.2	28.1	5.7	3.9	5.7	3.2
0	5.0	0 0 0 0	+ + + +	2.1	1.2	14.6	10.4	2.4	1.5	2.3	1.4
0	1.4	0 0 0 0	+ + 0 0	97.5	52.0	244.4	157.3	37.7	21.2	34.4	18.9
0	2.0	0 0 0 0	+ + 0 0	16.4	5.3	81.6	36.4	10.5	4.8	9.4	4.1
0	3.0	0 0 0 0	+ + 0 0	4.5	1.5	31.0	11.6	4.3	1.8	3.9	1.6
0	5.0	0 0 0 0	+ + 0 0	1.7	0.8	11.4	3.9	1.9	0.9	1.8	0.8
0	1.4	0 0 0 0	+ 0 0 0	75.9	24.9	227.1	119.8	27.8	13.0	28.8	15.3
0	2.0	0 0 0 0	+ 0 0 0	11.7	3.2	68.7	24.5	8.0	3.2	8.0	3.6
0	3.0	0 0 0 0	+ 0 0 0	3.6	1.3	25.5	7.8	3.5	1.5	3.5	1.6
0	5.0	0 0 0 0	+ 0 0 0	1.6	0.8	9.4	2.8	1.7	0.9	1.7	0.9

Table 9. SSATS for Some Control Chart Combinations with Regression Adjustment for Specific Shift Directions for $\boldsymbol{\mu}$ and $\boldsymbol{\sigma}$ when $p = 4$, $n = 1$, $\rho = 0$ or 0.9, the Charts for $\boldsymbol{\mu}$ have $\lambda = 0.02600$, and the Charts for $\boldsymbol{\sigma}$ have $\lambda = 0.11989$.

				ρ = 0				ρ = 0.9			
				UEZ	UEZ	MZ	MZ	UEA	UEA	MZ	MZ
				USZ	$UERZ^2$	SZ	M_2RZ^2	USA	$UERA^2$	SZ	M_2RA^2
δ	ψ	means	st dev	[1]	[2]	[3]	[4]	[5]	[6]	[7]	[8]
0	1.0	0000	0000	800.2	800.0	800.0	800.0	800.0	800.1	800.0	800.0
0.4	1.0	++++	0000	96.8	96.4	71.5	**71.2**	621.7	619.3	**71.5**	71.8
1.0	1.0	++++	0000	25.8	25.6	19.2	**18.9**	281.7	280.6	**19.2**	19.5
2.0	1.0	++++	0000	11.0	10.2	7.8	**7.1**	91.0	90.7	**7.8**	8.7
3.0	1.0	++++	0000	6.4	5.0	3.7	**3.1**	48.5	48.3	**3.7**	5.5
5.0	1.0	++++	0000	2.0	1.6	**0.7**	1.0	24.8	24.5	**0.7**	3.2
0.4	1.0	+++0	0000	90.3	89.9	71.5	**71.2**	71.3	**71.0**	71.5	**71.0**
1.0	1.0	+++0	0000	24.0	23.7	19.2	**18.9**	19.1	18.7	19.2	**18.6**
2.0	1.0	+++0	0000	10.2	9.1	7.8	**6.9**	7.8	**6.1**	7.8	**6.1**
3.0	1.0	+++0	0000	5.7	4.2	3.7	**2.9**	3.6	**2.4**	3.7	2.5
5.0	1.0	+++0	0000	1.5	1.3	**0.7**	0.9	**0.7**	0.8	**0.7**	0.8
0.4	1.0	++00	0000	81.1	80.8	71.5	**71.2**	84.4	84.1	71.5	**71.0**
1.0	1.0	++00	0000	21.6	21.2	19.2	**18.8**	23.2	22.8	19.2	**18.8**
2.0	1.0	++00	0000	9.0	7.6	7.8	**6.5**	9.8	8.7	7.8	**6.8**
3.0	1.0	++00	0000	4.7	3.2	3.7	**2.7**	5.4	4.0	3.7	**2.9**
5.0	1.0	++00	0000	1.0	1.0	**0.7**	0.8	1.5	1.3	**0.7**	0.9
0.4	1.0	+000	0000	66.5	**66.3**	71.5	71.1	64.5	**64.2**	71.5	71.0
1.0	1.0	+000	0000	17.9	**17.3**	19.2	18.5	17.6	**17.1**	19.2	18.4
2.0	1.0	+000	0000	7.1	**5.3**	7.8	5.7	7.0	**5.2**	7.8	5.6
3.0	1.0	+000	0000	3.1	**2.1**	3.7	2.3	3.0	**2.1**	3.7	2.2
5.0	1.0	+000	0000	**0.6**	0.7	0.7	0.7	**0.6**	0.7	0.7	0.7
0.4	1.0	++--	0000	96.6	96.2	71.5	**71.1**	82.8	82.5	71.5	**71.0**
1.0	1.0	++--	0000	25.8	25.6	19.2	**18.9**	22.8	22.5	19.2	**18.8**
2.0	1.0	++--	0000	11.0	10.2	7.8	**7.1**	9.6	8.6	7.8	**6.7**
3.0	1.0	++--	0000	6.4	4.9	3.7	**3.1**	5.3	3.9	3.7	**2.8**
5.0	1.0	++--	0000	2.0	1.6	**0.7**	1.0	1.4	1.3	**0.7**	0.9
0.4	1.0	+-00	0000	81.1	80.8	71.5	**71.2**	68.8	**68.5**	71.5	71.1
1.0	1.0	+-00	0000	21.5	21.2	19.2	**18.8**	18.9	**18.4**	19.2	18.5
2.0	1.0	+-00	0000	9.0	7.6	7.8	**6.5**	7.7	6.1	7.8	**5.8**
3.0	1.0	+-00	0000	4.7	3.2	3.7	**2.7**	3.6	2.5	3.7	**2.3**
5.0	1.0	+-00	0000	1.0	1.0	**0.7**	0.8	**0.7**	0.8	**0.7**	**0.7**
0	1.4	0000	++++	107.6	62.3	83.2	**37.7**	110.3	65.2	83.2	**47.3**
0	2.0	0000	++++	18.7	10.3	12.7	**6.8**	19.7	11.2	12.7	**8.5**
0	3.0	0000	++++	4.0	3.2	2.8	**2.4**	4.5	3.5	**2.8**	2.9
0	5.0	0000	++++	1.1	1.2	**0.9**	1.0	1.3	1.3	**0.9**	1.2
0	1.4	0000	++00	109.2	57.0	108.3	**47.6**	61.3	31.2	40.3	**26.4**
0	2.0	0000	++00	16.7	9.4	16.6	**8.3**	6.2	4.5	4.8	**3.9**
0	3.0	0000	++00	3.9	3.2	3.8	**2.9**	1.6	1.6	**1.5**	**1.5**
0	5.0	0000	++00	**1.3**	**1.3**	**1.3**	**1.3**	**0.8**	**0.8**	**0.8**	**0.8**
0	1.4	0000	+000	104.7	**50.5**	127.3	54.6	29.1	**15.1**	30.0	16.8
0	2.0	0000	+000	15.4	**9.1**	19.3	9.7	3.5	**3.0**	4.0	3.1
0	3.0	0000	+000	4.2	**3.4**	4.9	3.6	**1.4**	**1.4**	1.5	**1.4**
0	5.0	0000	+000	**1.7**	**1.7**	1.9	**1.7**	**0.8**	0.9	0.9	0.9

the means indicates that all means increased by the same amount (while producing the given value of δ), while "+ - 0 0" indicates that the first mean increased, the second mean decreased by the same amount, and the last two means did not change. Similarly, "+ + 0 0" for the standard deviations means that the first two standard deviations increased by the same amount (while producing the given value of ψ), and the last two standard deviations did not change.

From Table 8 we see that using regression adjustments gives dramatic improvements in performance for most shift directions. However, using regression adjustment gives dramatically worse performance in the case of a shift in $\boldsymbol{\mu}$ in the "+ + + +" direction (see also Hawkins (1991)).

Table 9 gives SSATS values for some chart combinations for some specific shift directions. The lowest SSATS for each shift is shown in bold. For a given value of δ or ψ, the shift direction does not seem to have a large effect, except for the case in which the shift in $\boldsymbol{\mu}$ is in the "+ + + +" direction and sets of univariate charts with regression adjusted variables are used. Recall from Table 8 that the M_2RA^2 chart does not perform well for a shift in $\boldsymbol{\mu}$ in the "+ + + +" direction, but, when the M_2RA^2 chart is used with the MZ chart, there is much less of a problem because the MZ chart is invariant to the direction of the shift in $\boldsymbol{\mu}$.

5 Conclusions

This paper investigates various combinations of control charts for multivariate monitoring of $\boldsymbol{\mu}$ and $\boldsymbol{\sigma}$. In general, using a chart based on squared deviations from target along with the chart for $\boldsymbol{\mu}$ gives better overall performance than using a Shewhart chart along with the chart for $\boldsymbol{\mu}$. Using sets of univariate charts results in a moderate loss of efficiency compared with using multivariate charts, assuming that regression adjustment of the variables is used in the sets of univariate charts when the variables are correlated.

As long as λ in the chart based on squared deviations from target is not too small, the squared deviations chart will be almost as effective as the Shewhart chart in detecting large shifts in $\boldsymbol{\mu}$. In this case, a relatively small value of λ can be used in the chart for $\boldsymbol{\mu}$ to provide fast detection of small shifts in $\boldsymbol{\mu}$. However, using a very small value of λ in the chart for $\boldsymbol{\mu}$ will result in some deterioration in the ability to detect intermediate shifts in $\boldsymbol{\mu}$, or small and intermediate shifts in $\boldsymbol{\mu}$ or $\boldsymbol{\sigma}$.

References

Hawkins, DM (1991) Multivariate Quality Control Based on Regression-Adjusted Variables. *Technometrics*, 33, 61-75.

Hawkins, DM (1993) Regression Adjustment for Variables in Multivariate Quality Control. *Journal of Quality Technology*, 25, 170-182.

Hotelling, H (1947) Multivariate Quality Control-Illustrated by the Air Testing of Sample Bombsights. In *Techniques of statistical Analysis* edited by C Eisenhart, MW Hastay, and WA Wallis, McGraw-Hill, New York, NY.

Lowry, CA, Woodall, WH, Champ, CW, Rigdon, SE (1992) A Multivariate Exponentially Weighted Moving Average Control Chart. *Technometrics*, 34, 46-53.

Lucas, JM, Saccucci, MS (1990) Exponentially Weighted Moving Average Control Schemes: Properties and Enhancements. *Technometrics*, 32, 1-12.

Reynolds, MR Jr, Cho, GY (2006) Multivariate Control Charts for Monitoring the Mean Vector and Covariance Matrix. *Journal of Quality Technology*, 38, 230-253.

Reynolds, MR Jr, Kim, K (2007) Multivariate Control Charts for Monitoring the Process Mean and Variability Using Sequential Sampling. *Sequential Analysis*, 26, 283-315.

Reynolds, MR Jr, Stoumbos, ZG (2001) Monitoring the Process Mean and Variance Using Individual Observations and Variable Sampling Intervals. *Journal of Quality Technology*, 33, 181-205.

Reynolds, MR Jr, Stoumbos, ZG (2004) Control Charts and the Efficient Allocation of Sampling Resources. *Technometrics*, 46, 200-214.

Reynolds, MR Jr, Stoumbos, ZG (2005) Should Exponentially Weighted Average and Cumulative Sum Charts be Used with Shewhart Limits?. *Technometrics*, 47, 409-424.

Reynolds, MR Jr, Stoumbos, ZG (2008) Combinations of Multivariate Shewhart and MEWMA Control Charts for Monitoring the Mean Vector and Covariance Matrix. *Journal of Quality Technology*, 40, 381-393.

Quality Control of Manufactured Surfaces

Bianca Maria Colosimo, Federica Mammarella, and Stefano Petrò

Dipartimento di Meccanica - Politecnico di Milano
Piazza Leonardo da Vinci 32, 20133 Milano, Italy
Email: biancamaria.colosimo@polimi.it

Summary. Recent literature on statistical process monitoring pointed out that the quality of products and processes can be often related to profiles, where the function relating a response to one or more location variables (in time or space) is the quality characteristic of interest. An important application of profile monitoring concerns geometric specifications of mechanical components, such as straightness, roundness or free-form tolerance. This paper presents a new approach aimed at extending the method proposed for profile monitoring to surface monitoring. In this case, a geometric specification (such as cylindricity, flatness, etc.) is assumed to characterize the machined surface. The proposed method is based on combining a Spatial Autoregressive Regression (SARX) model (i.e. a regression model with spatial autoregressive errors) to multivariate and univariate control charting. In this work, the approach is applied to a case study concerning surfaces obtained by turning and subject to cylindricity tolerance.

1 Introduction

During the last decades, technical drawings of mechanical components have been showing an increasing number of form tolerances (i.e., product specifications concerning profiles and surfaces). This increase can be probably ascribed to two main reasons. Firstly, a deeper knowledge of the relationships between the characteristics of profiles/surfaces and the functional properties of the produced items is now available. Secondly, the advent of modern measurement systems (such as Coordinate Measuring Machines - CMM) and computerized data analysis decisively reduced the complexity and the time required for checking form tolerances.

Form tolerances can concern simple shapes (such as straightness, roundness, cylindricity, flatness, etc.) or even complex ones (usually related to as free-forms) and are designed for defining a constraint on the final shape of the processed item. In fact, due to the natural variability of the material, the manufacturing and the measurement processes, the actual surface can deviate

H.-J. Lenz et al. (eds.), *Frontiers in Statistical Quality Control 9*,
DOI 10.1007/978-3-7908-2380-6_4, © Springer-Verlag Berlin Heidelberg 2010

from its ideal or nominal shape. The geometric tolerance is indeed designed for constraining this deviation, since it can affect the functional performances of the final produced item. In this framework, a statistical process monitoring procedure should quickly detect any deviation of the machined shape from the usual pattern obtained when the process is in its standard or in-control behavior. With this aim, Colosimo *et al.* [4] recently presented a procedure for monitoring a manufacturing process when the quality characteristic of interest is a profile, specifically a roundness or circular profile, obtained by turning. Further details on the roundness case study can be found in [3]. The approach proposed in [4] originated from recent literature on profile monitoring [15], where the "profile" is the functional relationship which links a response to one or more location variables and the main objective is to detect a change affecting the functional relationship, rather than the response itself. Approaches for profile monitoring have been presented for different applications ranging from calibration study [10] [11] [13], to signal analysis [6] [9].

The objective of this paper is to extend the method proposed in [4] for profile monitoring to quality control of manufactured surfaces, such as cylindricity, flatness, etc. Within the paper, a case study concerning cylindrical surfaces obtained by lathe-turning will be taken as reference. Regression models presented in the literature on cylindrical specifications [7] [16] will be considered as starting reference to identify the parametric models of the large-scale pattern characterizing all the cylindrical surfaces machined. The model presented in the literature will be further extended to represent the spatial correlation characterizing adjacent points on each machined surface. Therefore, as a first byproduct, this paper will present a novel parametric model of cylindrical surfaces able to represent both the large-scale and the small-scale (spatial correlation) characteristics of the machined surface. This model will make full use of Spatial Autoregressive Regression (SARX) models [5], used in spatial statistics to represent the spatial correlation in regular lattices. As suggested in the literature on profile monitoring [15], the analytical model describing the signature will be identified and combined with multivariate and univariate control charting for monitoring the estimated coefficients and the estimated noise variance, respectively. Therefore, a first monitoring approach will be developed to properly take into account both the large-scale and the small-scale characteristics of the machined surface.

A different monitoring procedure will instead focus on the large-scale behavior of the machined surface while neglecting the small-scale (spatial correlation) structure. Therefore regression will be simply used as modeling technique and the resulting residuals, even if spatially correlated, will be then treated as they were uncorrelated. In this second case, the monitoring approach will again use a multivariate control chart for monitoring the estimated coefficients of the regression model and a univariate control chart on the estimated variance of the (spatially correlated) noise sequence. This second competitor procedure will allow us to evaluate conditions under which the extra-effort required for modeling the spatial correlation structure is worthwhile.

Both the aforementioned approaches will be eventually compared with a simpler procedure which represents the industrial practice, where the information collected in the cloud of measured points is usually summarized in just one indicator representing the form error. In the case of cylindricity, this synthetic value will be referred to as out-of-cylindricity (OOC) and will represent the maximum deviation of the machined surface from the ideal cylindrical shape. Hence, a univariate control chart on the OOC will be the last competitor approach.

The performances of the three approaches for surface monitoring will be computed as the Average Run Length (ARL) to detect out-of-control states. These out-of-control states will be related to specific conditions which typically occur in manufacturing processes [16].

2 Cylindrical surfaces obtained by lathe-turning

Among different geometric specifications, cylindricity plays an important role when functionality relates to rotation or coupling of mechanical components. According to the standards [1] [8], the cylindricity is the condition of a surface of revolution in which all the points of the surface are equidistant from a common axis. The deviation of the actual shape from the ideal one is specified by the cylindricity error, which is the minimum radial distance between two coaxial cylinders which contain among them the actual surface. This error will be referred to as out-of-cylindricity (OOC) from now on. According to the ISO standard [8], the OOC error has to be computed in two steps. The first step consists in computing the *reference cylinder*, which is fitted on the actual data by using one of several possible criteria (the most common being the least-squares, the minimum zone, the minimum circumscribed and the maximum inscribed reference cylinders). Then, the deviation of the actual shape from the reference cylinder is computed and possibly modeled. Assuming a cylindrical coordinate system, let y_{nh} represent the radial deviation from this reference cylinder observed on the h-th item ($h = 1, ..., H$) when the axial height is z_n and the angular position (often referred to as the azimuthal coordinate) is θ_n ($n = 1, ..., N$). Figure 1 shows the cylindrical coordinate system as well as the representation of the cylindricity error OOC, which is the minimum radial distance between the two coaxial cylinders which contain among them the actual surface.

2.1 The large-scale regression model

In the case of cylindrical features, Henke *et al.* [7] presented an analytical model for aiding the interpretation or the relationship between manufacturing processes and the typical deviation from the ideal cylinder. The approach

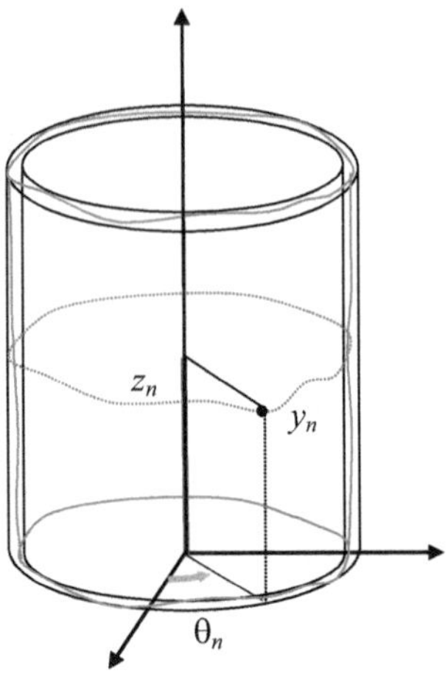

Fig. 1. The coordinate of the generic point observed

[7] combines Chebyshev polynomials and Fourier (periodic) functions for representing axial errors and circular errors, respectively. A similar approach is proposed by Zhang *et al.* [16], who used the Legendre polynomials instead of the Chebyshev ones to describe the deviation along the cylinder axis, while keeping the Fourier functions for describing the cross-section form errors. Figure 2 shows some typical errors observed on cylindrical components discussed in [7] and modeled by combining Chebyshev polynomials and Fourier functions.

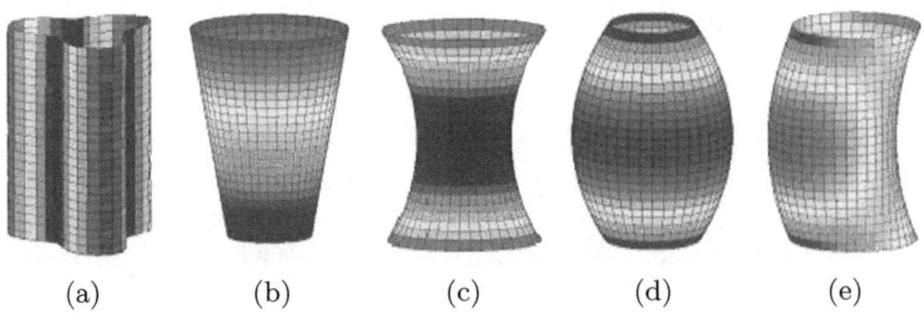

Fig. 2. Different types of form error for cylindrical surfaces, as classified by Henke *et al.* [7]. (a) Three-lobed; (b) taper; (c) hourglass; (d) barrel; (e) banana

According to the model in [7], the n-th value data observed on the h-th item can be expressed as:

$$y_{nh} = \mathbf{x}'_n \boldsymbol{\beta}_h + u_{nh} \quad , (n = 1, ..., N; h = 1, ..., H) \tag{1}$$

where $h = 1, ..., H$ is the index of the surface, $n = 1, ..., N$ is the index of the equally spaced observations on each surface, $\mathbf{x}'_n \boldsymbol{\beta}_h$ represents the *large-scale* model characterizing the h-th surface at the specific point n, and u_{nh} represents the noise term. In particular, $\mathbf{x}'_n = [\, x_{1n} \,...\, x_{kn} \,...\, x_{Kn} \,]$ is a (row) vector whose elements are K regressor variables which are assumed to be given and known functions of the coordinates of the n-th point; $\boldsymbol{\beta}_h = [\, \beta_{1h} \; \beta_{2h} \,...\, \beta_{Kh} \,]'$ represents a (column) vector whose elements are K unknown coefficients that have to be estimated for each surface h ($h = 1, 2, ..., H$). Table 1 summarizes the expression of the regressor function x_{kn} according to the model in [7]. Note that $T_i(\zeta_n)$ represents a second-type Chebyshev polynomial of order i, i.e., $T_0(\zeta_n) = 1$; $T_1(\zeta_n) = 2\zeta_n$; $T_2(\zeta_n) = 4\zeta_n^2 - 1$ and $\zeta_n = \frac{2(z_n - z_{min})}{z_{max} - z_{min}} - 1$. From Table 1 it is clear that some of the possible combinations of the Chebyshev polynomials and the periodic functions are not included as possible regressor functions. As examples, $T_0(\zeta_n)$–representing the (least-square) cylinder radius– or $T_0(\zeta_n)cos(\theta_n)$ and $T_0(\zeta_n)sin(\theta_n)$ –associated to the translation of the (least-square) cylinder axis– are not included in this table. In fact, the model in (1) is describing the *deviation* of the observed data from the ideal cylinder. Therefore, the ideal cylinder is firstly estimated and subtracted by the actual radii observed in order to compute the deviation [7].

k	Order of the Chebyshev polynomial	Order of the periodic (Fourier) component	x_{kn}
1	0	2	$T_0(\zeta_n)cos(2\theta_n)$
2	0	2	$T_0(\zeta_n)sin(2\theta_n)$
3	0	3	$T_0(\zeta_n)cos(3\theta_n)$
4	0	3	$T_0(\zeta_n)sin(3\theta_n)$
5	1	0	$T_1(\zeta_n)$
6	1	2	$T_1(\zeta_n)cos(2\theta_n)$
7	1	2	$T_1(\zeta_n)sin(2\theta_n)$
8	1	3	$T_1(\zeta_n)cos(3\theta_n)$
9	1	3	$T_1(\zeta_n)sin(3\theta_n)$
10	2	0	$T_2(\zeta_n)$
11	2	1	$T_2(\zeta_n)cos(\theta_n)$
12	2	1	$T_2(\zeta_n)sin(\theta_n)$
13	2	2	$T_2(\zeta_n)cos(2\theta_n)$
14	2	2	$T_2(\zeta_n)sin(2\theta_n)$
15	2	3	$T_2(\zeta_n)cos(3\theta_n)$
16	2	3	$T_2(\zeta_n)sin(3\theta_n)$

Table 1. The regressor functions x_{kn} as a function of the index k.

2.2 The small-scale model for the spatial correlated noise terms

With reference to the model in (1), the noise term u_{nh} observed at the n-th location of the h-th item is usually assumed to be uncorrelated and normally distributed. In particular, if the assumption of uncorrelated noise is rejected, models able to deal with spatial correlation structures should be used. In fact, traditional time-series models (in which past data are assumed to influence future ones) can not be used in the case of machined surfaces, since no specific direction of the influence can be identified for data observed on a surface. In this paper, we will use approaches taken from spatial statistics to deal with the issue of possibly correlated noise. In particular, we will assume that the grid of equally spaced points observed on the machined surface is kept fixed, as it were implemented on a coordinate measuring machine which automatically performs the required measurement path. Given this fixed grid of measurement locations, each point can be related to a different set of adjacent points. Therefore an hypothesis on the *adjacency*, *contiguity* or *spatial weight* matrices $\mathbf{W}^{(s)}$ $(s = 1, ..., p)$ has to be firstly considered. Let $w_{nm}^{(s)}$ represent the elements in row n of column m of the weight matrix for the s-th order neighbors, namely $\mathbf{W}^{(s)}$. In other words, $w_{nm}^{(1)}$ is set equal to 1 if the n-th point is assumed to be neighbor of the m-th point and 0 otherwise. Analogously, $w_{nm}^{(2)}$ is set equal to 1 if the n-th point is a neighbor of the original first-generation neighbors of the m-th point, and so on. Two traditional ways for defining these matrices are presented in the literature on spatial statistics. The first is the *rook-based* contiguity, where neighbors share a common border (Figure 3 on the left). The second is the *queen-based* contiguity, which defines neighbors as locations that share either a border or a vertex in their boundaries (Figure 3 on the right).

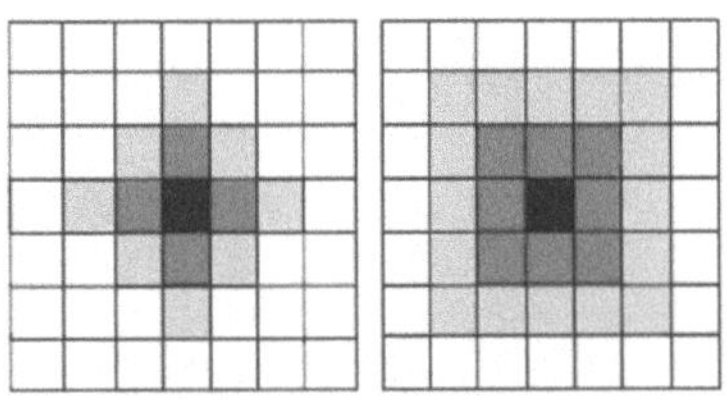

Fig. 3. Plot of the neighbors of the rook-based (on the left) and the queen-based (on the right) contiguity: the first-order and second-order neighbors of the center point (which is shown in black) are represented in darker and lighter gray, respectively.

Despite of the specific type of contiguity structure (rook or queen-based), a Spatial AutoRegressive (SAR) model of order p [5] can be used as reference model for the noise term of the h-th surface:

$$u_{nh} = \varrho_{1h} \sum_{m=1}^{N} w_{nm}^{(1)} u_{mh} + \varrho_{2h} \sum_{m=1}^{N} w_{nm}^{(2)} u_{mh} + ... + \varrho_{ph} \sum_{m=1}^{N} w_{nm}^{(p)} u_{mh} + \varepsilon_{nh} \quad (2)$$

where $\varepsilon_{nh} \sim (0, \sigma_h^2)$ is an uncorrelated noise error sequence ($n = 1, ..., N$), $\varrho_{1h}, ..., \varrho_{ph}$ are coefficients that have to be estimated, while $w_{nm}^{(s)}$ ($s = 1, ..., p$) is the generic element of the s-th order neighbor matrix $\mathbf{W}^{(s)}$. When the SAR(p) model of the noise error terms given in (2) is combined with the large-scale regression model given in (1), it determines a Spatial Autoregressive Regression model of order p or SARX(p) model [5]. Coefficients of the noise and large-scale models, given by the $\varrho'_{sh}s$ ($s = 1, ..., p$) and $\beta'_{kh}s$ ($k = 1, ..., K$), respectively, can be estimated by using the approach implemented in the Spatial Econometric Toolbox [12].

2.3 The parametric model for the case study

The case study refers to 100 C20 carbon steel cylinders, which were supplied in 30 mm diameter rolled bars and machined to a final diameter of 26 mm (cutting speed equal to 163 m/min, feed equal to 0.2 mm/rev). Two cutting steps of 1 mm depth each were required to arrive to the final diameter. Each machined surface was eventually measured by using a Coordinate Measuring Machine (CMM). A set of 68 generatrices was sampled on each turned specimen by continuous scanning. In particular a set of 61 points were measured on each of the 68 generatrices. Therefore, $61 \times 68 = 4148$ points were measured on each of the 100 machined surfaces. Figure 4 shows the actual shape of the form error observed in one of the machined items.

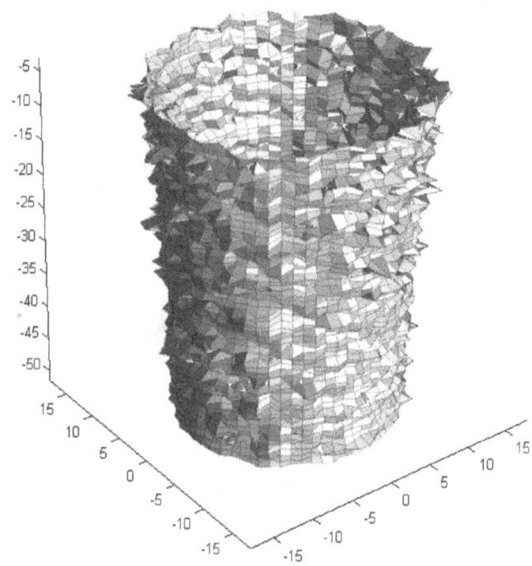

Fig. 4. The actual shape of one out of the 100 machined cylinders

With reference to this set of 100 surfaces, we assumed models in equations (1) and (2) as references for the large and small-scale patterns, respectively. All the regressor functions in Table 1 were initially considered for the large-scale model. The order p of the SARX model was increased in order to achieve uncorrelated residuals, as confirmed by the Moran's test [5]. According to this criterion, a SARX model of order $p = 2$ was deemed appropriate by using both the rook-based and the queen-based contiguity structures (Figure 3) for all the 100 items. Eventually, the rook-based model was preferred because of the reduced number of neighbors (i.e., its "parsimony").

Once identified the proper SARX model structure, a final check was performed to decide whether all the $K = 16$ regressors of Table 1 were effectively required for the large-scale model. In particular the (asymptotic) t-statistics for the SARX model's estimates were considered [12] and allowed us to conclude that all the regressor functions ranging from the 11-th to the 16-th are never required in any of the 100 surfaces for the large-scale model. Therefore, just the first 10 regressor functions in Table 1 were eventually considered in the final model.

Therefore, the model of the h-th cylindrical surfaces obtained by lathe-turning is a SARX(2) model, which can be synthetically represented in matrix form as:

$$\begin{aligned} \mathbf{y}_h &= \mathbf{X}\boldsymbol{\beta}_h + \mathbf{u}_h \quad , \\ \mathbf{u}_h &= \left(\varrho_{1h}\mathbf{W}^{(1)} + \varrho_{2h}\mathbf{W}^{(2)}\right)\mathbf{u}_h + \boldsymbol{\varepsilon}_h \quad , \end{aligned} \tag{3}$$

where $\mathbf{y}_h = [y_{1h} \dots y_{Nh}]'$ is the vector of data observed on the h-th item, $\mathbf{X}$ is the regressor matrix, whose n-th row is $\mathbf{x}'_n$, $\mathbf{u}_h$ and $\boldsymbol{\varepsilon}_h$ are the vectors of the spatially correlated and uncorrelated noise terms for the h-th surface, respectively. Eventually, $\mathbf{W}^{(1)}$ and $\mathbf{W}^{(2)}$ represent the first and second order adjacency matrices, respectively.

3 Control charts for surface monitoring

In order to detect out-of-control surfaces, three alternative approaches will be considered in this paper. The first approach is the simplest one and consists in computing the form error (out-of-cylindricity, OOC) value associated to each machined surface and in monitoring it with an individuals control chart. Similarly to control charts suggested in the literature on profile monitoring, the second and third approaches consist of a multivariate control chart for monitoring the vector of estimated parameters and a univariate control chart aimed at monitoring the estimated residuals variance. The main difference between these two approaches is in the in-control model assumption. In particular, the second procedure assumes that the manufacturing signature can be modeled using a SARX(2) model. Therefore this control charting is based

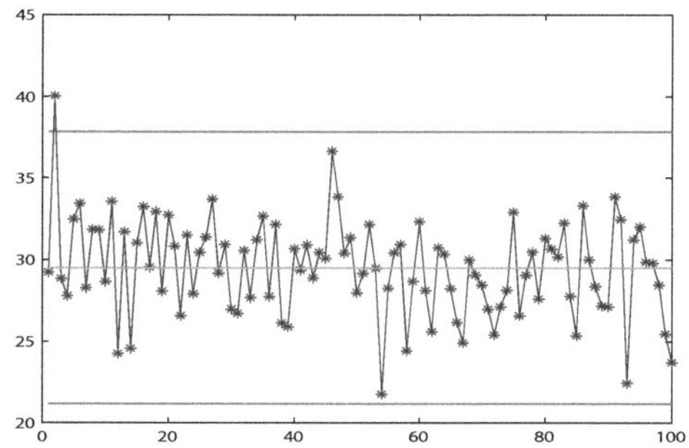

Fig. 5. Individuals control charts of the reciprocal of the OOC values.

on the actual model and thus, the resulting residuals will be effectively uncorrelated. The third approach will not include any spatial correlation structure into the model. Therefore, it will be referred to as a SARX(0) model. In this last approach we will be assuming that the relationship between adjacent measurements is neglected and hence only the regression model in (1) is considered as baseline model. Given that the model is a SARX(2), the residuals computed assuming a SARX(0) model will be obviously correlated but their variance will be estimated and monitored despite of this spatial correlation structure.

In all the three control charting approaches, a nominal false alarm probability $\alpha' = 0.01$ will be assumed. Since control charting for the SARX(2) and SARX(0) approaches consist of two control charts (a multivariate plus a univariate control charts), each of the two control charts will be set by assuming a false alarm probability $\alpha = 1 - \sqrt{1 - \alpha'}$.

3.1 Phase I of control charting

With reference to the sample of $H = 100$ machined surfaces, control charting can be firstly aimed at detecting possible out-of-control shapes in this set of items. In this case, the control charts have to be designed and retrospectively used to detect out-of-control surfaces, if any. A traditional individuals control chart on the OOC values was firstly designed. In particular, the least-square method was used to compute the OOC associated to each machined surface. Given the rejection of the normality assumption for this sequence of OOC values, a suitable transformation of these values was identified in order to achieve the normality of the transformed data. A value of $\lambda = -1$ was used in the power transformation, i.e. a standard Shewhart control chart for individuals was designed on the reciprocal of the OOC values. Figure 5 shows the individuals control charts obtained with these transformed data, given a false alarm probability equal to $\alpha = 1\%$. No assignable cause was found for the out-of-control sample (corresponding to the second surface) identified with this control chart.

A second monitoring procedure was designed by assuming the SARX(2) model in (3) as reference. In this case, the h-th machined surface was associated to: i) the vector of 12 estimated coefficients of the SARX(2) model $\mathbf{b}'_h = [\hat{\beta}_{1h} \dots \hat{\beta}_{10h}\ \hat{\varrho}_{1h}\ \hat{\varrho}_{2h}]$; and ii) the estimated variance $S^2_h = \hat{\sigma}^2_h$ of the (uncorrelated) residuals ε_h. Consequently, a multivariate T^2 control chart was designed for monitoring the mean of $\mathbf{b}_h$ $(h = 1, \dots, H)$. In particular, the recent approach of Williams *et al.* [14] was used in this design stage. A traditional univariate control chart for monitoring the estimated variance S^2_h of the residuals was designed too. Unfortunately, the normality assumption of SARX(2) model's residuals was rejected for most of the 100 machined surfaces. Therefore, traditional control limits of the Shewhart-type control chart for the variance were not applicable in this case. However, by applying a Box-Cox transformation to the sequence of estimated residual variances S^2_h for $h = 1, \dots, 100$, the suggested transformation was found to be the natural log. A traditional individuals control chart was then applied to the sequence of log-transformed variances. Figure 6 shows the T^2 and individuals control chart for the SARX(2)-based approach. Three out-of-control samples are now detected with this method. Since no assignable causes were found for these outlying surfaces, we did not delete any specific sample from the set of 100 reference ones.

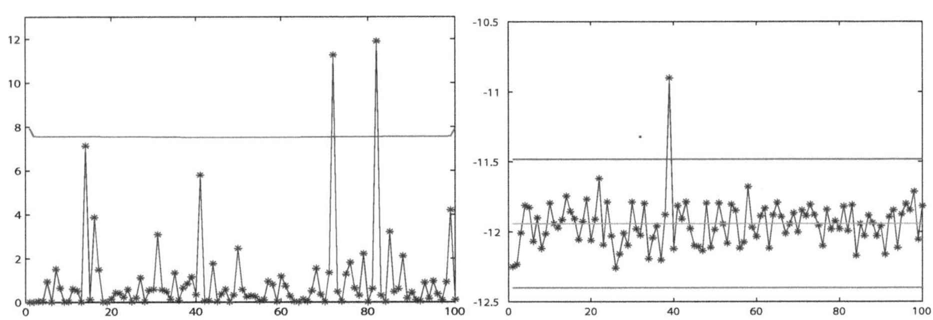

(a) (b)

Fig. 6. Phase I SARX(2)-based control charting for the 100 cylindrical surfaces. (a) T^2 control chart on the estimated coefficients (b) Shewhart control chart on the log-transformed estimated noise variance.

Eventually, the last approach was similar to the previous one but for the correlation structure of residuals, which was neglected in this last case. Specifically, the multivariate control chart was designed on the sub-vector $[\hat{\beta}_{1h} \dots \hat{\beta}_{10h}]$ of coefficients associated to the surface when just the large-scale model in (1) is considered. Therefore, the variance of residuals related to this SARX(0) model was computed for each of the machined surface, log-transformed to achieve normality and monitored via an individuals control chart. Figure 7 shows the SARX(0)-based Phase I control charting. It can be clearly observed that the results of this third approach are quite similar to the ones achieved with the SARX(2)-based one. On the contrary, these two last control charting procedures are different from the simplest OOC-based one showed in Figure 5, since the out-of-control signals issued by these last monitoring procedures (both the SARX(0) and the SARX(2) approaches) are different from the one issued by the first OOC control chart. This difference will be further investigated with reference to Phase II control charting in the simulation study presented in the next subsection.

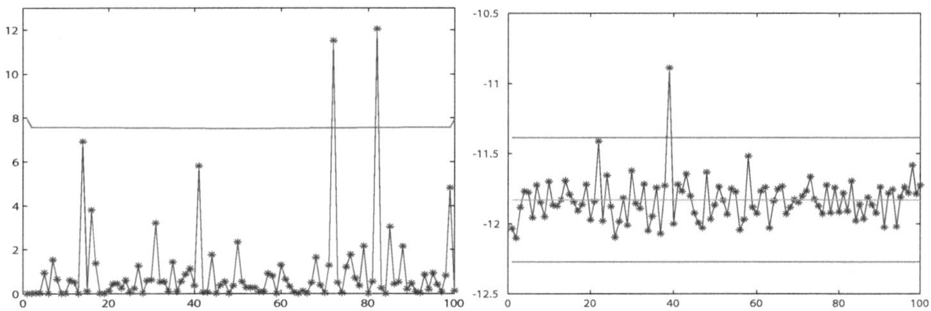

Fig. 7. Phase I SARX(0)-based control charting for the 100 cylindrical surfaces. (a) T^2 control chart on the estimated coefficients (b) Shewhart control chart on the log-transformed estimated noise variance.

3.2 Phase II of control charting: the run length performance

With reference to the control charts designed in the previous section, we used simulation to generate in-control and out-of-control surfaces. In particular, the SARX(2) model in (3) was taken as reference, assuming the unknown coefficients to be generated from a multinormal distribution with mean and covariance matrix estimated on the set of 100 samples used in Phase I. According to this model, a new set of 20000 new cylindrical surfaces were randomly generated for properly tuning the three monitoring procedures to achieve exactly the same in-control average run lengths (ARL) performances. In particular, for all the monitoring procedures, the control charts were tuned to obtain a nominal value of the false alarm probability equal to 1%, for all the three approaches. Figure 8 shows the 95 % confidence intervals on the in-control ARL obtained by simulating in-control surfaces until a set of 1000 run lengths (i.e., 1000 false alarms) were available for all the three monitoring procedures.

In order to generate a set of possible out-of-control conditions, we considered form errors presented in the literature [2], [7], [16] as results of technological problems which are common in lathe-turning. In particular, the following set of out-of-control conditions were simulated:

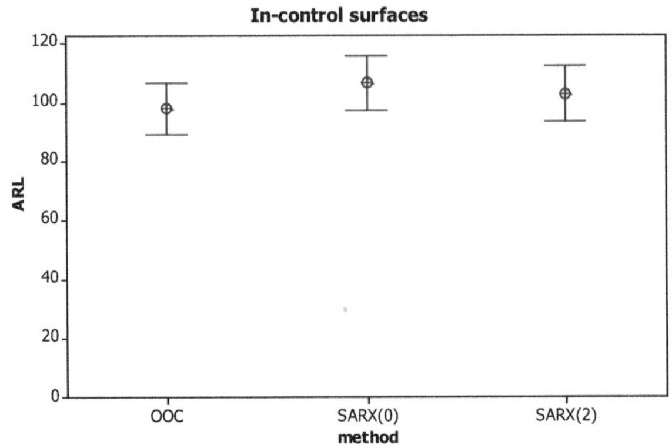

Fig. 8. 95% confidencce interval on the Average Run Length (on 1000 instances of run lengths) for in-control surfaces in Phase II.

1. **Increase of tapering** (Figure 2 (b)), which can be due to an increased inflection of the workpiece axis while machining. This out-of-control condition can be obtained by generating the in-control surfaces and then multiplying the coefficient of the 5-th regressor function in Table 1 by a factor δ_1, since this regressor function is responsible of the conical shape. In particular, we assumed $\delta_1 = 1.3$.
2. **Change of the trilobed error.** The in-control surfaces were already characterized by this type of error, which was probably due to a spindle motion error because its effect was increasing while moving far from the the spindle. On the other side, a trilobed error can be further due to an excessive force of the clamping fixture and, in this case, its effect reduces while moving far from the spindle. We assumed that a possible out of control can be due to the excessive clamping force and we modeled it by multiplying the 8-th regressor function in Table 1 (namely, $T_1(\zeta_n)cos(3\theta_n)$) by $\delta_3 = 1 - 0.001 = 0.999$.
3. **Half frequency spindle motion error**, which can be due to wear on one ball bearing spindle or to whirling in a hydrodynamic bearing [2]. This out-of-control condition can be modeled by introducing a spurious harmonic at frequency 1/2 upr to the baseline model, i.e. adding $\delta_2 \left[cos(0.5\theta_n) + T_1(\zeta_n)cos(0.5\theta_n)\right]$, to the set of regressor functions in Table 1. In this case we assumed $\delta_2 = 0.001$.
4. **Four-lobed spindle motion error.** This error was not originally present in the in-control model but can be due to a further type of spindle motion error, according to the literature on surfaces obtained by turning [2]. This condition was simulated by adding $\delta_4 \left[cos(4\theta_n) + T_1(\zeta_n)cos(4\theta_n)\right]$ to the in-control surfaces. In particular, we used $\delta_4 = 0.002$.
5. **Change of the spatial correlation structure.** As previously shown, the in-control model is a SARX model of order 2. We assumed that, due to

some changes in the properties of the material machined or in the machining conditions, the spatial correlation reduces to a first-order contiguity structure. In other words, the order of the SARX model is assumed to decrease to 1 (i.e., $\varrho_{2h} = 0$, $\forall h$) in this out-of-control condition.

6. **Increase of the uncorrelated residuals' variance.** Because of possible changes in the material properties or machining conditions, we assumed that a possible out-of-control shape can be due to an increase in the variance σ_h^2 of the residuals $\varepsilon'_{nh}s$. In particular, an increase equal to 5% was considered in this case.

For each of the aforementioned out-of-control conditions, we went on simulating surfaces until all the monitoring procedures allowed us to collect the same number of 1000 run lengths. Then, we computed a 95% confidence intervals on the ARLs obtained with each procedure in each of the aforementioned out-of-control conditions. Figure 9 shows the results of this performance comparison study. It can be observed that both the SARX(0) and the SARX(2) approaches outperform the industrial practice of monitoring the OOC value only. These results are consistent with the ones shown by Colostimo *et al.* for roundness profiles [4] and can be intuitively explained considering that the industrial approach reduces all the information contained in the cloud of measured points in just a single variable that reflects the out-of-control shape with some delay. An even more interesting result is that the advantage of properly modeling the spatial correlation structure –thus adopting a SARX(2) model instead of the simpler SARX(0)– can be observed just when the change affects somehow the noise. In fact, the SARX(2)-based approach outperforms the SARX(0)-based one just in the last two scenarios, where the correlation structure or the noise variance is assumed to change with time (note that an increase of 5% in the variance is ordinarily very difficult to detect but the SARX(2)-based approach shows very good performance, possibly because of the large number of measurements it is based on).

4 Conclusions

In this paper the method for monitoring profile data developed in [4] was extended to three-dimensional forms (i.e. surfaces). The approach proposed combines a regression model with spatial correlated noise to univariate and multivariate control charting and is applied to data on cylindricity surfaces obtained by turning. The simulation study showed that the results found in [4] for roundness profiles are confirmed when surfaces are machined instead of profiles. In particular, the main results is that approaches based on monitoring the coefficients of the parametric model which describes the systematic pattern of machined surface outperforms the industrial practice, which is based on monitoring a synthetic indicator of form error. Furthermore, the simulation study showed that the extra-effort required for modeling the spatial correlation

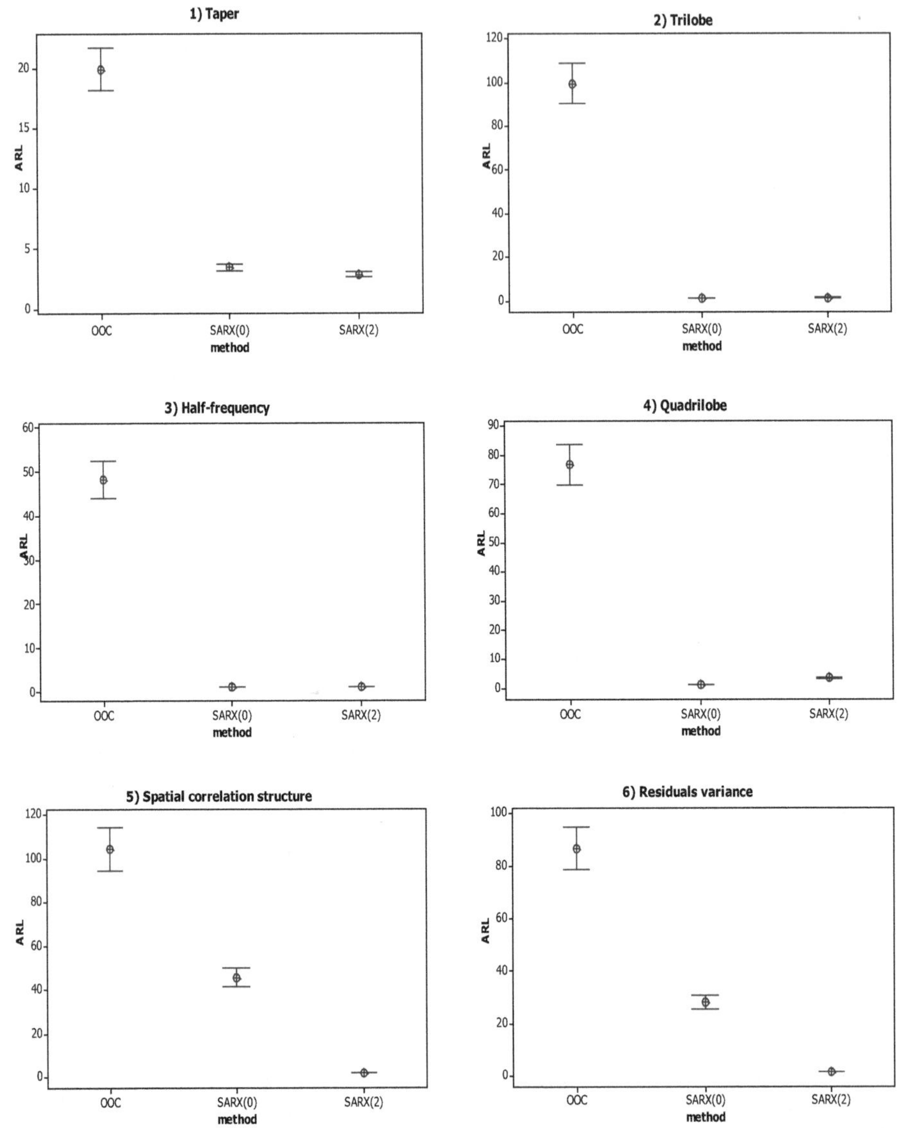

Fig. 9. Effect of different out-of-control conditions on the ARL performance of the different control charting methods.

can be worthwhile just when the small-scale (e.g., correlation structure and/or noise variance) characteristics of the machined surface are of interest.

Although a specific geometry was used, we point out that any three-dimension shape can be monitored with an approach similar to the one proposed, given that the regression terms and the spatial weight matrices should be appropriately selected. Therefore, the approach presented can be easily extended to different geometric specifications characterizing discrete part manufacturing in the automotive and aeronautical productions.

Acknowledgements.- This paper was partially funded by the Ministry of Education, University and Research of Italy (MIUR).

References

1. ASME Y14.5-M-1994 "Dimensioning and Tolerancing".
2. Cho, N.W. and Tu, J.F. (2001). "Roundness Modeling of Machined Parts for Tolerance Analysis". *Precision Engineering*, 25, pp. 35–47.
3. Colosimo, B. M. and Pacella, M. (2007). "On the Use of Principal Component Analysis to Identify Systematic Patterns in Roundness Profiles". *Quality & Reliability Engineering International*, 23(6), pp. 707–725.
4. Colosimo, B. M., Pacella, M. and Semeraro Q. (2008). "Statistical Process Control for Geometric Specifications: On the Monitoring of Roundness Profiles". *Journal of Quality Technology*, 40(1), pp. 1–18.
5. Cressie, N.A.C. (1993). *Statistics for Spatial Data, Revised Edition.* John Wiley & Sons, New York, NY.
6. Ding, Y., Zeng, L. and Zhou, S. (2006). "Phase I Analysis for Monitoring Nonlinear Profiles in Manufacturing Processes". *Journal of Quality Technology*, 38, pp. 199–216.
7. Henke, R.P., Summerhays, K.D., Baldwin, J.M., Cassou, R.M. and Brown, C.W. (1999). "Methods for Evaluation of Systematic Geometric Deviations in Machined Parts and Their Relationships to Process Variables". *Precision Engineering*, 23, pp. 273–292.
8. ISO 1101:2004 - Geometrical Product Specification (GPS) - Cylindricity (ISO/TS 12180:2003).
9. Jin, J. and Shi, J. (2001). "Automatic Feature Extraction of Waveform Signals for In-Process Diagnostic Performance Improvement", *Journal of Intelligent Manufacturing*, 12, pp. 257–268.
10. Kang, L. and Albin, S.L. (2000). "On-Line Monitoring When the Process Yields a Linear Profile". *Journal of Quality Technology*, 32, pp. 418–426.
11. Kim, K., Mahmoud, M.A. and Woodall, W.H. (2003). "On the Monitoring of Linear Profiles". *Journal of Quality Technology*, 35, pp. 317–328.
12. LeSage, J.P. (1999). "The Theory and Practice of Spatial Econometrics" Department of Economics University of Toledo, available at http://www.spatial-econometrics.com/.
13. Mahmoud, M.A. and Woodall, W.H. (2004). "Phase I Analysis of Linear Profiles with Calibration Applications". *Technometrics*, 46, pp. 380–391.
14. Williams, J.D., Woodall, W.H., Birch, J.B. and Sullivan J.H. (2006)."Distribution of Hotellings T^2 Statistic Based on the Successive Differences Estimator". *Journal of Quality Technology*, 38, pp. 217–229.
15. Woodall, W.H., Spitzner, D.J., Montgomery, D.C. and Gupta, S. (2004). "Using Control Charts to Monitor Process and Product Quality Profiles". *Journal of Quality Technology*, 36, pp. 309–320.

16. Zhang, X.D., Zhang, C., Wang, B. and Feng, S.C. (2005). "Unified Functional Approach for Precision Cylindrical Components". *International Journal of Production Research*, 43, pp. 25–47.

Statistical Process Control for Semiconductor Manufacturing Processes

Masanobu Higashide[1], Ken Nishina[2], Hironobu Kawamura[2], and Naru Ishii[2]

[1]NEC Electronics Corp, [2]Nagoya Institute of Technology

Summary. This paper considers statistical process control (SPC) for the semiconductor manufacuturing industry, where automatic process adjustment and process maintenance are widely used. However, SPC has been developed in parts industry an, thus, application of SPC to chemical processes such as those in the semiconductor manufacuturinghas not been systematically investigated

Two case studies are presented; one is an example for process adjustment and the other is an example for process maintenance. Either of them is based on control charts and it is discussed how to take into account process autocorrelation.

1 Introduction

Statistical process control (SPC) is used to control and reduce process variation by means of statistical methods based on information about the product quality produced by the process. It began with the control charts proposed by Shewhart in the 1920s, and afterward SPC spread with a focus on the parts industry (see Box and Kramer [1]). Control charts are used for an economic and efficient detection of the out-of-control state of processes.

A semiconductor manufacturing process is different than a mechanical machining process. The semiconductor manufacturing process is centered on processes that use chemical reactions. Semiconductor products are characterized by rapid changes in both improvement and deterioration of quality. Moreover, the quality requirement become stricter year by year, and at the same time the lifecycle of a single product is getting shorter. As a result, the development, trial manufacturing, and mass production cycles get shorter and shorter. Therefore, process stability has a large impact on corporate earnings. Generally, process stability is demanded right away with the start of mass production.

In conventional SPC, after detection of an out-of-control state actions are undertaken to eliminate the assignable cause. However, in the semiconductor manufacturing it is not uncommon that variation in a certain process affects product quality some 100 steps downstream, i.e., days or weeks later. There are many causes of variation and the cause-and-effect relationships between processes

H.-J. Lenz et al. (eds.), *Frontiers in Statistical Quality Control 9*,
DOI 10.1007/978-3-7908-2380-6_5,

are complex. Moreover, a semiconductor manufacturing process may be affected by barometric pressure and other external factors. Under these conditions process adjustment and process maintenance are used for limiting process variation and, therefore, it is of crucial importance to controll the functioning of SPC.

In the present study, we propose a method for controlling the functioning of automatic process adjustment and process maintenance based on two case studies

2 SPC fot Semiconductor Manufacturing

2.1 Process Model

The major characteristics of a semiconductor manufacturing process are as follow:

a. External factors (environments, materials) may have an immediate impact on the process.

b. Process variation is caused by a large number of different factors, and cause-and-effect relationships are complex.

c. There is autocorrelation.

d. The range of effects from process variation is so large.

Typical external factors are barometric pressure, outside temperature, and many of the materials used during processing. Because the manufacturing process for semiconductors is driven by chemical reactions, there are effects from the many factors that cause variation in the state of the chemical reaction or the properties of the reaction products. As mentioned above, these factors are of many different types with effects not limited to a single process. The cause-and-effect relationships are very complex and the impact of a cause may show up an effect only dozens or hundred process steps after its occurrence.

Another awkward feature of processes including chemical reactions is that they are heavily autocorreleted. Reaction by-products accumulate in the reaction chamber and surrounding area, and the amount of accumulated by-product increases together with the cumulative reaction time. The volume of the reaction chamber and the flow of the raw material gas are changed by the accumulation of by-product, which leads to a change in the state of the chemical reaction. This gives rise to autocorrelation.

Referring to the above characteristics, Kawamura et al. [2] proposed the control model shown in Fig. 1. The model regers to the characteristic y and admits the adjustment factor w used in tuning, plus variation factors z and x and error factors δ and ε, which affect y and w. The effects on the characteristic and the tuning are outlined in the figure. Automatic adjustment is included in tuning. The time is denoted by t and the double arrows represent the time axes. Generally, the process $y(t)$'s is autocorrelated. Both factor z and x affect the characteristic y. In contrast to factor x, factor z affects also the tuning precision.

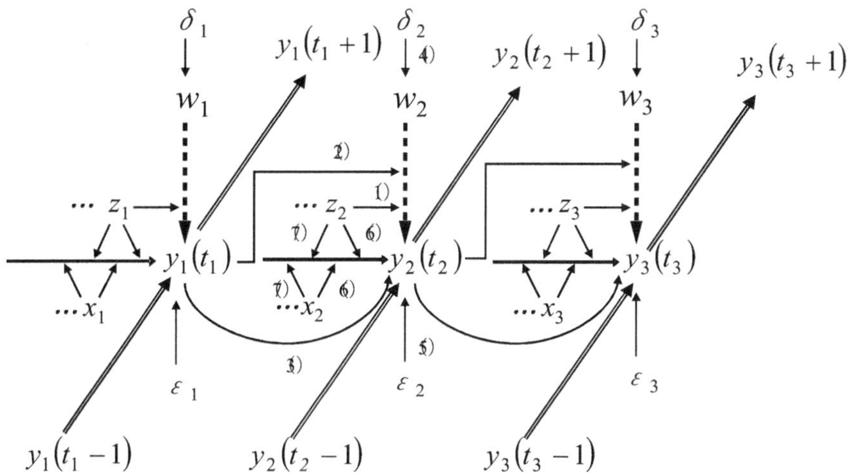

Fig. 1. SPC model for a semiconductor manufacturing process

There are seven effects ((1) - (7)) in Fig. 1, which are explained with respect to control characteristic y_2 as follows:

(1) Effect of factor z_2 of characteristic y_2 on tuning precision (interaction between factor z_2 and adjustment factor w_2).

(2) Effect of characteristic y_1 of the preceding process on the tuning precision of the following process (interaction between characteristic y_1 of the preceding process and adjustment factor w_2).

(3) Effect of preceding characteristic y_1 on the characteristic y_2 of the following process.

(4) Effect of error δ on the level setting in tuning.

(5) Effect of residual error ε on the correction formula that is the foundation for tuning.

(6) Effect of factors z_2 and x_2 on between-batch variation of y_2.

(7) Effect of factors z_2 and x_2 on within-batch variation of y_2.

It should be noted that there is not only correlation between subsequent process steps, but also between more distant process steps.

2.2 Process adjustment and process maintenance

As mentioned earlier, in semiconductor manufacturing processes by-products of chemical reactions accumulate in the reaction chamber with ongoing reaction time. The between-batch variation (process mean shift) and the within-batch variation are generated by the effects of these by-products. The amount of reaction by-products accumulated depends increases with time t.

The accumulation of reaction by-products first causes between-batch variation. This is the effect described by effect (6) in Fig. 1. This variation is adjusted with adjustment factor w so as to maintain a constant process mean.

With continuing time t, the amount of accumulated by-product increases, parts of equipment deteriorate, and within-batch variation begins to increase. This is the effect (7) described in Fig. 1. This within-batch variation cannot be removed with adjustment factor w. Therefore, the process is stopped and refreshed by removing by-products or replacing degraded parts, which is subsumed as process maintenance. Since process maintenance requires stoppage of the line, there is a need to keep it to a minimum, without adversely effect product quality or process yield.

3 Process adjustment

3.1 Process adjustment and the need to control it

Conventionally SPC is used to detect and eliminate assignable causes; that is, to reduce the effects (6) and (7) in Fig. 1. If the effects (6) and (7) are not considerably, the characteristic y_1 will vary only slightly and effect (3) will be small too. If variation of characteristic y_2 small, then there is no need for process adjustment. However, as mentioned earlier, semiconductor manufacturing processes have complex cause-and-effect relations and, moroever, there are narrow limits to process analysis and often it is impossible to specify assignable causes. For this reason, reducing process variation by process adjustment is more effective than attempting to eliminate variation factors.

Reducing process variation with the use of process adjustment requires that the adjustment will work with high reliability. Four effects, (1), (2), (4), and (5) in Fig.1 affect the correct functioning of the process adjustment. In conducting process adjustment, correction formula $y = f(w)$ is used to approximate the relationship between characteristic y and adjustment factor w, as oulined in Fig. 2. Effects (4) and (5) are determined at the time of design of the correction formula.

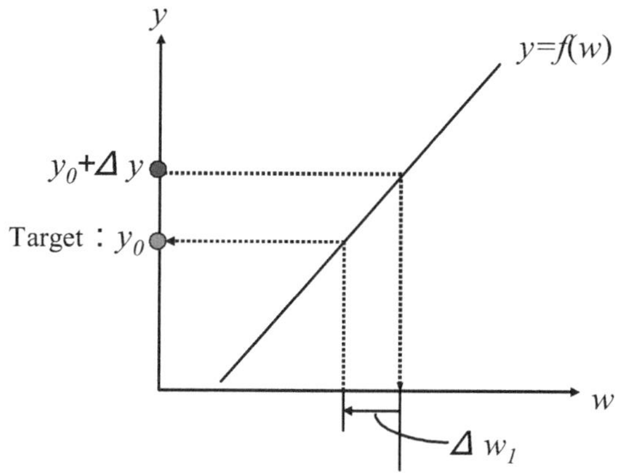

Fig. 2. Process adjustment using correction formula

By means of the correction formula in process adjustment the amount of adjustment is determined. As shown in Fig. 2, in cases when the characteristic y is by Δy larger than target the value y_0, the correction formula is used and an appropriate adjustment volume Δw_1 is obtained. However, the relation between characteristic y and adjustment factor w is affected by some factors ((1) and (2) in Fig.1). Therefore, it becomes necessary to control the factors that interact with the adjustment factor. The problem then is to select an appropritae control characteristic.

Process adjustment that is done automatically is called automatic process control (APC). It should be noted that APC may introduce additional process autocorrelation making the use of conventional control charts questionable. One way to solve the related control problem, is to use control charts for the residuals as proposed, for example, by Montgomery and Friedman [3]. Thus, we try to apply the residual X control charts to control process adjustment.

3.2 The process of the case study

As an example for explaining the control of process adjustment, the process of formation by chemical vapor deposition (CVD) of the polysilicon is used that becomes the electrical wiring for semiconductor elements. In this process, there is the process of forming a thin polysilicon film on a single-crystal silicon wafer. This process is outlined in Fig. 3. The wafer is introduced into a reaction chamber called a silicon carbide tube, and heated to several hundred degrees under reduced pressure, after which polysilicon raw material gas is flowed in. A polysilicon film then forms through a chemical reaction by heat. In this process, multiple lots made up of single lots with a maximum of 25 silicon wafers are processed simultaneously. This work unit is called a batch. The thickness of the polysilicon film formed is measured using a special film thickness measurement wafer inserted between the product wafers. The position where the film measurement wafer is inserted is constant, and films are always monitored in the same place.

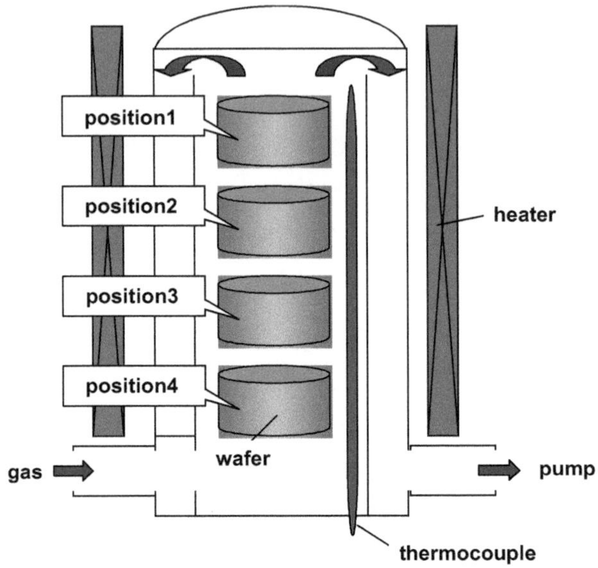

Fig. 3. Schema of general low pressure CVD apparatus [2]

A maximum of four lots can be inserted simultaneously, and one batch of a maximum of 100 product wafers can be processed simultaneously. It is also possible to process fewer than four lots per batch, and the number of lots processed in a single batch varies. The product lot insertion position is determined as follows according to the number of lots processed (Fig. 3):

One lot processing: processing position 2.

Two lots processing: processing position 2, 3.

Three lots processing: processing position 1, 2, 3.

Four lots processing: processing position 1, 2, 3, 4.

Film measurement wafers are inserted immediately above or immediately below product lots, and processing is conducted simultaneously with the product wafers. The film thickness is measured after processing with a film thickness gauge. Therefore, except when simultaneously processing four lots, the film immediately under processing position 4 is not measured and there is a missing value.

From Fig. 3, we see that the raw material gas is introduced from the lower part of the reaction chamber and flows upward implying that the film is produced more abundantly in the lower part of the reaction chamber, and the by-products also accumulate more in the lower part of the chamber. It was mentioned above that between-batch and within-batch variation is generated by the accumulated reaction by-products. Thus, the lower part of the reaction chamber, where there is a greater amount of by-products, has greater between-batch variation than the upper part.

The thickness of a film formed with a low pressure polysilicon deposition equipment differs with the degree of integration and intended purpose of the device, but the thickness is less than several hundred nanometers. The film variation, which is the quality demanded in this process, differs depending on the device, but needs to be kept roughly at 2–5%.

Process adjustment in the process under discussion has been done automatically using the film thickness value of the most recent four lots and film deposition time when those lots were being worked. Film deposition time is the time for the occurrence of the chemical reaction that causes the film to deposit. This is adjustment factor w in Fig. 1. The correction formula shown below is used in automatic process adjustment.

$$T_t = \frac{1}{a} \times \left(X_0 - \overline{X}_{t-1}\right) \times b + \overline{T}_{t-1} \tag{1}$$

Here, T_t is the deposition time of the film of the next batch calculated from the correction formula, X_0 is the target thickness of the film, $\overline{X}_{t-1}$ is the mean film thickness value for the most recent four lots up to the previous batch, $\overline{T}_{t-1}$ is the mean film thickness deposition time for the most recent four lots up to the previous batch, a is a correction coefficient, b is a relaxation coefficient, and t is time.

3.3 Residual X control chart based on a time series model

Controlling the function of process adjustment means to stabilize the relationship between the characteristic and the adjustment factor. The problems are effects (1) and (2) in Fig. 1. Therefore, the deposition rate, that is, the film thickness of the wafer divided by the film deposition time (corresponding to the slope of the correction formula shown in Fig. 2) should be considered as the control characteristic. However, the deposition rate is autocorrelated, because of the automatic adjustment using equation (1). Therefore, residual X control charts based on the time series model mentioned in 3.1 are used.

An autoregressive integrated moving average (ARIMA (p, d, q)) model of Equation (2) including an autoregressive parameter (p), difference (d), and moving average parameter (q) (see Box & Jenkins [4]) is applied to the deposition rate as the time series model.

$$y_t = (1 - \phi_1 - \cdots - \phi_p)\mu + \phi_1 y_{t-1} + \cdots + \phi_p y_{t-p} + \varepsilon_t - \theta_1 \varepsilon_{t-1} - \theta_2 \varepsilon_{t-2} - \cdots - \theta_q \varepsilon_{t-q} \quad (2)$$

$$y_t = z_t - z_{t-l}$$

Here, z_t is the observed value at time t, φ_j ($j = 1,2,\ldots,p$) is the autoregressive parameter, θ_j is the moving average parameter, μ is the average of y_t, and ε_t is the white noise at time t.

Fig. 4 shows the deposition rate at process position 2 with no missing values. The variation displayed in Fig. 4 shows no unusual patterns. Because a straight line of the correction system indicated in Fig. 2 is kept, it is possible to use Eq. 1 just as it is, to the entire range in Fig. 4.

Several ARIMA models are considered, and as a result of model selection based on the Akaike Information Criterion, an ARIMA (1, 1, 0) model is selected.

The formula for the ARIMA (1, 1, 0) model in "period a" in Fig. 4 is shown below:

$$\hat{y}_t = -0.334 y_{t-1} + 0.014, \quad y_t = z_t - z_{t-1} \quad (3)$$

The residual e_t is given as $e_t = y_t - \hat{y}_t$.

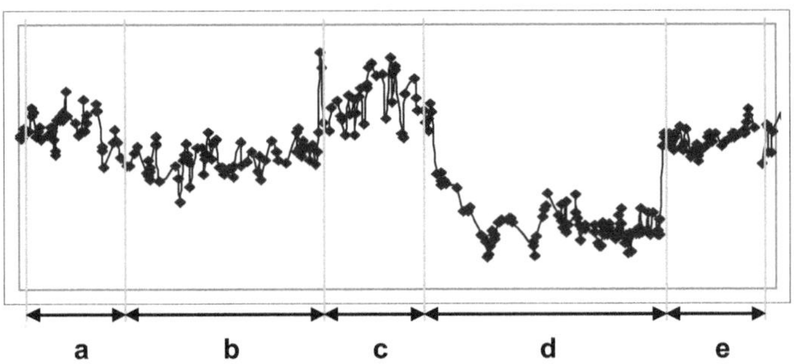

Fig. 4. Trend in deposition rate

Fig. 5 displays the residual X control chart for "period a" in Fig. 4. The upper control limit (UCL) and the lower control limit (LCL) are obtained by multiplying the standard deviation of the residual based on the ARIMA model of "period a" by 3. The control limits calculated for "period a" are extended, and investigations of other periods in Fig. 4 are conducted. The residual X control charts for "period c" and "period d" are shown in Fig. 6 and Fig. 7, respectively. It can be seen that five points are beyond the control limits for "period c" in Fig. 6. On the other hand, the X control chart in Fig. 7 for "period d" indicates that the control process is in control. The X control charts indicate the in control state in "period b" and "period e", like in "period d". Thus, process adjustment in these periods to work correctly.

However, the control chart in Fig. 6 for "period c" indicates that it is out-of-control. In "period c" there is a lower frequency of simultaneous processing of four lots than in the other periods, and there are many instances in which film thickness cannot be measured at processing position 4. Unfortunately, the variation in film thickness at position 4 is the largest. It may be that the missing values at processing position 4 cause the out-of-control state. Even in cases, when the number of processed lots in one batch is less than 4, measurement at all processing positions should be performed.

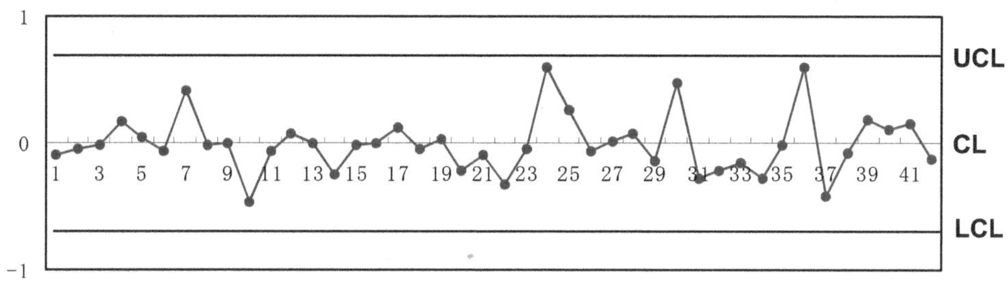

Fig. 5. Residual X control chart for "period a"

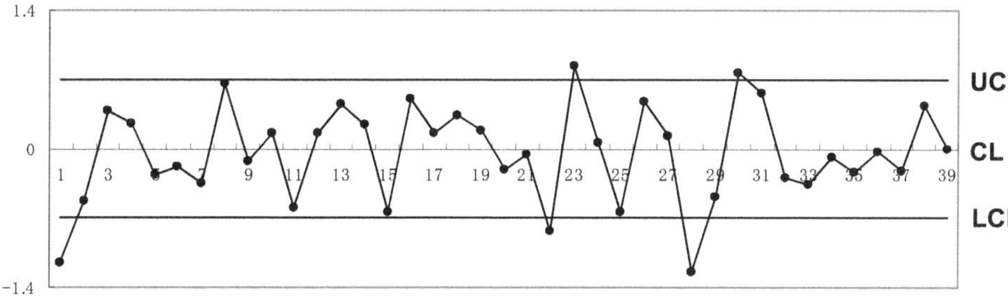

Fig. 6. Residual X control chart for "period c"

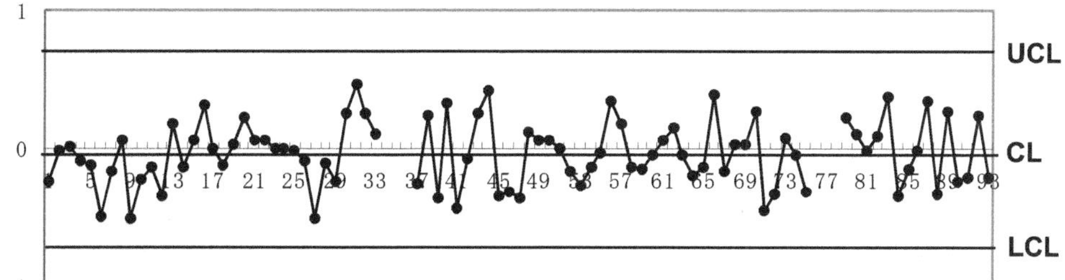

Fig. 7. Residual X control chart for "period d"

4 Process maintenance

4.1 Process maintenance and the need to control it

Accumulating reaction by-products and/or deteriorating parts may affect not only between-batch variation but also within-batch variation. Process adjustment is done to stabilize the process mean, that is, to reduce the between-batch variation; however, process adjustment can not control within-batch variation.

Therefore, process maintenance must complement process adjustment. Process maintenance aims at eliminating reaction by-products and replacing degraded parts to restore the process. Process maintenance is not only effective in restoring the between-batch conditions, but also the within-batch conditions.

However, for each maintenance action the process has to be stopped and, therefore, it is maintenance should be performed only as much as necessary, which leads to the problem to fix the maintenance times.

The second problem refers to which degree the process conditions shall be improved by process maintenance. Moreover, if maintenance is conducted wrongly, it may lead to an out-of-control state. Thus, a maintenance failures can be cause great damage.

We consider the maintenance times and the maintenance results and propose for them an effective control method.

4.2 Outline of the process in the case study

We consider a CMP (Chemical Mechanical Polishing) process as a case study for process maintenance. The CMP process is currently the most common planarizing technique in the semiconductor manufacturing process. The surface of a semiconductor wafer is made to contact a polishing pad while a polishing compound called slurry, a liquid containing silica powder, is added to the process. The CMP process is based on chemical and mechanical properties, but it is unknown, whichof them is dominant. A schematic diagram of a CMP apparatus is shown in Fig. 8.

The lots in a semiconductor manufacturing process are made up of a maximum of 25 wafers. The film thickness of 1 to 4 wafers is measured before and after CMP. In this study, however, measurements are taken from one wafer per lot only, because the within lot variation is very small. Measurements are made at the nine points of a wafer shown in Fig. 9.

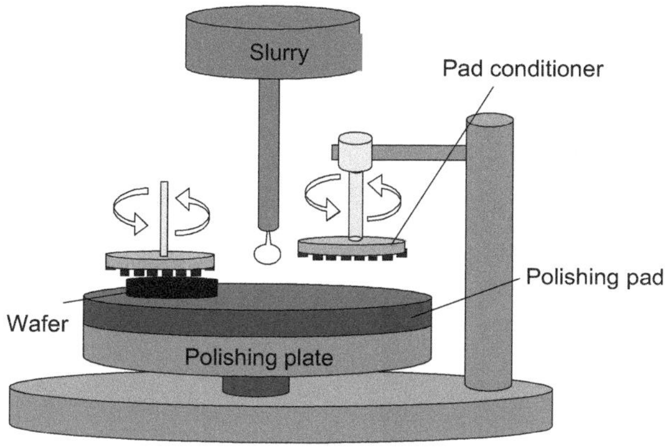

Fig. 8. Simple illustration of CMP process

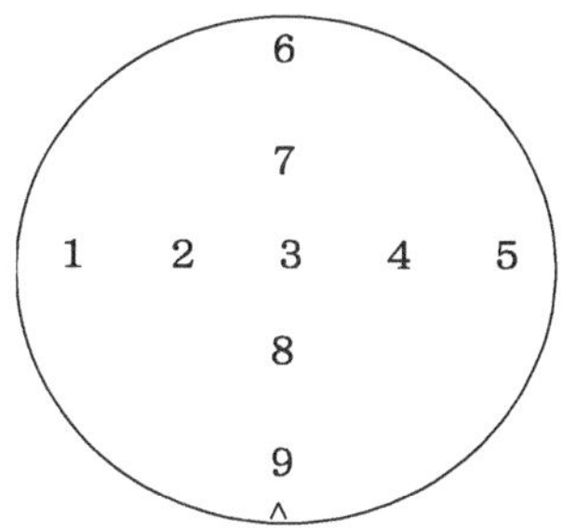

Fig. 9. Film thickness measurement points on wafer

4.3 Control chart using principal component analysis

During the CMP process, the film thickness constitutes the control characzteristic, which is measured before and after polishing. A change in the film thickness pattern on the surface is seen before and after maintenance. We tried to quantify the change of pattern and analyze the relation between the change pattern and process maintenance.

Fig. 10 shows two frequently observed patterns of the remaining film thickness. The problem is to derive an appropriate statistic to quantify the differences in these patterns.

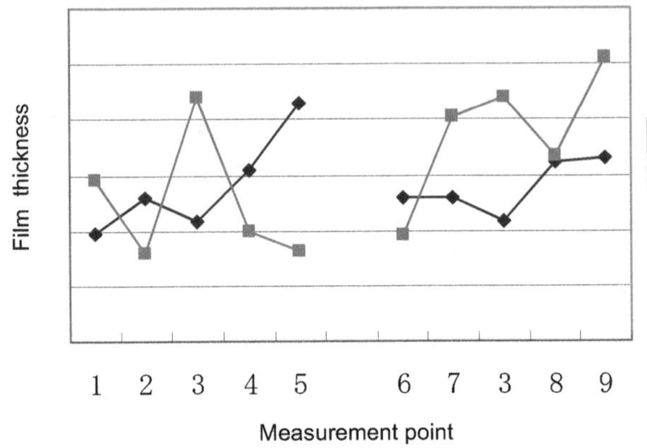

Fig. 10. Typical film patterns

Table 1. Principal component analysis after double centralized transformation

eigenvalue	83489.72	26026.51	15146.84
contribution	53.75	16.76	9.75
cumulative contribution	53.75	70.50	80.25
eigenvector			
point 1	-0.267	0.115	0.320
point 2	0.155	0.147	0.174
point 3	-0.424	0.519	-0.116
point 4	0.086	0.398	0.071
point 5	0.627	0.115	0.198
point 6	0.253	-0.354	0.063
point 7	-0.342	-0.577	0.407
point 8	0.220	-0.186	-0.647
point 9	-0.308	-0.177	-0.470

For deriving an appropriate statistics, we used Principal Component Analysis for quantifying the remaining film pattern. Double centralized transformation y_{ij} (see Equation (4)) is useful as preprocessing to identify wafer with different remaining film pattern. PCA from a sum of products matrix of y_{ij} is applied.

$$y_{ij} = z_{ij} - z_{i\cdot} - z_{\cdot j} + z_{\cdot\cdot} \quad (4)$$

z_{ij}: remaining film thickness,

$$z_{i\cdot} = \frac{\sum_{j=1}^{9} z_{ij}}{9}, \quad z_{\cdot j} = \frac{\sum_{i} z_{ij}}{sample\ size}, \quad z_{\cdot\cdot}: \text{total average}$$

The results of the analysis are shown in Table 1. The first principal component has an explanatory power of 53%, although double centralized transformation is conducted. In addition, it is confirmed that the first eigenvector shown in Table 1 is reproduced in other periods. From this result, the first principal component scores are utilized as the statistic to represent the pattern. Fig. 11 shows the first principal component scores as a time series. The dashed lines in Fig. 11 stand for the maintenance time points.

From Fig. 11, it can be seen that between two subsequent maintenance times the first principal component score gradually rises. After maintenance the score changes greatly, shifting from a large value to a small value.

Thus, for CMP process maintenance the change in remaining film pattern can be detected by using the first principal component score after double centralized transformation. Thus, by monitoring the first principal component score before and after maintenance, it is possible to control the effect of process maintenance.

However, we can not say so far whether monitoring the proposed statistic alone is adequate to control process maintenance. Besides, the number of particles on a wafer should be monitored for controlling process maintenance.

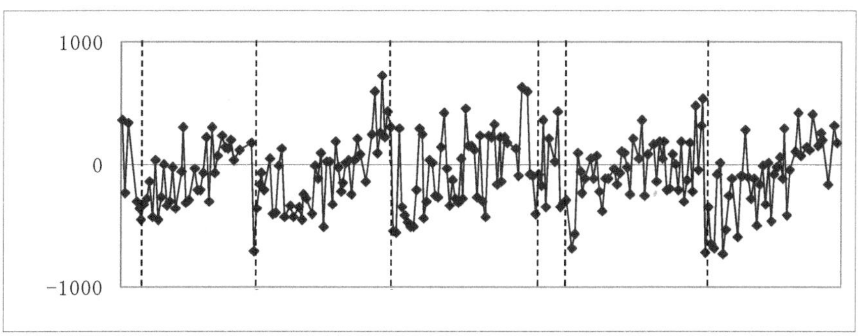

Fig. 11. Trend in the first principal component score

5 Conclusive remarks

Statistical process control has been developed focussing on the parts industry. Consequently, it is questionable whether the methods may be applied to the chemical process industry. In the present study we considered the semiconductor manufacturing process and proposed a statistical control method, which integrates SPC and APC.

Two case studies were analysed and led to the following conclusions with respect to SPC for the semiconductor manufacturing process:

(a) Process adjustment and process maintenance are the keys to preserve process stability; therefore, effective control of them is required.

(b) Residual *X* control charts based on a time series model is useful to control the process adjustment system.

(c) Monitoring within-batch variation is useful to control process maintenance.

It should be noted that these conclusive remarks do not only apply to semiconductor manufacturing processes. Therefore, it is planned to extend the research on other chemical processes.

References

1. Box, G. E. P. and Kramer, T. (1992), "Statistical Process Monitoring and Feedback Adjustment - A Discussion", Technometrics, Vol. 34, 251 - 285.

2. Kawamura, H., Nishina, K., Higashide, M., and Shimazu, Y. (2008), "Integrating SPC and APC in Semiconductor Manufacturing Process", *Quality*, Vol. 38, No. 3, 99 - 107 (in Japanese).

3. Montgomery, D. C., and Friedman, D. J. (1989), "Some Statistical Process Control Methods for Autocorrelated Data", *Journal of Quality Technology*, Vol. 23, 179 - 193.

4. Box, G. E. P., Jenkins, G. M., and Reinsel, G. C. (1994), *Time Series Analysis, Forecasting, and Control*, 3rd edition, Prentice-Hall, Englewood Cliffs, NJ.

The MAX-CUSUM Chart

Smiley W. Cheng[1] and Keoagile Thaga[2]

[1]Department of Statistics, University of Manitoba, Winnipeg, Manitoba, Canada, R3T 2N2

[2]Department of Statistics, University of Botswana, Private Bag UB00705, Gaborone, Botswana

Summary. Control charts have been widely used in industries to monitor process quality. We usually use two control charts to monitor the process. One chart is used for monitoring process mean and another for monitoring process variability, when dealing with variables data. A single Cumulative Sum (CUSUM) control chart capable of detecting changes in both mean and standard deviation, referred to as the Max-CUSUM chart is proposed. This chart is based on standardizing the sample means and standard deviations. This chart uses only one plotting variable to monitor both parameters. The proposed chart is compared with other recently developed single charts. Comparisons are based on the average run lengths. The Max-CUSUM chart detects small shifts in the mean and standard deviation quicker than the Max-EWMA chart and the Max chart. This makes the Max-CUSUM chart more applicable in modern production process where high quality goods are produced with very low fraction of nonconforming products and there is high demand for good quality.

1 Introduction

Control charts are basic and most powerful tools in statistical process control (SPC) and are widely used for monitoring quality characteristics of production processes. The first types of control charts were developed by Shewhart in the 1920s and ever since, several new charts have been developed in an effort to improve their capability to quickly detect a shift of the process from a target value. The statistical control chart, generally with 3σ action limits and 2σ warning limits, is the longest established statistical form of graphical tool. The control chart statistics are plotted by simply plotting time on the horizontal axis and a quality characteristic on the vertical axis. A quality characteristic is regarded to be in an in-control state if the statistic falls within the action limits of the chart and out-of-control if the statistic plots outside the action limits.

The disadvantage of the Shewhart control charts is that they only use the information about the process contained in the last plotted point and

H.-J. Lenz et al. (eds.), *Frontiers in Statistical Quality Control 9*,
DOI 10.1007/978-3-7908-2380-6_6, © Springer-Verlag Berlin Heidelberg 2010

information given by the entire sequence of points. This has led to various additional rules about runs of points above the mean. Attempts to incorporate the information from several successive results have resulted in charts based on some form of weighted mean of past results. In particular the Arithmetic Running Mean has been used in some instances by Ewan (1963).

One of the charts developed as an improvement to the Shewhart chart is the cumulative sum (CUSUM) chart developed by Page in 1954. This technique plots the cumulative sums of deviations of the sample values from a target value against time. The Shewhart control chart is effective if the magnitude of the shift is 1.5σ to 2σ or larger (Montgomery (2001)). The CUSUM charts are highly recommended by Marquardt (1984) for use in the U.S industry since they can detect small changes in the distribution of a quality characteristic and thus maintain tight control over a process. An important feature of the CUSUM chart is that it incorporates all the information in the sequence of sample values by plotting the cumulative sums of the deviations of the sample values from a target value. If there is no assignable cause of variation, the two-sided CUSUM chart is a random walk with a drift zero. If the process statistic shifts, a trend will develop either upwards or downwards depending on the direction of the shift and thus if a trend develops, a search for the assignable causes of variation should be taken. The magnitude of a change can be determined from the slope of the CUSUM chart and a point at which a change first occurred is the point where a trend first developed. The ability to show a point at which changes in the process mean began makes the CUSUM chart a viable chart and also helps to quickly diagnose the cause of changes in the process. There are two ways to represent CUSUM charts: the tabular (or algorithmic) CUSUM chart and the V-mask form of the CUSUM chart. We shall discuss the tabular CUSUM chart in this article.

The CUSUM chart for the change in the mean and the standard deviation for variables data have been extensively studied and two separate plots generated to assess for a shift from the targeted values. Constructing individual charts for the mean and standard deviation is very cumbersome and sometimes tedious and Hawkins (1993) suggested plotting the two statistics on the same plot using different plotting variables. This produces a chart that is somewhat complicated to interpret and is congested with many plotting points on the same chart. Other charts have been developed with an effort to propose single charts to monitor both the mean and the standard deviation of the process such as those suggested by Cheng and Spiring (1998), Domangue and Patch (1991), Chao and Cheng (1996), Chen and Cheng (1998) and Cheng and Xie (1999).

The major objective of this paper is to develop a single CUSUM chart that simultaneously monitors both the process mean and variability by using a single plotting variable. This chart is capable of quickly detecting both small and large shifts in the process mean and/or standard deviation and is also capable of handling cases of varying sample sizes.

2 The New Control Chart

Let $X_i = X_{i1}, \ldots, X_{in}, i = 1, 2, \ldots$ denote a sequence of samples of size n taken on a quality characteristic X. It is assumed that, for each i, $X_{i1}, \ldots, X_{in}$ are independent and identically distributed observations following a normal distribution with means and standard deviations possibly depending on i, where i indicates the ith group. Let μ_0 and σ_0 be the nominal process mean and standard deviation previously established. Assume that the process parameters μ and σ can be expressed as $\mu = \mu_0 + a\sigma_0$ and $\sigma = b\sigma_0$, where $a = 0$ and $b = 1$ when the process is in control, otherwise the process has changed due to some assignable causes. Then a represents a shift in the process mean and b a shift in the process standard deviation and $b > 0$.

Let $\bar{X}_i = (X_{i1} + \ldots + X_{in})/n$ and $S_i^2 = \frac{\sum_{j=1}^{n} (X_{ij} - \bar{X}_i)^2}{n-1}$ be the mean and variance for the ith sample respectively. The sample mean $\bar{X}_i$ and sample variance S_i^2 are the uniformly minimum variance unbiased estimators for the corresponding population parameters. These statistics are also independently distributed as do the sample values. These two statistics follow different distributions. The CUSUM charts for the mean and standard deviation are based on $\bar{X}_i$ and S_i respectively.

To develop a single CUSUM chart, we define the following statistics:

$$Z_i = \sqrt{n}\frac{(\bar{X}_i - \mu_0)}{\sigma_0} \tag{1}$$

$$Y_i = \Phi^{-1}\left\{H\left[\frac{(n-1)S_i^2}{\sigma_0^2}; n-1\right]\right\}, \tag{2}$$

where $\Phi(z) = P(Z \leq z)$, for $Z \sim$N (0, 1) the standard normal distribution. Φ^{-1} is the inverse of the standard normal cumulative distribution function, and $H(w; p) = P(W \leq w|p)$ for $W \sim \chi_p^2$ the chi-square distribution with p degrees of freedom. The functions Z_i and Y_i are independent and when the process variance is at its nominal value, Y_i follows the standard normal distribution. The CUSUM statistics based on Z_i and Y_i are given by

$$C_i^+ = \max[0, Z_i - k + C_{i-1}^+], \tag{3}$$

$$C_i^- = \max[0, -Z_i - k + C_{i-1}^-], \tag{4}$$

and

$$S_i^+ = \max[0, Y_i - v + S_{i-1}^+], \tag{5}$$

$$S_i^- = \max[0, -Y_i - v + S_{i-1}^-], \tag{6}$$

respectively, where C_0 and S_0 are starting points. Because Z_i and Y_i follow the same distribution, a new statistic for the single control chart can be defined as

$$M_i = \max[C_i^+, C_i^-, S_i^+, S_i^-]. \tag{7}$$

The statistic M_i will be large when the process mean has drifted away from μ_0 and/or when the process standard deviation has drifted away from σ_0. Small values of M_i indicate that the process is in statistical control. Since M_i's are non-negative, they are compared with the upper decision interval only.

The average run length (ARL) of a control chart is often used as the sole measure of performance of the chart. The ARL of the chart is the average number of points that must be plotted before a point plots above or below the decision interval. If this happens, an out-of-control signal is issued and a search for an assignable cause(s) of variation must be mounted. A chart is considered to be more efficient if its ARL is smaller than those of all other competing charts when the process is out of control and the largest when the process is in control.

The out-of-control signal is issued when either the mean or the standard deviation or both have shifted from their target values. Therefore the plan (the sample size and control limits) is chosen so that the ARL is large, when the process is in control and small when the process is out of control. Cox (1999) suggested that the criteria for a good chart are acceptable risks of incorrect actions, expected average quality levels reaching the customer and expected average inspection loads. Therefore the in-control ARL should be chosen so as to minimize the frequency of false alarms and to ensure adequate response times to genuine shifts.

For a predetermined in-control ARL, for quickly detecting shifts in the mean and variability, an optimal combination of h and k is determined which will minimize the out-of-control ARL for a specified change in the mean and standard deviation, where h is the decision interval and k is the reference value of the chart. The proposed chart is sensitive to changes in both mean and standard deviation when there is an increase in the standard deviation and is less sensitive when the standard deviation shifts downwards. This phenomenon has been observed for other charts based on the standardized values Domangue and Patch (1991).

3 Design of a Max-CUSUM chart

We use the statistic M_i to construct a new control chart. Because M_i is the maximum of four statistics, we call this new chart the Max-CUSUM chart. Monte Carlo simulation is used to compute the in control ARL for our Max-CUSUM chart. For a given in-control ARL, and a shift for the mean and/or standard deviation intended to be detected by the chart, the reference value k is computed as half the shift. For these values (ARL, k), the value of the decision interval (h) follows. For various changes in the process mean and/or standard deviation, each ARL value is also obtained by using 10 000 simulations.

Table 1 gives the combinations of k and h for an in-control ARL fixed at 250. We assume that the process starts in an in-control state with mean zero ($\mu_0 = 0$) and standard deviation of one ($\sigma_0 = 1$) and thus the initial value of the CUSUM statistic is set at zero. For example if one wants to have in-control ARL of 250 and to guard against $3\sigma_0$ increase in the mean and $1.25\sigma_0$ increase in the standard deviation, i.e., $a = 3$ and $b = 1.25$, the optimal parameter values are $h = 1.215$ and $k = 1.500$. These shifts can be detected on the second sample, i.e., the ARL is approximately two. A good feature of the Max-CUSUM chart is that smaller shifts in the process mean are detected much faster than in the single Shewhart chart (Max chart) as seen in the next section.

Table 1 shows that small values of k with large values of h result in quick detection of small shifts in mean and/or standard deviations. If one wants to guard against $3\sigma_0$ increase in the mean and $3\sigma_0$ increase in the standard deviation, the value of $h = 1.220$ and the value of $k = 1.500$. But for a $1\sigma_0$ increase in mean and $1.25\sigma_0$ increase in standard deviation, $h = 4.051$ and the value of k decreases to $k = 0.500$.

The Max-CUSUM scheme is sensitive to both small and large shifts in both mean and standard deviation. A $0.25\sigma_0$ increase in the process mean reduces the ARL from 250 to about 53 and a $1.25\sigma_0$ increase in the process standard deviation with a $0.25\sigma_0$increase in the process mean reduces the ARL from 250 to about 41 runs. If both parameters increase by large values, the ARL is reduced to 2. Thus the increase will be detected within the second sample. For example, a $3\sigma_0$ increase in both parameters will be detected within the second sample.

Another alternative method of assessing the performance of the CUSUM chart is to fix the values of h and k and calculate the ARL's for various shifts in the mean and/or standard deviation. This is displayed in Table 2. The value of $k = 0.5$ and thus we want to detect a $1\sigma_0$ shift in the mean and $h = 4.051$. This combination gives an in-control ARL = 250. From Table 2 it can be concluded that, even when the chart is designed to detect a $1\sigma_0$ shift in the process, it is sensitive to both small and large shifts in the mean and/or standard deviation.

4 Comparison with other Procedures

In this section, the performance of the Max-CUSUM chart is compared with those of several other charts used for quality monitoring. Most of the CUSUM charts developed are designed to monitor the mean and standard deviation separately, even the combined CUSUM charts developed monitor these parameters separately in the same plots. This is done by plotting the charts using different plotting variables for the means and standard deviations, and then calculating ARLs separately for each parameter. The ARL for the chart will be taken as the minimum of the two. The new chart (Max-CUSUM) is

Table 1. (k, h) combinations and the corresponding ARL for the Max-CUSUM chart with $\text{ARL}_0 = 250$

		$ARL_0 = 250$							
		a							
b	Para-meter	0.00	0.25	0.50	1.00	1.50	2.00	2.50	3.00
1.00	h	4.051	8.572	6.161	4.051	2.981	2.103	1.554	1.220
	k	0.500	0.125	0.250	0.500	0.750	1.000	1.250	1.500
	ARL	250.21	53.21	22.04	7.99	4.56	2.76	1.96	1.50
0.5	h	4.051	8.572	6.161	4.051	2.981	2.103	1.554	1.220
	k	0.500	0.125	0.250	0.500	0.750	1.000	1.250	1.500
	ARL	91.30	67.33	25.00	8.24	4.43	2.62	1.84	1.30
1.25	h	4.051	8.572	6.161	4.051	2.981	2.103	1.554	1.220
	k	0.500	0.125	0.250	0.500	0.750	1.000	1.250	1.500
	ARL	82.42	41.01	18.98	7.40	4.21	2.73	1.91	1.47
1.50	h	4.051	8.572	6.161	4.051	2.981	2.103	1.554	1.220
	k	0.500	0.125	0.250	0.500	0.750	1.000	1.250	1.500
	ARL	41.84	34.92	16.37	6.93	4.11	2.72	1.89	1.43
2.00	h	4.051	8.572	6.161	4.051	2.981	2.103	1.554	1.220
	k	0.500	0.125	0.250	0.500	0.750	1.000	1.250	1.500
	ARL	18.81	23.97	12.75	5.95	3.77	2.61	1.87	1.43
2.50	h	4.051	8.572	6.161	4.051	2.981	2.103	1.554	1.220
	k	0.500	0.125	0.250	0.500	0.750	1.000	1.250	1.500
	ARL	11.78	18.16	10.22	5.11	3.37	2.55	1.86	1.42
3.00	h	4.051	8.572	6.161	4.051	2.981	2.103	1.554	1.220
	k	0.500	0.125	0.250	0.500	0.750	1.000	1.250	1.500
	ARL	8.63	14.32	8.49	4.53	3.15	2.47	1.84	1.39
4.00	h	4.051	8.572	6.161	4.051	2.981	2.103	1.554	1.220
	k	0.500	0.125	0.250	0.500	0.750	1.000	1.250	1.500
	ARL	5.85	10.16	6.38	3.80	2.87	2.36	1.81	1.37

Table 2. ARL's for the Max-CUSUM chart with $h = 4.051$ and $k = 0.500$

	a							
b	0.00	0.25	0.50	1.00	1.50	2.00	2.50	3.00
1.00	250.21	69.66	29.33	7.99	4.95	3.44	2.68	2.24
1.25	82.42	36.97	19.58	7.40	4.93	3.44	2.64	2.21
1.50	41.84	24.08	15.20	6.93	4.86	3.42	2.61	2.17
2.00	18.81	13.72	10.39	5.95	4.59	3.37	2.59	2.12
2.50	11.78	9.55	7.88	5.11	4.26	3.33	2.55	2.04
3.00	8.63	7.40	6.41	4.53	3.96	3.28	2.47	1.94
4.00	5.85	5.31	4.84	3.80	3.49	3.25	2.44	1.88

compared with the omnibus CUSUM chart proposed by Domanque and Patch (1991), the Max chart by Chen and Cheng (1998) and the Max-EWMA chart by Cheng and Xie (1999).

Table 4 shows the ARL's for the Max-CUSUM chart and the omnibus CUSUM chart developed by Domangue and Patch (1991) for shifts shown in Table 3. For various changes in the mean and/or standard deviation, we have calculated the ARL's for the Max CUSUM chart and compared them with those given by Domangue and Patch (1991) in Table 4. The Max-CUSUM chart performs better than the omnibus CUSUM chart for all shifts since its ARL's are smaller than those of the omnibus chart. The Max-CUSUM chart is also easy to plot and read as compared to the omnibus CUSUM chart since it plots only one plotting variable for each sample.

Table 3. Level of shifts in mean and standard deviation considered

Label	μ	σ
S_1	0.75	1.0
S_2	1.5	1.0
S_3	0	1.2
S_4	0	1.4
S_5	0.75	1.3
S_6	1.0	1.2

Table 4. ARL's of the Max-CUSUM chart and the omnibus CUSUM chart

$k = 1$ $h = 1.279$ $\alpha = 0.5$ $n = 1$						
Scheme	S_1	S_2	S_3	S_4	S_5	S_6
Omnibus CUSUM	37.0	7.0	50.4	21.5	15.7	13.0
Max-CUSUM	9.2	3.1	26.1	15.5	6.3	5.0

In Table 5 we compare the Max-CUSUM chart with the Max chart. The Max-CUSUM chart is more sensitive for small shifts in the mean than the Max chart and there is no significant difference in the performance of these charts at larger shifts even though the Max chart has slightly lower ARLs for large shifts. This is a major improvement in the CUSUM scheme as existing CUSUM charts are less sensitive to large shifts in the process mean and/or standard deviation.

Table 6 shows the performance of the Max-CUSUM chart and Max-EWMA chart for in-control ARL = 250. Both charts are sensitive to small and large shifts in the mean and/or standard deviation with the Max-EWMA chart performing better than the Max- CUSUM chart for both small and large shifts. These two charts use only one plotting variable for each sample and have good procedures of indicating the source and direction of shifts in the process.

5 Charting Procedures

The charting procedure of a Max-CUSUM chart is similar to that of the standard upper CUSUM chart. The successive CUSUM values, M_i's are plotted against the sample numbers. If a point plots below the decision interval, the process is said to be in statistical control and the point is plotted as a dot point. An out-of-control signal is given if any point plots above the decision interval and is plotted as one of the characters defined below. The Max-CUSUM chart is a combination of two two-sided standard CUSUM charts. The following procedure is followed in building the CUSUM chart:

1. Specify the following parameters; h, k, δ and the in-control or target value of the mean μ_0 and the nominal value of the standard deviation σ_0.
2. If μ_0 is unknown, use the sample grand average $\overline{\overline{X}}$ of the data to estimate it, where $\overline{\overline{X}} = (\bar{X}_1 + \cdots + \bar{X}_m)/m$. If σ_0 is unknown, use $\overline{R}/d_2$ or $\overline{S}/c_4$ to estimate it, where $\overline{R} = (R_1 + \cdots + R_m)/m$ is the average of the sample ranges and $\overline{S} = (S_1 + \cdots + S_m)/m$ is the average of the sample standard deviations, and d_2 and c_4 are statistically determined constants.
3. For each sample compute Z_i and Y_i.
4. To detect specified changes in the process mean and standard deviation, choose an optimal (h, k) combination and calculate C_i^+, C_i^-, S_i^+ and S_i^-.
5. Compute the M_i's and compare them with h, the decision interval.
6. Denote the sample points with a dot and plot them against the sample number if $M_i \leq h$.
7. If any of the M_i's is greater than h, the following plotting characters should be used to show the direction as well as the statistic(s) that is plotting above the decision interval:
   a) If $C_i^+ > h$, plot $C+$. This shows an increase in the process mean.
   b) If $C_i^- > h$, plot $C-$. This indicates a decrease in the process mean.
   c) If $S_i^+ > h$, plot $S+$. This shows an increase in the process standard deviation.
   d) If $S_i^- > h$, plot $S-$. This shows a decrease in the process standard deviation.
   e) If both $C_i^+ > h$ and $S_i^+ > h$, plot $B++$. This indicates an increase in both the mean and the standard deviation of the process.
   f) If $C_i^+ > h$ and $S_i^- > h$, plot $B+-$. This indicates an increase in the mean and a decrease in the standard deviation of the process.
   g) If $C_i^- > h$ and $S_i^+ > h$, plot $B-+$. This indicates a decrease in the mean and an increase in the standard deviation of the process.
   h) If $C_i^- > h$ and $S_i^- > h$, plot $B--$. This shows a decrease in both the mean and the standard deviation of the process.

Table 5. ARL for Max-CUSUM chart and the Max chart

$ARL_0 = 250$							$n = 4$					
	Max-CUSUM						Max Chart					
	a						a					
b	0.00	0.25	0.50	1.00	2.00	3.00	0.00	0.25	0.50	1.00	2.00	3.00
1.00	250.21	69.66	29.33	7.99	3.44	2.24	250.0	143.8	49.3	7.2	1.2	1.0
1.25	82.42	36.97	19.58	7.40	3.44	2.21	34.3	27.2	15.9	4.9	1.3	1.0
1.50	41.84	24.08	15.20	6.93	3.42	2.17	9.8	8.9	6.9	3.5	1.3	1.0
2.00	18.81	13.72	10.39	5.95	3.37	2.04	2.9	2.8	2.6	2.1	1.3	1.1
3.00	8.63	7.40	6.41	4.53	3.28	1.94	1.4	1.4	1.4	1.3	1.2	1.1

Table 6. ARL for Max-CUSUM chart and the Max-EWMA chart

$ARL_0 = 250$												
	Max-CUSUM						Max-EWMA					
	a						a					
b	0.00	0.25	0.50	1.00	2.00	3.00	0.00	0.25	0.50	1.00	2.00	3.00
1.00	250.21	69.66	29.33	7.99	3.44	2.24	250.0	24.6	8.6	2.9	1.1	1.0
1.25	82.42	36.97	19.58	7.40	3.44	2.21	17.8	12.3	7.1	2.9	1.2	1.0
1.50	41.84	24.08	15.20	6.93	3.42	2.17	6.3	5.7	4.5	2.5	1.2	1.0
2.00	18.81	13.72	10.39	5.95	3.37	2.04	2.5	2.5	2.3	1.8	1.2	1.0
3.00	8.63	7.40	6.41	4.53	3.28	1.94	1.7	1.6	1.6	1.5	1.2	1.1

8. Investigate the cause(s) of the shift for each out-of-control point in the chart and carry out the remedial measures needed to bring the process back into an in-control state.

6 An Example

A Max-CUSUM chart is applied to real data obtained from DeVor, Chang and Sutherland (1992). The data is for measurements of the inside diameter of the cylinder bores in an engine block. The measurements are made to 1/10,000 of an inch. Samples of size $n = 5$ are taken roughly every half hour, and the first 35 samples are given in Table 7. The actual measurements are of the form 3.5205, 3.5202, 3.5204 and so on. The entries given in Table 7 provide the last three digits in the measurements.

Suppose based on past experience, an operator wants to detect a 1σ shift in the mean, that is $a = 1$ and a 2σ shift in the standard deviation, that is $b = 2$ with an in-control ARL $= 250$, the corresponding decision interval from Table 1 is $h = 2.475$ and the reference value is $k = 0.500$. The chart is developed as follows: The nominal mean μ_0 is estimated by $\overline{\overline{X}}$ and σ_0 is estimated by $\bar{S}/c_4$. The sample produced the following estimates $\overline{\overline{X}} = 200.25$ and $\bar{S}/c_4 = 3.31$.

The Max-CUSUM chart in Figure 1 which plots all the 35 observations shows that several points plot above the decision interval. Sample number 6 shows an increase in the standard deviation. After this sample, samples 7 and 8 also plot above the decision interval. However these points show that the standard deviation is decreasing towards the in-control region. Due to very high value of the CUSUM statistic for the standard deviation in sample 6, the successive cumulative values at samples 7 and 8 show a value above the decision interval even though the standard deviation values corresponding to these samples are in control. We therefore investigate the cause of higher variability at sample number 6. According to DeVor, Chang and Sutherland (1992), this sample was taken when the regular operator was absent, and a relief, inexperienced operator was in charge of the production line and thus could have affected the process.

Sample number 11 also plots above the decision interval. This point corresponds to an increase in the mean. This corresponds to a sample taken at 1:00 P.M. when production had just resumed after lunch break. The machines were shut down at lunch time for tool changing and thus these items were produced when the machines were still cold. Once the machines warmed up, the process settled to an in-control state. This shows that the shift in the mean was caused by the machine tune-up problem.

Sample 16 also plots above the decision interval; this shift shows an increase in the standard deviation. According to DeVor, Chang and Sutherland (1992), this sample corresponds to a time when an inexperienced operator was in control of the process. In addition to the above mentioned points which also

Table 7. Cylinder diameter data

Sample	X_{i1}	X_{i2}	X_{i3}	X_{i4}	X_{i5}
1	205	202	204	207	205
2	202	196	201	198	202
3	201	202	199	197	196
4	205	203	196	201	197
5	199	196	201	200	195
6	203	198	192	217	196
7	202	202	198	203	202
8	197	196	196	200	204
9	199	200	204	196	202
10	202	196	204	195	197
11	205	204	202	208	205
12	200	201	199	200	201
13	205	196	201	197	198
14	202	199	200	198	200
15	200	200	201	205	201
16	201	187	209	202	200
17	202	202	204	198	203
18	201	198	204	201	201
19	207	206	194	197	201
20	200	204	198	199	199
21	203	200	204	199	200
22	196	203	197	201	194
23	197	199	203	200	196
24	201	197	196	199	207
25	204	196	201	199	197
26	206	206	199	200	203
27	204	203	199	199	197
28	199	201	201	194	200
29	201	196	197	204	200
30	203	206	201	196	201
31	203	197	199	197	201
32	197	194	199	200	199
33	200	201	200	197	200
34	199	199	201	201	201
35	200	204	197	197	199

plotted above the control limit in the Shewhart chart, Max chart and EWMA chart, sample 34 plots above the decision interval. This point corresponds to a decrease in the standard deviation. The Shewhart S chart plotted this value close to the lower control limit but within the acceptable area. Table 1 show that the Max-CUSUM chart is very sensitive to small shift and thus signals for this small decrease in the standard deviation.

When these four samples are removed from the data, new estimates for the mean and standard deviation were computed, giving the following; $\overline{\overline{X}} =$

200.08 and $\bar{S}/c_4 = 3.02$. The revised chart is shown in Figure 2. The chart plots only one point above the decision interval. This point corresponding to sample 1 shows an increase in the mean. It corresponds to a sample that was taken at 8:00 A.M., roughly the start up of the production line in the morning, when the machine was cold. Once the machine warmed up, the production returns to an in-control state.

When sample 1 is removed from the data, we re-calculate the estimates and obtain $\overline{\overline{X}} = 199.93$ and $\bar{S}/c_4 = 3.06$. The Max-CUSUM chart for this new data is shown in Figure 3. All the points plot within the decision interval showing that the process is in-control.

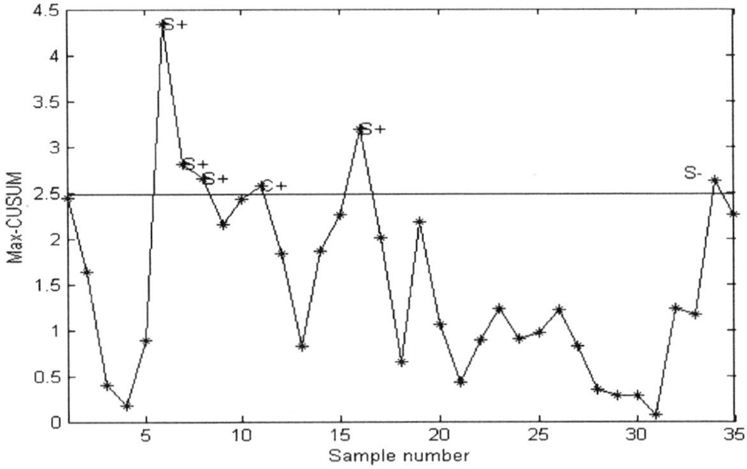

Fig. 1. The first Max-CUSUM control chart for the cylinder diameter data

7 Conclusion

The ARL for this chart reduces as the shift increases. One disadvantage of the standard CUSUM chart is that it does not quickly detect a large increase in the process parameters and thus is not recommended for large increase in both mean and variability.

A good feature of the Max-CUSUM chart developed here is its ability to quickly detect both small and large changes in both the process mean and the process variability. Another advantage of the Max-CUSUM is that we are able to monitor both the process center and spread by looking at one chart. The performance of the proposed Max-CUSUM is very competitive in comparison with the Max chart and the Max-EWMA chart.

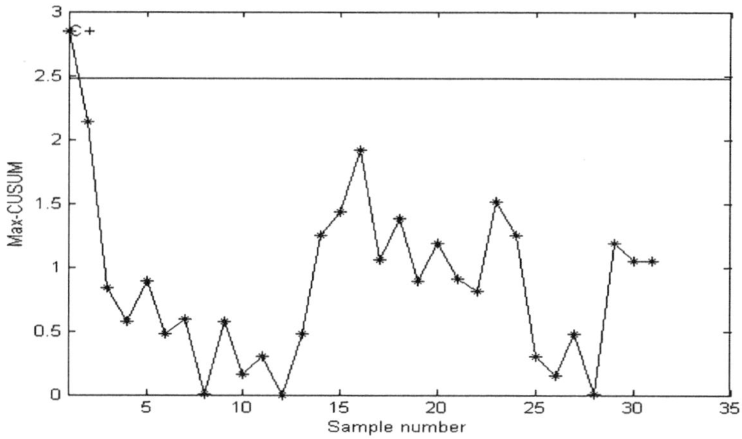

Fig. 2. The second Max-CUSUM control chart for the cylinder diameter data

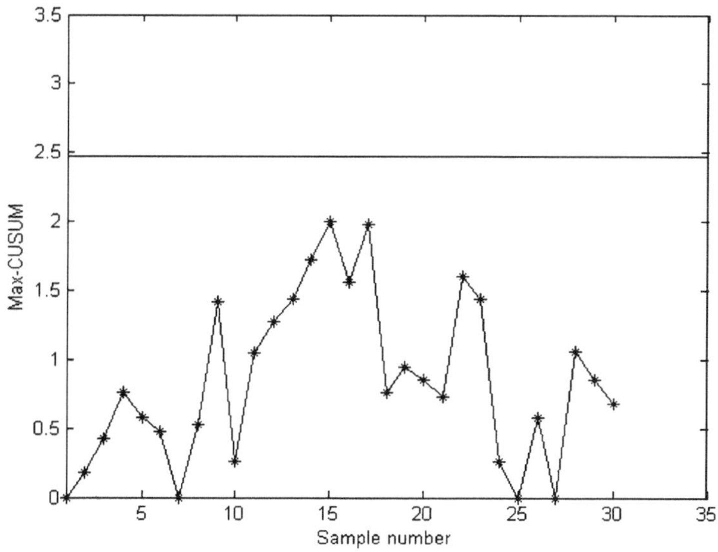

Fig. 3. The third Max-CUSUM control chart for the cylinder diameter data

References

1. Brook, D. and Evans, D. A. (1972). An Approach to the Probability Distribution of CUSUM Run Length. *Biometrika* **59**, 539549.
2. Champ, C. W. and Woodall, W. H. (1987). Exact Results for Shewhart Control Charts With Supplementary Runs Rules. *Technometrics*, **29**, 393-399.
3. Chao, M. T. and Cheng, S. W. (1996). Semicircle Control Chart for

Variables Data. *Quality Engineering*, **8(3)**, 441-446.
4. Chen, G. and Cheng, S. W. (1998). Max-chart: Combining X-Bar Chart and S Chart. *Statistica Sinica* **8**, 263-271.
5. Cheng, S. W. and Spiring, F. A. (1990). An Alternative Variable Control Chart: The Univariate and MultivariateCase. *Statistica Sinica*, **8**, 273-287.
6. Cox, M. A. A. (1999). Toward the Implementation of a Universal Control Chart and Estimation of its Average Run Length Using a Spreadsheet. *Quality Engineering*, **11**, 511-536.
7. Domangue, R. and Patch, S. C. (1991). Some Omnibus Exponentially Weighted Moving Average Statistical Process Monitoring Schemes. *Technometrics* **33**, 299-313.
8. DeVor, R. E., Chang, T. and Sutherland, J. W. (1992). *Statistical Quality Design and Control.* Macmillan, New York.
9. Ewan, W. D. (1963). When and How to Use CUSUM Charts. *Technometrics*, **5**, 1-22.
10. Gan, F. F. (1993). The Run Length Distribution of a Cumulative Sum Control Chart. *Journal of Quality Technology* **25**, 205-215.
11. Hawkins, D. M. (1981). A CUSUM for Scale Parameter. *Journal of Quality Technology* **13**, 228-231.
12. Hawkins, D. M. (1992). A Fast Approximation for Average Run Length of CUSUM Control Charts. *Journal of Quality Technology* **24**, 37-43.
13. Hawkins, D. M. (1993). Cumulative Sum Control Charting: An Underutilized SPC Tool. *Quality Engineering*, **5**, 463-477.
Hawkins, D. M. and Olwell, D .H. (1998). *Cumulative Sum Charts and Charting for Quality Improvement.* Springer, New York.
14. Lucas, J. M. (1982). Combined Shewhart-CUSUM Quality Control Schemes. *Journal of Quality Technology* **14**, 51-59.
15. Lucas, J. M. and Crosier, R. B. (2000). Fast Initial Response for CUSUM Quality Control Scheme: Give Your CUSUM a Head Start. *Technometrics*, **42**, 102-107.
16. Lucas, J. M. and Saccucci, M. S. (1990). Exponentially Weighted Moving Average Control Schemes: Properties and Enhancements. *Technometrics*, **32**, 1-12.
17. Marquardt, D. W. (1984). New Technical and Educational Directions for Managing Product Quality. *The American Statistician*, **38**, 8-14.
18. Montgomery, D. C. (2001). *Introduction to Statistical Quality Control.* 4th Edition, John Wiley & Sons, Inc., New York.
Page, E. S. (1954). Continuous Inspection Schemes. *Biometrika*, **41**, 100-115.
19. Xie, H. (1999). Contribution to Qualimetry. PhD. thesis, University of Manitoba, Winnipeg, Canada.

Sequential Signals on a Control Chart Based on Nonparametric Statistical Tests

Olgierd Hryniewicz and Anna Szediw

Systems Research Institute, ul. Newelska 6, 01-447 Warsaw, Poland
{hryniewi, szediw}@ibspan.waw.pl

Summary. The existence of dependencies between consecutive observations of a process makes the usage of SPC tools much more complicated. In order to avoid unnecessary costs we need to have simple tools for the discrimination between correlated and uncorrelated process data. In the paper we propose a new control chart based on Kendall's tau statistic which can be used for this purpose. In case of normally distributed observations with dependence of an autoregressive type the proposed Kendall control chart is nearly as good as a well known autocorrelation chart, but outperforms this chart when these basic assumptions are not fulfilled.

1 Introduction

Statistical process control (SPC) is a collection of methods for achieving continuous improvement in quality. This objective is accomplished by continuous monitoring of the process under study in order to quickly detect the occurrence of assignable causes. The Shewhart control chart, and other control charts - like CUSUM, MAV, and EWMA - are the most popular SPC methods used to detect whether observed process is under control. Their classical, and widely known, versions are designed under the assumption that process measurements are described by independent and identically distributed random variables. In the majority of practical cases these assumptions are fulfilled at least approximately. However, there exist production processes where consecutive observations are correlated. This phenomenon can be frequently observed in chemical processes, and many other continuous production processes (see Wardell et al. [25] or Alwan and Roberts [2] for examples). The presence of correlations between consecutive measurements should be taken into account during the design of control charts. This need was noticed in the 1970s, see for example the papers by Johnson and Bagshaw [9] and by Vasilopoulos and Stamboulis [24], but the real outburst of papers related to this problem took place in the late 1980s and in the 1990s.

There exist several approaches for dealing with serially correlated SPC data. First approach, historically the oldest one, consists in adjusting con-

H.-J. Lenz et al. (eds.), *Frontiers in Statistical Quality Control 9*,
DOI 10.1007/978-3-7908-2380-6_7, © Springer-Verlag Berlin Heidelberg 2010

trol limits of classical control charts. This approach was used, for example, in papers [11],[12],[13],[18],[19],[23],[24],[30]. The second approach, which is represented by a seminal paper by Alwan and Roberts [1], profits from the knowledge of the correlation structure of a measured process. Alwan and Roberts [1] propose to chart so called residuals, i.e. differences between actual observations and their predicted, in accordance with a prespecified mathematical model, values. This approach has its roots in the statistical analysis of time series used in automatic control, originated by the famous book of Box and Jenkins [3]. Among many papers on control charts for residuals we can mention e.g. the following: [11],[12],[13],[16],[19]. Lu and Reynolds [11],[12],[13] compared control charts for original observations (with corrected limits) with control charts for residuals. They have shown that the best practical results may be obtained while using simultaneously EWMA (or CUSUM) modified charts for original observations and Shewhart charts for residuals. According to the third approach, introduced in the area of SPC by Yourstone and Montgomery [28],[29], a process is monitored by charting statistics, such as the coefficient of serial correlation, that reflect the correlation structure of the monitored process. Another original approach of this type was proposed in the paper by Jiang et al. [8] who proposed a new type of a chart for monitoring of autocorrelated data an ARMA chart.

SPC procedures for autocorrelated data have been usually proposed for charting individual observations that are typical for continuous production processes. Relatively few papers have been proposed for the analysis of SPC procedures in case of autocorrelation within the sample. A good overview of those papers together with interesting original results can be found in the paper by Knoth et al. [10]. Another important problem which has attracted only few authors is related to the control of short-run processes. Some interesting results in this area can be found in papers [21],[27].

The number of papers devoted to the problem of charting autocorrelated processes is quite large. Therefore the readers are encouraged to look at those papers for further references. The review of first papers devoted to the problem of SPC with correlated data can be found, for example, in the aforementioned paper by Wardell et al. [25], and in a short overview paper by Woodall and Faltin [26]. A good overview of the papers published in 1990s can be found in [10],[12].

While discussing different SPC methods used for the analysis of correlated data we have to take into account their efficiency and as it was pointed out by Lucas [14] in the discussion of [25], simplicity. The results obtained by Wardell et al. [25] for the case of Auto Regressive Moving Average (ARMA) time series that describe correlated measurements have shown that Alwan Roberts type control charts for residuals in certain cases of positive correlation may be outperformed even by a classical Shewhart control chart with unmodified control limits. However, in other cases the application of these rather complicated procedures (their usage requires the software for the analysis of time series) may be quite useful. Also the results obtained by Timmer et al. [22]

show that in the case of Auto Regressive AR(1) processes the application of simple control charts based on the serial autocorrelation coefficient as the only SPC tools may be not effective.

While dealing with correlated data we cannot rely, even in the case of classical control charts, on the methods used for the estimation of their parameters in case independent observations. Some corrections are necessary, as it was mentioned e.g. in the paper by Maragah and Woodall [15]. Another problem with the application of the procedures designed to control autocorrelated data is the knowledge of the structure of correlation. In the majority of papers it is assumed that the type of the stochastic process that describes the process data is known. Moreover, it is also assumed that the parameters of this stochastic process are also known. However, Lu and Reynolds [12],[13] have shown that precise estimation of such parameters requires at least hundreds of observations. Taking into account that all computations required for designing and running SPC procedures for autocorrelated data are not easy for an average practitioner, the problem arises then: how to verify the hypothesis of correlation in a simple way? The answer to this question is very important, as it indicates the amount of possible future difficulties with running SPC procedures. It is quite obvious that practitioners would like to avoid these problems as it can be only possible. The simplest solution to this problem is to use the serial autocorrelation control chart, as it was proposed in [28],[29]. However, the coefficient of serial autocorrelation performs well for processes with Gaussian random error components. For more general processes specialists in time series analysis suggest to apply nonparametric statistical tests. An interesting review of such tests can be found in the paper by Hallin and Mélard [7]. Unfortunately, the majority of those tests are either complicated or unsuitable for process control. However, one of the recent papers [5] on the application of Kendall's tau statistics for testing serial dependence seems to be promising in the context of SPC. We consider this possibility in this paper.

In the second section of the paper we present some basic information on the Kendall's tau statistic when it is used for the analysis of autocorrelated data. Using basic properties of this statistic we propose a relatively simple control chart based on this statistic. Statistical properties of this control chart have been investigated using Monte Carlo simulations. In the third section of the paper we present the results of Monte Carlo experiments in the case when two consecutive observations are described by a two-dimensional normal distribution, i.e. in the case of a simple autoregressive model. We compare the behavior of our chart with the behavior of a chart based on the serial autocorrelation coefficient. In the fourth section we analyze the results of simulations when the dependence structure is more complicated. We consider the case when two consecutive observations are described by a two-dimensional copula with different marginal probability distribution functions. In the last section of the paper we formulate some conclusions, and propose the directions for further investigations.

2 Control chart based on Kendall's tau statistic for serially autocorrelated data

Many of problems experienced when applying traditional SPC to monitoring processes are caused by the violation of the basic assumption of statistical independence of consecutive observations. However, in practice this condition is very often not fulfilled, and consecutive observations are correlated. It should be stressed that some small disturbances of independence conditions may be natural and even desirable (e.g. in case of the existence of favourable trends of process parameters). However, in the majority of cases autocorrelation of process parameters should be considered either as an obstacle in monitoring the process or even as unwanted feature, when it increases process variation. In such situation special statistical methods for detecting dependencies (autocorrelations) between consecutive process observations are strongly recommended. For this purpose we propose to use the Kendall's τ statistic, which is a fundamental statistical measure of association.

Let $Z_1, Z_2, \ldots, Z_n$ denote a random sample of n consecutive process observations and (X_i, Y_i), where $X_i = Z_i$ and $Y_i = Z_{i+1}$ for $i = 1, 2, \ldots, n-1$ is a bivariate random vector. Then, the Kendall's τ sample statistic measuring the association between random variables X and Y is given by the following formula

$$\tau_n = \frac{4}{n-1} \sum_{i=1}^{n-1} V_i - 1, \tag{1}$$

where

$$V_i = \frac{card\{(X_j, Y_j) : X_j < X_i, Y_j < Y_i\}}{n-2}, i = 1, \ldots, n-1. \tag{2}$$

In terms of the original observations Kendall's tau can be represented as a function of the number of disconcordances M, i.e. the number of pairs (Z_i, Z_{i+1}) and (Z_j, Z_{j+1}) that satisfy either $Z_i < Z_j$ and $Z_{i+1} > Z_{j+1}$ or $Z_i > Z_j$ and $Z_{i+1} < Z_{j+1}$. In these terms we have

$$\tau_n = 1 - \frac{4M}{(n-1)(n-2)}, \tag{3}$$

where

$$M = \sum_{i=1}^{n-1} \sum_{j=1}^{n-1} I(Z_i < Z_j, Z_{i+1} > Z_{j+1}), \tag{4}$$

and $I(A)$ represents the indicator function of the set A. In case of mutually independent pairs of observations (X_i, Y_i), $i = 1, 2, \ldots, n-1$ the probability distribution of (1) is well known. However, in case of time series, even in the case of mutual independence of $Z_1, Z_2, \ldots, Z_n$ pairs of observations (X_i, Y_i) become dependent, and the probability distribution of τ_n for small values of n has been obtained only recently [5]. Ferguson et al. [5] obtained precise probabilities $P_n(M \leq m)$ for $n = 3, \ldots, 10$, and approximate probabilities for $n > 10$. In Table 1 we present the probabilities of $\tau_n \geq \tau_{crit}$ for some selected small values of n. This type of presentation is more useful for the discussion of the applicability of Kendall's tau in SPC. First of all, from Table 1 it is clearly seen that except for the case $n = 6$ and $\tau_{crit} = 1$ it is not possible to construct a one-sided statistical test of independence against the alternative of positive dependence with the same probability of false alarms as in the case of a Shewhart control chart. By the way, this exceptional case is equivalent to a well known supplementary pattern test signal on a classical control chart: "six observations in a row are either increasing or decreasing". Moreover, all critical values of τ_n that are equal to one correspond to sequential signals of the type "n observations in a row are either increasing or decreasing", and for the values τ_{crit} that are close to one it is possible to formulate similar pattern rules. Thus, for small values of n it is in principle impossible to make precise comparisons of control charts based on Kendall's tau with classical three-sigma Shewhart control charts. Close investigation of the probability distribution of M presented in [5] shows that due to a discrete nature of the Kendall's tau this situation is the same in case of a two-sided test and also similar even for larger values of n.

Table 1. Critical values for Kendall's tau statistic in presence of dependence between pairs of observations

n=6		n=7		n=8		n=9	
τ_{crit}	$P(\tau_n \geq \tau_{crit})$	τ_{crit}	$P(\tau_n \geq \tau_{crit})$	τ_{crit}	$P(\tau_n \geq \tau_{crit})$	τ_{crit}	$P(\tau_n \geq \tau_{crit})$
1	0.00267	1	0.00042	1	0.00006	1	0.00001
0.8	0.00834	0.866	0.00119	0.904	0.00014	0.857	0.00007
0.6	0.03056	0.733	0.00477	0.809	0.00069	0.785	0.00021
		0.600	0.01356	0.714	0.00178	0.714	0.00071
				0.619	0.00565	0.642	0.00185
						0.571	0.00514

Consecutive values of τ_n are dependent even for independent original observations. Therefore, it is rather difficult to obtain the values of ARLs. In Table 2 we present such values, each based on over 1 million simulations in case of small ARLs and over 10 000 simulations for very large ARLs, obtained for the case of mutual independence of normally distributed observations, when the alarm signal is generated when $\tau_n \geq \tau_{crit}$. The results presented in Table 2 confirm our claim that the construction of a test having statistical properties similar to the properties of a Shewhart control chart is hardly possible. Due to a discrete character of τ_n ARLs in case of independence are either very large, and this suggests poor discrimination power of the test, or rather low, resulting in a high rate of false alarms.

Bearing in mind the requirement of simplicity we can now propose a Shewhart type control chart based on τ_n with control limits of the following form:

$$LCL = \max\left(E(\tau_n) - k\sigma(\tau_n), -1\right), \tag{5}$$

$$UCL = \min\left(E(\tau_n) + k\sigma(\tau_n), 1\right), \tag{6}$$

where LCL and UCL are the lower and upper limit, respectively. In the remaining part of this paper we will name it *the Kendall control chart.* To calculate the limits of the Kendall control chart we use the following formulae for the expected value and the variance of τ_n given in [5]:

$$E(\tau_n) = -\frac{2}{3(n-1)}, n \geq 3, \tag{7}$$

$$V(\tau_n) = \frac{20n^3 - 74n^2 + 54n + 148}{45(n-1)^2(n-2)^2}, n \geq 4. \tag{8}$$

It is worth noting that for small values of n the probability distribution of τ_n is not symmetric. Therefore, the properties of the proposed control chart for testing independence of consecutive observations from a process may be improved by using control lines that are asymmetric around the expected value of τ_n. However, for sake of simplicity, in this paper we will not consider this possibility. The properties of the proposed control chart are investigated in the next section of the paper.

Table 2. Values of ARL of Kendall's test for independent observations

n=6		n=7		n=8		n=9	
τ_{crit}	ARL	τ_{crit}	ARL	τ_{crit}	ARL	τ_{crit}	ARL
1	422.0	1	2885.8	1	22586.0	1	> 150000.0
0.8	151.4	0.866	1013.5	0.904	7799.0	0.857	15395.0
0.6	47.5	0.733	270.6	0.809	1891.2	0.785	5618.0
		0.600	106.5	0.714	705.9	0.714	1784.7
				0.619	246.0	0.642	724.9
						0.571	291.4

3 Properties of the Kendall control chart in case of dependencies described by a multivariate normal distribution

In order to analyze basic properties of the proposed Kendall control chart let us consider the simplest case when two consecutive observations are described by a bivariate normal distribution. Let $X_i = Z_i$ and $Y_i = Z_{i+1}$, $i = 1, 2, \ldots, n-1$ denote the random variables describing two consecutive observations in a sample of size n. We want to model the stochastic dependence between them. In order to do it we assume that the join probability distribution of the random vector (X, Y) is the bivariate normal distribution with the following probability distribution

$$f(x,y) = \frac{1}{2\pi\sigma_X\sigma_Y\sqrt{1-\rho^2}} \exp{(-Q)}, \tag{9}$$

where

$$Q = \frac{1}{1-\rho^2}\left\{\frac{(x-m_X)^2}{\sigma_X^2} - 2\rho\frac{(x-m_X)(y-m_Y)}{\sigma_X\sigma_Y} + \frac{(y-m_Y)^2}{\sigma_Y^2}\right\}, \tag{10}$$

and parameter ρ is the coefficient of correlation. If the variables X and Y are independent, then we have $\rho = 0$.

The conditional cumulative distribution function of Y given $X = x$ is the normal distribution with the mean $m_Y + \rho(\sigma_X/\sigma_Y)(x - m_X)$ and the variance of $\sigma_Y\sqrt{1-\rho^2}$. In a particular case when $m_X = m_Y = 0$ and $\sigma_X = \sigma_Y = 1$ it is the normal distribution with mean of $x\rho$ and variance of $\sqrt{1-\rho^2}$. Thus, the proposed model is the well known autoregression model.

The basic characteristic that describes the performance of control charts is the Average Run Length (ARL). ARL is calculated as the average number of samples (or individual observations) plotted on a control chart up to and including the point that gives rise to a decision that a special cause is present. In Table 3 we present the results of simulation (each entry of the table is calculated as the average from one million simulation runs) for different sample sizes (numbers of considered consecutive points) n, and different values of the correlation coefficient ρ.

The results presented in Table 3 reveal that a simple Kendall control chart with a simple to remember three-sigma decision rule, and a small sample size n, is not a good tool for finding dependencies between consecutive observations. Obviously, the common value $k = 3$ cannot be used for all values of n. Moreover, the discrimination ability of the Kendall control chart for the sample sizes n smaller than 10 seems to be insufficient, even if we decrease the value of k. For larger value of n the situation looks better, but the discrimination power of this simple Kendall chart is still insufficient.

Now, let us analyze the behavior of a simple autocorrelation chart in similar circumstances. To design this chart we assume that the expected value

Table 3. Values of ARL of the Kendall control chart with k=3 for observations described by a bivariate normal distribution

				n			
ρ	6	7	8	9	10	20	50
0.8	47.36	63.71	67.81	50.56	44.96	26.67	50.11
0.5	90.94	165.44	221.86	165.35	154.87	66.41	59.94
0.2	140.82	416.60	702.13	610.14	689.44	473.52	328.82
0.1	146.23	520.36	1156.77	859.95	1593.03	1026.06	1058.76
0	141.90	595.74	1623.15	1043.36	1497.27	1502.60	2597.57
-0.1	129.64	607.00	2003.14	1020.18	1597.13	1011.92	1053.87
-0.2	111.47	577.85	2286.24	796.90	1257.70	465.59	327.97
-0.5	56.18	220.44	952.73	194.62	247.45	61.30	58.82
-0.8	23.97	53.98	145.83	41.04	43.43	23.78	50.05

of the plotted statistic and its variance are equal to their asymptotic values. Hence, we set $E(\rho_n) = 0$, and for the calculation of the variance we use an approximate simple formula proposed by Moran [17]

$$V(\rho_n) = \frac{n-1}{n(n+2)}. \tag{11}$$

Now, let us define control limits of this simple autocorrelation chart as $\pm k\sigma(\rho_n)$, and set $k = 3$. The values of ARLs for different values of n and ρ are presented in Table 4.

Table 4. Values of ARL of the autocorrelation control chart with k=3 for observations described by a bivariate normal distribution

				n			
ρ	6	7	8	9	10	20	50
0.8	2713.4	88.7	39.0	30.0	26.5	24.0	50.0
0.5	1539.3	251.7	115.8	87.6	75.7	51.7	57.0
0.2	732.8	796.3	494.7	415.5	386.1	352.9	301.1
0.1	400.4	991.1	793.0	720.3	693.6	822.7	1018.9
0	400.6	1004.8	1183.0	1190.2	1143.7	1427.0	2797.7
-0.1	287.3	834.3	1454.2	1662.3	1485.7	1121.7	1093.6
-0.2	200.1	590.9	1326.9	1677.6	1253.8	499.2	319.6
-0.5	61.3	133.9	266.9	297.2	185.3	57.8	56.8
-0.8	19.5	28.3	38.0	37.6	29.3	22.8	50.0

The results given in Table 4 look rather surprisingly. We would expect that for the assumed model of dependence the behaviour of the autocorrelation chart should be much better than the behaviour of the Kendall chart which is based on a nonparametric statistic. Surprisingly though, the behaviour of a simple autocorrelation chart does not seem much better. For small values of n the coefficient of autocorrelation is obviously biased, and the simple Moran's approximation may influence the results in a negative way. The direct comparison of the both charts using the data given in Table 3 and Table 4 is, of course, impossible. In order to make this comparison relevant we have calibrated both charts for $n = 10$ and $n = 50$ by choosing the proper value k such that the in-control ARLs are nearly the same. In case of $n = 10$ for the Kendall chart we set k to 2.7, and for the autocorrelation chart we set k to 2.65. In case of $n = 50$ for the Kendall chart we set k to 2.2, and for the autocorrelation chart we set k to 2.16. The results of the comparison (each value based on 10^5 simulations) are presented in Table 5 and Figure 1.

The results presented in Table 5 confirm our previous finding that a simply designed autocorrelation chart (without a correction for bias) for small values of n does not perform better than the Kendall chart. For large value of n, such as $n = 50$, surprisingly though, for the assumed dependence model the autocorrelation chart does not perform better, as it is expected to do. The effect of bias is still observed, and this results in better discrimination of negative dependence, and visibly worse discrimination for positive dependence. It is also worth noticing that fast detection of small correlations requires samples even much larger than $n = 50$.

Table 5. Values of ARL for equivalent Kendall and autocorrelation control chart

	n=10		n=50	
ρ	Kendall	autocorrelation	Kendall	autocorrelation
0.8	30.9	19.8	50	50
0.5	84.3	43.3	52.1	51.9
0.3	189.1	97.9	73.9	75.8
0.2	271.2	156.8	117.5	127.4
0.1	344.6	249.4	226.6	259.1
0	351.2	351.9	350.7	350.9
-0.1	286.4	385.6	226.3	189.8
-0.2	200.6	305.8	117.1	99.3
-0.3	132.2	193.3	73.3	66
-0.5	55.7	67.0	51.8	50.9
-0.8	19.4	19.4	50	50

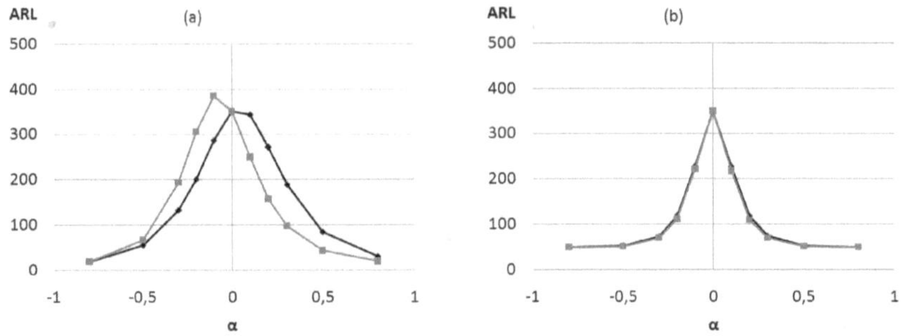

Fig. 1. ARL comparison for Kendall control chart (black line) and autocorrelation control chart (grey line) (a) $n = 10$, (b) $n = 50$

4 Properties of the Kendall control chart in case of other models of dependence

Let us consider now the sensitivity of the proposed Kendall chart and the simple autocorrelation chart to the change of some basic assumptions. First, we check how changes the performance of these charts when we assume that the consecutive observations in a sample are not independent. To describe diffrent kinds of dependencies between them we need model to model dependencies which are different from autoregression.

For the bivariate normal distribution the correlation coefficient completely defines the dependence structure between random variables. However, it is worth to notice that the random vector (X, Y) can be described by any two-dimensional probability distribution function. In such case, the information given by a correlation coefficient may be not sufficient to define the dependence structure between random variables. Therefore, to fully capture this structure one may consider another type of dependence described by a so called copula. The copula contains all of the information on the nature of the dependence between random variables. The joint cumulative distribution function $F_{12}(x, y)$ of any pair (X, Y) of continuous random variables may be written in the form $F_{12}(x, y) = C(F_1(x), F_2(y))$, $x, y \in R$, where $F_1(x)$ and $F_2(y)$ are the marginal cumulative distribution functions of X and Y, respectively, and $C : [0, 1]^2 \rightarrow [0, 1]$ is a copula. In this paper, to model another type of dependence between random variables we use, as examples, the Farlie-Gumbel-Morgenstern (FGM) copula (C_{FGM}), Plackett copula (C_P) and Frank copula (C_F). Using the probability integral transformations $u_1 = F_1(x)$, $u_2 = F_2(y)$, where u_1, u_2 have a uniform distribution on the interval $[0, 1]$, we can write the particular copulas as

$$C_{FGM}(u_1, u_2) = u_1 u_2 \{1 + \alpha(1 - u_1)(1 - u_2)\}, \tag{12}$$

$$C_P(u_1,u_2) = \begin{cases} u_1u_2, & \alpha = 1, \\ \frac{[1+(\alpha-1)(u_1+u_2)]-\sqrt{[1+(\alpha-1)(u_1+u_2)]^2-4u_1u_2\alpha(\alpha-1)}}{2(\alpha-1)}, & \alpha \in R_+ \setminus \{1\}, \end{cases} \tag{13}$$

$$C_F(u_1,u_2) = -\frac{1}{\alpha}\ln\left(1 + \frac{(e^{-\alpha u_1}-1)(e^{-\alpha u_2}-1)}{e^{-\alpha}-1}\right), \quad \alpha \in R \setminus 0. \tag{14}$$

The parameter α in above formulas for C_{FGM}, C_P and C_F describes the power and the direction of association between X and Y. The variables X and Y are independent if and only if $\alpha = 0, \alpha = 1, \alpha \cong 0$, respectively for C_{FGM}, C_P, C_F.

When $F_1(x)$ and $F_2(y)$ are the univariate cumulative distribution functions of the normal distribution, the marginal distributions of the FGM copula are normal, but the structure of dependence is different than in the case of the bivariate normal distribution. For example, when $\alpha \in [-1,1]$, there exists a limit on the coefficient of correlation, namely $\rho \leq \alpha/\pi$ (see [20]). The similar fact is also true for Plackett and Frank copulas (see [4], [6] for more details).

The Kendall control chart is based on a nonparametric statistic. Thus, its performance should not depend not only upon the type of a marginal distribution of observations, but upon the type of dependence as well. Using Monte Carlo simulations we investiagte the performance of the Kendall and autocorrelation charts. We assume that the dependence between pairs of observations Z_i, Z_{i+1}, $i = 1,\ldots,n-1$ in a sample of size n is described by a copula and we generate the random numbers sample. If the Kendall τ statistic calculated for this sample does not fall outside of the limits established for the Kendall chart, we move to the next process observation, i.e. we genarate random number Z_{n+1} and we calculate the Kendall τ for the sample $Z_1' = Z_2,\ldots,Z_n' = Z_{n+1}$. We repeat this step as long as the Kendall τ falls outside the limits. The number of process observations Z_i plotted on the control chart up to and including the last observation in a sample for which the Kendall τ is outside the limits defines ARL.

In Tables 6-7 and Figures 2-4 we present the average ARL values, each based on 10^5 simulations, obtained in case of normally distributed observations, where the dependence between consecutive observations is described by FGM, Plackett and Frank copula.

In Table 6 there are given the ARL values for the Kendall chart in case of FGM copula. On the basis of these results we can observe that the Kendall chart for FGM copula with normal marginal distributions behaves similarly as in case of bivariate normal distributions (some slight differences may come from two different applied generators of random numbers). This fact arises from nonparametric character of Kendall's tau statistic. It confirms fact that the performance of the Kendall control chart based on a nonparametric statistic should not be dependent on the type of dependence. Obviously, results for autocorrelation control chart for simple autoregressive model (see Table 4) and the same chart for FGM copula (see Table 7) do not confirm this fact and they are differ from each other in a meaningful way.

Table 6. ARL for the Kendall control chart with $k = 3$ for observations described by FGM copula with normal marginal distributions

		n						
α	ρ	6	7	8	9	10	20	50
1	0.318	128.83	297.60	470.94	371.06	389.99	191.19	112.55
0.942	0.3	129.50	315.19	510.86	395.31	420.79	216.42	127.04
0.628	0.2	136.43	402.49	729.07	578.95	651.87	443.14	288.60
0.314	0.1	145.80	490.94	1049.37	808.29	1016.84	969.05	955.37
0	0	144.74	578.02	1506.35	1009.02	1475.79	1481.33	2593.94
-0.314	-0.1	132.16	593.57	2101.18	994.87	1637.61	950.73	962.21
-0.628	-0.2	115.96	553.11	2469.10	809.24	1327.59	444.13	287.80
-0.942	-0.3	91.57	459.41	2325.11	566.49	875.34	213.47	124.05
-1	-0.318	92.87	438.73	2238.41	538.43	816.31	188.24	110.40

Table 7. ARL for the autocorrelation control chart with $k = 3$ for observations described by FGM copula with normal marginal distributions

		n						
α	ρ	6	7	8	9	10	20	50
1	0.318	1048.79	561.54	287.43	225.15	202.95	153.47	115.72
0.942	0.3	1021.87	592.45	311.75	246.39	222.83	170.98	129.73
0.628	0.2	813.13	794.85	472.06	397.84	373.21	348.94	303.60
0.314	0.1	566.77	961.89	750.39	670.50	657.14	793.17	999.54
0	0	392.76	998.18	1137.59	1155.81	1143.13	1459.37	2805.56
-0.314	-0.1	261.16	771.92	1443.81	1706.57	1512.51	1083.89	1079.53
-0.628	-0.2	177.08	531.78	1263.12	1661.35	1233.48	493.23	321.80
-0.942	-0.3	124.47	344.45	842.86	1102.55	736.39	226.63	133.82
-1	-0.318	116.06	317.42	772.82	1003.40	661.75	198.12	117.73

Now, we compare both control charts applied for FGM copula. In order to do it, in case of $n = 10$ for Kendall chart we set k to 2.71 and for autocorrelation chart we set k to 2.73. In case of $n = 50$ we set k to 2.17 for both charts. The results of those experiments are presented in Figure 2

We have repeated the analogical research for Plackett and Frank copulas and we have obtained similar results. It is worth to notice that in case of these two copulas for extreme or high values of parameter α, ARL are the minimum and equal to the sample size. Analysing results presented in Figures 2-4 we get the conclusion that the autocorrelation control chart does not perform better than the Kendall chart based on a nonparametric statistic.

It seems to be interesting to make the comparison of the Kendall and autocorrelations control charts in case of non-normal distributions. To verify the influence of character of distribution on the performance of the considered charts we make the analogical research under assumption that marginal

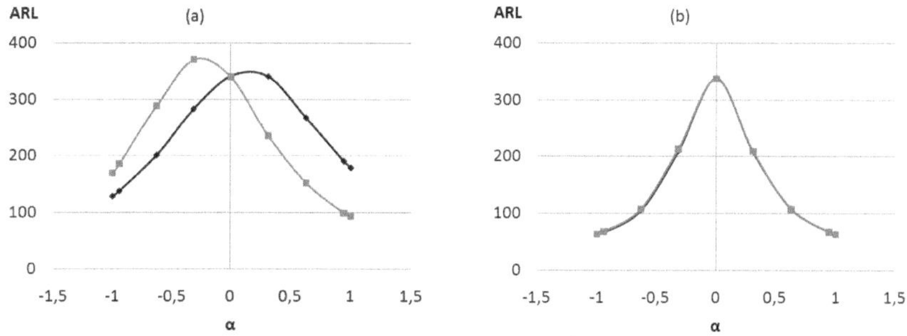

Fig. 2. ARL comparison for the Kendall chart (black line) and autocorrelation chart (grey line) for FGM copula with normal marginal distributions (a) $n = 10$, (b) $n = 50$

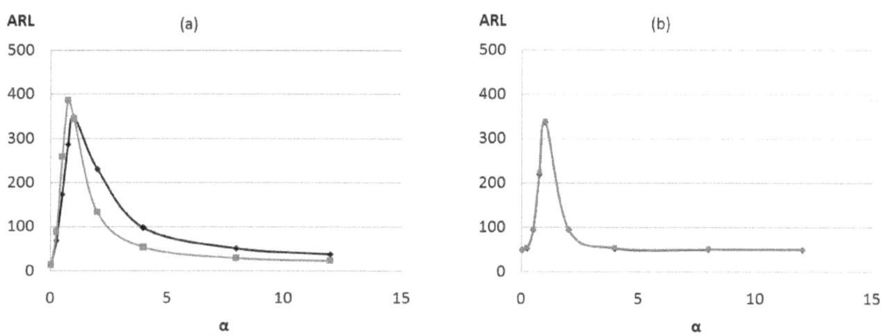

Fig. 3. ARL comparison for the Kendall chart (black line) and autocorrelation chart (grey line) for Plackett copula with normal marginal distributions (a) $n = 10$, (b) $n = 50$

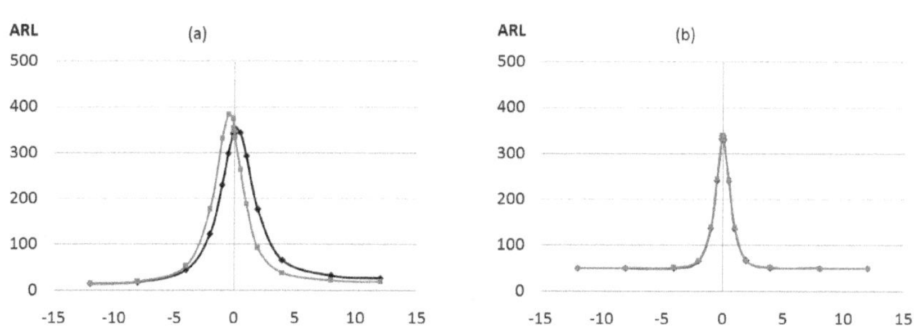

Fig. 4. ARL comparison for Kendall control chart (black line) and autocorrelation control chart (grey line) for Frank copula with normal marginal distributions (a) $n = 10$, (b) $n = 50$

distribution function is not normal. We consider two cases, first - we do not know that the distribution of our data is not normal, and second - we know the character of a non-normal distribution function of our data. In the last case we modify the width of the autocorrelation control chart and set a proper value of parameter k to get nearly the same ARLs for this chart and the Kendall chart in case of independence. The results obtained for FGM, Plackett and Frank copulas with exponential and uniform marginal distributions are presented in Figures 5-10.

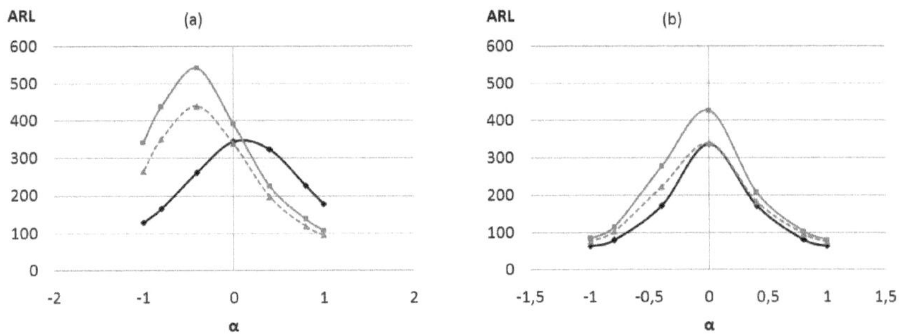

Fig. 5. ARL comparison for Kendall control chart (black line), autocorrelation control chart (grey line) and modified autocorrelation chart (grey dotted line) for FGM copula with exponential distributions (a) $n = 10$, (b) $n = 50$

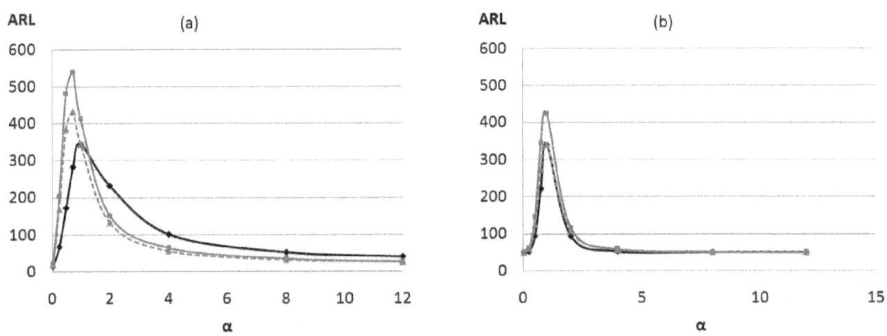

Fig. 6. ARL comparison for Kendall control chart (black line), autocorrelation control chart (grey line) and modified autocorrelation chart (grey dotted line) for Plackett copula with exponential marginal distributions (a) $n = 10$, (b) $n = 50$

Let's assume that we do not know the type of marginal distribution and we use the autocorrelation chart with determined for normal distribution value of $k = 2.73$. Then, if the marginal distributions are really non-normal, we see

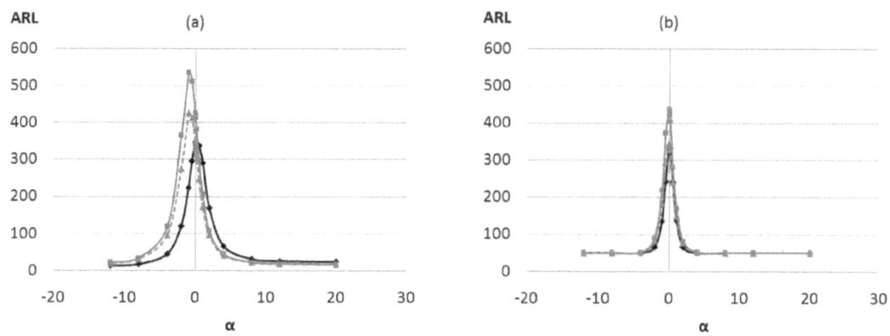

Fig. 7. ARL comparison for Kendall control chart (black line), autocorrelation control chart (grey line) and modified autocorrelation chart (grey dotted line) for Frank copula with exponential marginal distributions (a) $n = 10$, (b) $n = 50$

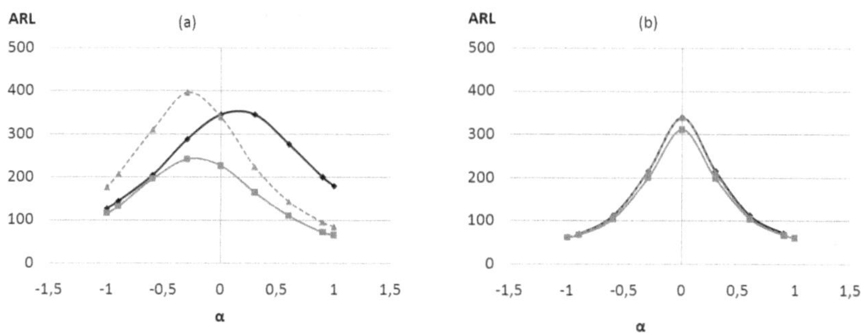

Fig. 8. ARL comparison for Kendall control chart (black line), autocorrelation control chart (grey line) and modified autocorrelation chart (grey dotted line) for FGM copula with uniform marginal distributions (a) $n = 10$, (b) $n = 50$

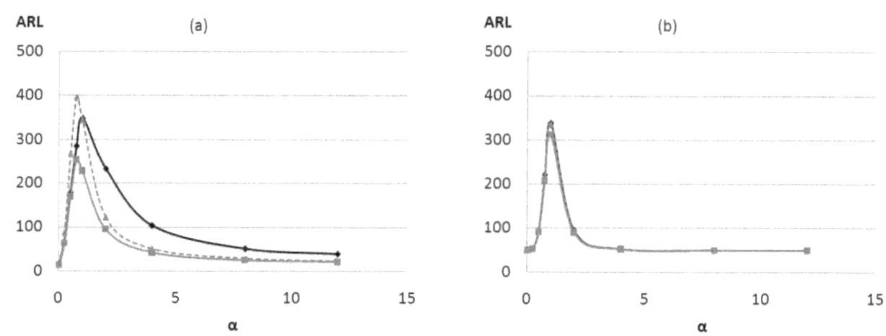

Fig. 9. ARL comparison for Kendall control chart (black line), autocorrelation control chart (grey line) and modified autocorrelation chart (grey dotted line) for Plackett copula with uniform marginal distributions (a) $n = 10$, (b) $n = 50$

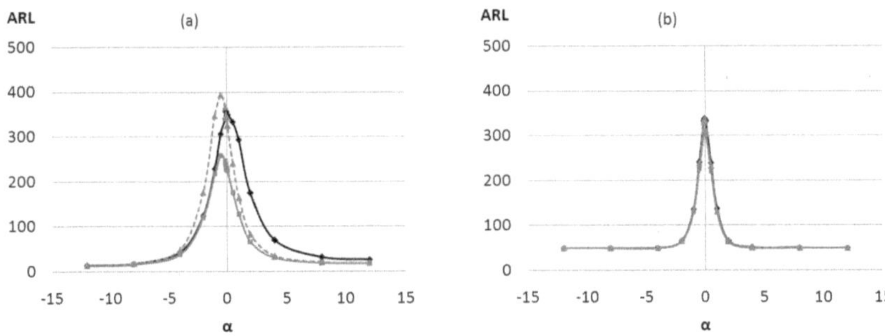

Fig. 10. ARL comparison for Kendall control chart (black line), autocorrelation control chart (grey line) and modified autocorrelation chart (grey dotted line) for Frank copula with uniform marginal distributions (a) $n = 10$, (b) $n = 50$

that ARL values are different from ARL values obtained in case of normal marginal distributions. So, in such case we have to set a suitable value of k, but for that purpose we need the information about our distribution. For $n = 10$, in case of positive dependence the ARL values are lower or nearly the same and in case of negative dependence they are greater than ARL values for Kendall chart. Whereas for $n = 50$ the Kendall control chart is not worse than autocorrelation control chart. In case of uni-form marginals it works almost like the autocorrelation chart, but in case of exponential marginal distributions it works better.

So, now let's assume that we know the type of our marginal distributions. Then we can set a proper value of k. In our research we set $k = 2.68$ and $k = 2.85$, respectively, for exponential and uniform distributions. Let's notice that even if we profit from this knowledge the autocorrelation chart does not perform better then the Kendall chart.

The results given in Figures 5-10 show, how sensitive is the autocorrelation chart to the assumption of the underlying distribution. The compared distributions have been chosen deliberately so different in order to magnify the differences. For small sample sizes n, in both cases of a heavy-tailed distribution (uniform) and a skewed distribution (exponential), the rate of false alarms is unacceptable high. Therefore, the autocorrelation chart should be specially tailored in case of different probability distribution of the plotted observations. This conclusion is hardly unexpected, but the observed differences in the behavior of the autocorrelation control chart in case of different distributions are very significant from a practical point of view. So, we should apply the autocorrelation chart very carefully if we do not know that our distribution is really the normal distribution. On the other hand, if we know the type of our distribution and can set a proper value of k, we do not obtain the control chart with better performance than Kendall control chart.

5 Conclusions

Mutual dependencies (correlations) between consecutive observations of processes may influence properties of SPC procedures in a dramatic way. This phenomenon has attracted the attention of many researchers for the last over twenty years. Many new or modified SPC tools have been pro-posed for dealing with this problem. However, their usage requires addi-tional skills, specialized software and is usually much more complicated than in the case of classical tools used for mutually independent observa-tions. Thus, from a practical point of view, it is very important to identify situations when an additional treatment of data is really necessary. In other words, there is a need to have a simple SPC tool, like a Shewhart control chart, which could be used for the detection of dependencies (autocorrela-tions) between observations of processes.

In this paper we have proposed such a tool based on a nonparametric Kendall's tau statistic. This tool is similar to a Shewhart control chart with plotted values of the Kendall's tau. We compared this new tool to a known autocorrelation chart. The results of the comparison show that in the case of dependencies described by autoregressive processes with a normal error component the new tool performs nearly as well as the autocorrelation chart. However, when the assumption of multivariate normality of consecu-tive observations is not fulfilled, the newly proposed Kendall control chart performs much better due to its distribution-free character.

The Kendall chart, in its simplest "three-sigma" form, does not perform well for small and moderate sample sizes. It has unnecessarily high values of ARL in case of independence, and hence, a very low rate of false alarms. Unwanted consequences of this feature are the high values of ARL in case of the existence of weak dependencies between observations. This situation can be improved by changing the limits of a control chart, but such im-provements may not be sufficient for the effective detection of weak de-pendencies. For such cases large sample sizes are required, and this un-pleasant from a practical point of view situation does not depend upon a statistical tool used for the detection of autocorrelation.

As it was mentioned above, it seems to be impossible to propose a very simple design (like e.g. using a "three-sigma" rule) of the Kendall control chart. Additional research is required in order to work-out guidance how to design an effective chart for different practical situations. Special attention should be paid to situations, when the type of dependence differs from a simple autoregressive process. There is also a need to investigate properties of the Kendall chart in case of shifts in a process mean value, and for other types of process deterioration. Preliminary investigations, not reported in this paper, suggest, however, that the Kendall control chart is not a good tool for the detection of such deteriorations.

References

1. L.C. Alwan and H.V. Roberts. Time-series modeling for statistical process control. *Journal of Business & Economic Statistics*, 6:87–95, 1988.
2. L.C. Alwan and H.V. Roberts. The problem of misplaced control limits. *Journal of the Royal Statistical Society Series C (Applied Statistics)*, 44:269–306, 1995.
3. G.E.P. Box, G.M. Jenkins, and G.C. Reinsel. *Time Series Analysis, Forecasting and Control*. Prentice-Hall, Englewood Cliffs, 3 edition, 1994.
4. D. Conway. Plackett family of distributions. In Campbell B. Kotz, S.and Read, N. Balakrishnan, and B. Vidakovic, editors, *Encyclopedia of Statistical Sciences*, pages 6164–6168. Wiley, New York, 2006.
5. T.S. Ferguson, C. Genest, and M. Hallin. Kendall's tau for serial dependence. *The Canadian Journal of Statistics*, 28:587–604, 2000.
6. C. Genest. Frank's family of bivariate distributions. *Biometrika*, 74:549–555, 1987.
7. M. Hallin and G. Mélard. Rank-based tests for randomness against first-order serial dependence. *Journal of the American Statistical Association*, 83:1117–1128, 1988.
8. W. Jiang, K.L. Tsui, and W.H. Woodall. The new SPC monitoring method: The ARMA chart. *Technomerics*, 42:399–410, 2000.
9. R.A. Johnson and M. Bagshaw. The effect of serial correlation on the peformance of CUSUM tests. *Technomerics*, 16:103–112, 1974.
10. S. Knoth, W. Schmid, and A. Schone. Simultaneous Shewhart-type charts for the mean and the variance of a time series. In H.J. Lenz and P.T. Wilrich, editors, *Frontiers in Statistical Quality Control VI*, pages 61–79. Physica Verlag, Heidelberg, 2001.
11. C.W. Lu and M.R. Reynolds Jr. Control charts for monitoring the mean and variance of autocorrelated processes. *Journal of Quality Technology*, 31:259–274, 1999.
12. C.W. Lu and M.R. Reynolds Jr. EWMA control charts for monitoring the mean of autocorrelated processes. *Journal of Quality Technology*, 31:166–188, 1999.
13. C.W. Lu and M.R. Reynolds Jr. CUSUM charts for monitoring an autocorrelated process. *Journal of Quality Technology*, 33:316–334, 2001.
14. J.M. Lucas. Discussion to the paper ,,Run-length distributions of special-cause control charts for correlated processes" by Wardell et al. *Technometrics*, 36:17–19, 2001.
15. H.D. Maragah and W.H. Woodall. The effect of autocorrelation on the retrospective X-chart. *Journal of Statistical Computation and Simulation*, 40:29–42, 1992.
16. D.C. Montgomery and C.M. Mastrangelo. Some statistical process control methods for autocorrelated data. *Journal of Quality Technology*, 23:179–193, 1991.
17. P.A.P. Moran. Some theorems on time series I. *Biometrika*, 34:281–291, 1947.
18. W. Schmid. On the run length of a Shewhart control chart for correlated data. *Statistical Papers*, 36:111–130, 1995.
19. W. Schmid. On EWMA charts for time series. In H.J. Lenz and P.T. Wilrich, editors, *Frontiers in Statistical Quality Control V*, pages 115–137. Physica Verlag, Heidelberg, 1997.
20. W. R. Schucany, W. C. Parr, and J. E. Boyer. Correlation structure in Farlie-Gumbel-Morgenstern distributions. *Biometrika*, 65:650–653, 1978.

21. A. Snoussi, M. El Ghourabi, and M. Limam. On SPC for short run autocorrelated data. *Communications in Statistics Simulation and Computation*, 34:219–234, 2005.
22. D.H. Timmer, J. Pignatello Jr, and M. Longnecker. The development and evaluation of CUSUM-based control charts for an AR(1) process. *IIE Transactions*, 30:525–534, 1998.
23. L.N. VanBrackle III and M.R. Reynolds Jr. EWMA and CUSUM control charts in presence of correlations. *Communications in Statistics Simulation and Computation*, 26:979–1008, 1997.
24. A.V. Vasilopoulos and A.P. Stamboulis. Modification of control limits in the presence of correlation. *Journal of Quality Technology*, 10:20–30, 1978.
25. D.G. Wardell, H. Moskowitz, and R.D. Plante. Run-length distributions of special-cause control charts for correlated processes. *Technometrics*, 36:3–17, 1994.
26. W.H. Woodall and F.W. Faltin. Autocorrelated data and SPC. *ASQC Statistics Division Newsletter*, 13:18–21, 1993.
27. C.M. Wright, D.E. Booth, and M.Y. Hu. Joint estimation: SPC method for short-run autocorrelated data. *Journal of Quality Technology*, 33:365–378, 2001.
28. S.A. Yourstone and D.C. Montgomery. A time-series approach to discrete real-time process quality control. *Quality and Reliability Engineering International*, 5:309–317, 1989.
29. S.A. Yourstone and D.C. Montgomery. Detection of process upsets sample autocorrelation control chart and group autocorrelation control chart applications. *Quality and Reliability Engineering International*, 7:133–140, 1991.
30. N.F. Zhang. Statistical control chart for stationary process data. *Technometrics*, 40:24–38, 1998.

On the Application of SPC in Finance

Vasyl Golosnoy[a], Iryna Okhrin[b], Sergiy Ragulin[b], and Wolfgang Schmid[b] *

a. Institute of Statistics and Econometrics, University of Kiel, Germany
b. Department of Statistics, European University Viadrina, Frankfurt (Oder), Germany

Summary. A financial analyst is interested in a fast on-line detection of changes in the optimal portfolio composition. Although this is a typical sequential problem the majority of papers in financial literature ignores this fact and handles it in a non-sequential way. This paper deals with the problem of monitoring the weights of the global minimum variance portfolio (GMVP).

We consider several control charts based on the estimated GMVP weights as well as on other closely related characteristic processes. Different types of EWMA and CUSUM control schemes are applied for our purpose. The behavior of the schemes is investigated within an extensive Monte Carlo simulation study. The average run length criterion serves as a comparison measure for the discussed charts.

1 Introduction

During the last decades the amount of money invested in risky assets has increased dramatically. Since Markowitz (1952) portfolio theory recommends to hold not a single asset but a well-diversified portfolio. On the one hand an investor is interested to have a portfolio composition providing a large expected return, on the other hand he wants to avoid risks associated with investment decisions. An optimal portfolio is often defined as a form of trade-off between the expected return and the variance of the portfolio.

A buy and hold portfolio strategy is one of the conventional investment recommendations (Michaud (1998)). During the holding period, however, newly incoming information may change the optimal portfolio weights. Then the portfolio should be revised and adjusted to the new market situation. There is an overwhelming evidence about the presence of structural breaks in the

* Correspondence to: Wolfgang Schmid, Department of Statistics, European University Viadrina, Grosse Scharrnstrasse 59, 15230 Frankfurt (Oder), Germany, schmid@euv-frankfurt-o.de

H.-J. Lenz et al. (eds.), *Frontiers in Statistical Quality Control 9*,
DOI 10.1007/978-3-7908-2380-6_8,

parameters of the asset return distribution (cf. Banerjee and Urga (2005)) which may cause alterations in the optimal portfolio proportions. Since the investor has to decide about the optimal portfolio composition at each point in time a sequential problem is present. Such questions are treated within statistical process control (SPC). The main monitoring tool of SPC is called control chart (e.g., Montgomery (2005)).

The optimal portfolio weights form the process of interest in our case, because the investor requires only these quantities for all decisions concerning the wealth allocation. A Markowitz-type portfolio investor obtains the optimal portfolio weights as a solution of a mean-variance optimization task. Since there are significant difficulties in estimating and forecasting mean asset returns (Merton (1980)), it is often reasonable to construct a portfolio by minimizing the portfolio variance. Here we are dealing with the surveillance of the weights of the global minimum variance portfolio (GMVP). The GMVP is the portfolio with the smallest attainable variance. It requires only the knowledge of the covariance matrix of the asset returns. Thus the GMVP weights do not suffer from estimation risk in mean asset returns. Golosnoy and Schmid (2007) derived several exponentially weighted control charts for monitoring the GMVP weights and their first differences. The developed procedures, however, fail to detect quickly some important types of changes (cf. Golosnoy et al. (2007)).

In this paper we consider two additional characteristics which resemble the weight and difference processes. These quantities are obtained as approximations to the estimated weights and the differences at lag 1. We apply different types of exponentially weighted moving average (EWMA) charts and cumulative sum (CUSUM) control charts to these characteristics. Both the multivariate EWMA scheme (Lowry et al. (1992)) and the EWMA chart based on the Mahalanobis distance are used. The CUSUM schemes developed in spirit of Pignatiello and Runger (1990) are applied as well. Since all these control charts have been initially introduced for monitoring the mean vector of an independent sample of multivariate normally distributed variables, it is not possible to exploit these procedures directly. We adopt the charts for our purpose following the approach of Bodnar and Schmid (2007).

The discussed control charts are investigated with respect to their ability to detect changes in the GMVP weights quickly. The control limits of all charts are calibrated to have the same prespecified in-control average run length (ARL). The in-control ARL measures the average number of observations until a chart gives a false signal. Using the determined control limits, the charts are compared with each other for various types of changes. The best chart provides the smallest out-of-control ARLs. Since there are no explicit formulas for the ARLs in the present case both in- and out-of-control results are calculated in a Monte Carlo simulation study. No single chart appears to be uniformly better. The obtained evidence allows to recommend a combination of the appropriate procedures for detecting important types of changes in the GMVP weights.

The paper is organized as follows. Section 2.1 provides the required results from portfolio theory. The surveillance of the weights and their first differences is discussed in Section 2.2, while in Section 2.3 approximations to these quantities are monitored. EWMA and CUSUM type charts are introduced in Sections 3 and 4, respectively. A comparison of all charts within a Monte Carlo simulations study is given in Section 5. Section 6 completes the paper.

2 Monitoring Optimal Portfolio Composition

2.1 Portfolio Problem

The considered portfolio consists of k risky assets. Let $\mathbf{X}$ denote the vector of asset returns with $E(\mathbf{X}) = \boldsymbol{\mu}$ and $Cov(\mathbf{X}) = \boldsymbol{\Sigma}$. The matrix $\boldsymbol{\Sigma}$ is assumed to be positive definite. Moreover, $\boldsymbol{w} = (w_1, ..., w_k)'$ denotes the vector of portfolio weights. Thus w_i gives the fraction of the investor's wealth invested in the i-th asset. The portfolio return is given by $X_P = \boldsymbol{w}'\mathbf{X}$. The classical Markowitz analysis is based on finding the portfolio proportions $\boldsymbol{w}$, providing the optimal trade-off between the mean portfolio return $E(X_P) = \boldsymbol{\mu}'\boldsymbol{w}$ and the portfolio variance $Var(X_P) = \boldsymbol{w}'\boldsymbol{\Sigma}\boldsymbol{w}$. The investor should choose the portfolio by maximizing the mean-variance objective function $EU(\boldsymbol{w})$:

$$\max_{\boldsymbol{w}} \left[EU(\boldsymbol{w}) = \boldsymbol{\mu}'\boldsymbol{w} - \frac{\gamma}{2}\boldsymbol{w}'\boldsymbol{\Sigma}\boldsymbol{w} \right],$$

where $\gamma > 0$ denotes the risk aversion coefficient.

The optimal portfolio weights depend on both the forecasts of the mean vector $\boldsymbol{\mu}$ and the covariance matrix $\boldsymbol{\Sigma}$. The matrix $\boldsymbol{\Sigma}$ could be precisely estimated and forecasted. However, there are considerable difficulties in estimating and forecasting the mean returns $\boldsymbol{\mu}$ precisely (cf. Best and Grauer (1991)). For this reason the GMVP suggests an attractive choice for practical investment decisions. The GMVP denotes the portfolio with the smallest attainable variance under the constraint $\mathbf{w}'\mathbf{1} = 1$, where $\mathbf{1} = (1 ... 1)'$ denotes the k-dimensional vector of units. The vector of the GMVP weights $\mathbf{w}$ is given by

$$\mathbf{w} = \frac{\boldsymbol{\Sigma}^{-1}\mathbf{1}}{\mathbf{1}'\boldsymbol{\Sigma}^{-1}\mathbf{1}}.$$

It corresponds to the optimal mean-variance portfolio with $\gamma = \infty$. Note that the GMVP composition $\mathbf{w}$ allows short sellings, i.e. some portfolio weights may be negative. Although an individual investor cannot construct such a portfolio, short positions for some assets should not be a problem for a bank or investment fund. The short-sellings mechanism is discussed in detail in Farrell (1997).

The covariance matrix $\boldsymbol{\Sigma}$ is unknown in practice and should be estimated. It is supposed that a sample of returns is available. Since the parameters, here

$\boldsymbol{\mu}$ and $\boldsymbol{\Sigma}$, may change over time we use rolling window estimators based on the last n observations $X_{t-n+1}, ..., X_t$ at each time point t

$$\hat{\boldsymbol{\Sigma}}_{t,n} = \frac{1}{n-1} \sum_{j=t-n+1}^{t} (\mathbf{X}_j - \hat{\boldsymbol{\mu}}_{t,n})(\mathbf{X}_j - \hat{\boldsymbol{\mu}}_{t,n})', \qquad \hat{\boldsymbol{\mu}}_{t,n} = \frac{1}{n} \sum_{v=t-n+1}^{t} \mathbf{X}_v.$$

Replacing $\boldsymbol{\Sigma}$ by the sample estimator $\hat{\boldsymbol{\Sigma}}_{t,n}$ we obtain an estimator of the GMVP weights at time t, given by $\hat{\mathbf{w}}_{t,n} = \hat{\boldsymbol{\Sigma}}_{t,n}^{-1}\mathbf{1}/\mathbf{1}'\hat{\boldsymbol{\Sigma}}_{t,n}^{-1}\mathbf{1}$.

Assuming that the returns follow an independent and normally distributed random process the estimated GMPV weights are multivariate t-distributed with the moments (Okhrin and Schmid (2006))

$$E(\hat{\mathbf{w}}_{t,n}) = \mathbf{w}, \qquad Cov(\hat{\mathbf{w}}_{t,n}) = \boldsymbol{\Omega} = \frac{1}{n-k-1}\left(\boldsymbol{\Sigma}^{-1} - \frac{\boldsymbol{\Sigma}^{-1}\mathbf{1}\mathbf{1}'\boldsymbol{\Sigma}^{-1}}{\mathbf{1}'\boldsymbol{\Sigma}^{-1}\mathbf{1}}\right)/\mathbf{1}'\boldsymbol{\Sigma}^{-1}\mathbf{1}.$$

2.2 Surveillance of the Portfolio Weights

Empirical evidence reports the presence of clusters in the second moment of the asset returns (cf. Engle (1982)). The locally constant volatility model (cf. Hsu et al., 1974, Härdle et al., 2003) suggests that the covariance matrix $\boldsymbol{\Sigma}$ stays unchanged between two consecutive change points. The timepoints of changes are unknown. The locally constant volatility model allows to mimic clusters, heavy tails and other stylized facts frequently observed in the distribution of risky asset returns (Rachev and Mittnik (2000)). The level changes in the covariance matrix may alter the optimal portfolio proportions. For the investor it is of great importance to detect the changes in the portfolio weights as soon as possible.

Hereafter the returns $\mathbf{X}_t$ are assumed to be independent and normally distributed. The monitoring phase starts at time point $t = 1$. For the period $t \leq 0$, it is assumed that there is no change in the process. This means that for $t \leq 0$ the returns $\mathbf{X}_t$ are k-dimensional identically normally distributed random variables with $\mathbf{X}_t \sim \mathcal{N}_k(\boldsymbol{\mu}, \boldsymbol{\Sigma})$. Thus the vector $\hat{\mathbf{w}}_{t,n}$ for $t \leq 0$ follows a k-dimensional t-distribution, i.e. $\hat{\mathbf{w}}_{t,n} \sim t_k(\mathbf{w}, \boldsymbol{\Omega})$ for $t \leq 0$. The notation E_0, Cov_0, etc. denotes that the characteristics are calculated assuming that $\mathbf{X}_t \sim \mathcal{N}_k(\boldsymbol{\mu}, \boldsymbol{\Sigma})$ for all $t \in I\!N$. Since $E_0(\hat{\mathbf{w}}_{t,n}) = \mathbf{w}$, we aim to detect deviations from the value $\mathbf{w}$. The observed process is considered to be in control if $E(\hat{\mathbf{w}}_{t,n}) = \mathbf{w}$ holds for all $t \geq 1$. Otherwise, the observed process is denoted to be out of control.

Golosnoy and Schmid (2007) suggest to monitor the mean behavior of two characteristics of the weight process. The first approach is directly based on the process of the estimated weights $\{\hat{\mathbf{w}}_{t,n}\}$. The second one considers the process of the first differences $\{\boldsymbol{\Delta}_{t,n}\}$, defined as $\boldsymbol{\Delta}_{t,n} = \hat{\mathbf{w}}_{t,n} - \hat{\mathbf{w}}_{t-1,n}$. Taking first differences aims to reduce the high dependency in the estimated optimal weights. The EWMA control schemes of Golosnoy and Schmid (2007) are applied to these processes. However, they have shown poor detection ability for some out-of-control situations. This evidence can be explained by the non-standard exact distributions of both weights and differences.

2.3 Alternative Characteristics

These two alternative processes are proposed to be monitored. They resemble the weights and the differences but their exact distributions can easily be calculated. The processes ground on the following result, proven by Golosnoy et al. (2008).

Theorem 1. *Assume that in the in-control state $\{\mathbf{X}_t\}$ is a sequence of independent and normally distributed k-dimensional random vectors with mean $\boldsymbol{\mu}$ and covariance matrix $\boldsymbol{\Sigma}$, which is assumed to be positive definite. Then it holds in the in-control state that as $n \to \infty$*

$$n\,\boldsymbol{\Delta}_{t,n} - \mathbf{p}_{t,n} \xrightarrow{p} 0, \qquad \textit{where}$$

$$\mathbf{p}_{t,n} = -\,\mathbf{Q}\ \left((\mathbf{X}_t - \boldsymbol{\mu})(\mathbf{X}_t - \boldsymbol{\mu})' \ - \ (\mathbf{X}_{t-n} - \boldsymbol{\mu})(\mathbf{X}_{t-n} - \boldsymbol{\mu})' \right)\ \mathbf{w}\,.$$

Using this result we analyze the alternative processes from the SPC viewpoint.

An alternative to the difference process

The analysis of the process $\{\mathbf{p}_{t,n}\}$ is more suitable than that of $\{\boldsymbol{\Delta}_{t,n}\}$ since the exact distribution of $\mathbf{p}_{t,n}$ can easily be derived. Suppose that $\mathbf{X}_t \sim \mathcal{N}_k(\boldsymbol{\mu}, \boldsymbol{\Sigma})$ for $t \leq 0$ and $\mathbf{X}_t \sim \mathcal{N}_k(\boldsymbol{\mu}, \boldsymbol{\Sigma}_1)$ for $t \geq 1$. We use the notation E_1, Cov_1 for the characteristics calculated with respect to this model, i.e. if the process is out of control. Then the out-of-control mean and covariance matrix of $\mathbf{p}_{t,n}$ for $t \geq 1$ are given by

$$E_1(\mathbf{p}_{t,n}) = \begin{cases} \mathbf{0} & \text{for} \quad t \geq n+1 \\ -\mathbf{Q}\boldsymbol{\Sigma}_1\mathbf{w} & \text{for} \quad 1 \leq t \leq n \end{cases},$$

$$Cov_1(\mathbf{p}_{t,n}) = \begin{cases} 2\mathbf{Q}\,(\boldsymbol{\Sigma}_1\mathbf{w}\mathbf{w}'\boldsymbol{\Sigma}_1 + (\mathbf{w}'\boldsymbol{\Sigma}_1\mathbf{w})\boldsymbol{\Sigma}_1)\,\mathbf{Q} & \text{for } t \geq n+1 \\ \mathbf{Q}\,(\boldsymbol{\Sigma}_1\mathbf{w}\mathbf{w}'\boldsymbol{\Sigma}_1 + (\mathbf{w}'\boldsymbol{\Sigma}_1\mathbf{w})\boldsymbol{\Sigma}_1)\,\mathbf{Q} + \frac{\mathbf{Q}}{\mathbf{1}'\boldsymbol{\Sigma}^{-1}\mathbf{1}} & \text{for } 1 \leq t \leq n \end{cases}.$$

This result shows that the change in $\boldsymbol{\Sigma}$ has no influence on the mean behavior of $\mathbf{p}_{t,n}$ for $t \geq n+1$. This may lead to the undesirable inertia phenomenon if a chart gives no signal within the first n time periods.

An alternative to the weight process

In order to avoid the undesirable inertia behavior of the $\{\mathbf{p}_{t,n}\}$ process we consider the characteristic

$$\mathbf{q}_t = -\mathbf{Q}[(\mathbf{X}_t - \boldsymbol{\mu})(\mathbf{X}_t - \boldsymbol{\mu})' - \boldsymbol{\Sigma}]\mathbf{w} = -\mathbf{Q}(\mathbf{X}_t - \boldsymbol{\mu})(\mathbf{X}_t - \boldsymbol{\mu})'\mathbf{w} \,. \qquad (1)$$

Note that $\mathbf{q}_t$ does not depend on n and that $\mathbf{p}_{t,n} = \mathbf{q}_t - \mathbf{q}_{t-n}$. This property is very advantageous, since the choice of n significantly influences stochastic properties of the weights and differences. It holds that for $t \geq 1$

$$E_1(\mathbf{q}_t) = -\mathbf{Q}\, \boldsymbol{\Sigma}_1\, \mathbf{w},$$
$$Cov_1(\mathbf{q}_t) = \mathbf{Q}\, (\boldsymbol{\Sigma}_1 \mathbf{w}\mathbf{w}' \boldsymbol{\Sigma}_1 + (\mathbf{w}' \boldsymbol{\Sigma}_1 \mathbf{w}) \boldsymbol{\Sigma}_1)\, \mathbf{Q}.$$

Thus the change influences the mean behavior of $\mathbf{q}_t$ for all $t \geq 1$.

3 Control Charts Based on Exponential Smoothing

Though the characteristics $\hat{\mathbf{w}}_{t,n}$, $\boldsymbol{\Delta}_{t,n}$, $\mathbf{p}_{t,n}$ and $\mathbf{q}_t$ are k-dimensional vectors, the sums of their elements are equal to some constant, e.g. $\hat{\mathbf{w}}_{t,n}'\, \mathbf{1} = 1$. Consequently, their covariance matrices are not regular. The rank of all covariance matrices is $k-1$. For that reason in the following we consider the corresponding vectors of the first $k-1$ elements, in general denoted by $\mathbf{T}_t$.

3.1 Charts Based on a Multivariate EWMA Recursion

A multivariate EWMA recursion (Lowry et al. (1992)) is applied to each component of the characteristic $\mathbf{T}_t$. This leads to

$$\mathbf{Z}_t = (\mathbf{I} - \mathbf{R})\mathbf{Z}_{t-1} + \mathbf{R}\, \mathbf{T}_t, \quad t \geq 1, \qquad \mathbf{Z}_0 = E_0(\mathbf{T}_t)\,, \qquad (2)$$

where $\mathbf{I}$ is the $(k-1) \times (k-1)$ identity matrix and $\mathbf{R} = diag(r_1, ..., r_{k-1})$ is a $(k-1) \times (k-1)$ diagonal matrix with diagonal elements $0 < r_i \leq 1$ for $i \in \{1, ..., k-1\}$. The quantities r_i are smoothing parameters describing the influence of past observations. The smaller the parameter r_i the larger the influence of the past. If r_i is chosen equal to one then no previous observations of $T_{i,t}$ are taken into account.

The control chart gives a signal at time t if $(\mathbf{Z}_t - E_0(\mathbf{Z}_t))'[Cov_0(\mathbf{Z}_t)]^{-1}(\mathbf{Z}_t - E_0(\mathbf{Z}_t))$ is sufficiently large. The advantage of this approach is that each element has its own smoothing factor r_i which makes the scheme flexible. However, the calculation of the covariance matrix for $\mathbf{Z}_t$ remains complicated.

3.2 Charts Based on a Univariate EWMA Recursion

An alternative to the multivariate EWMA scheme is to apply a univariate EWMA recursion. Here we first transform $\mathbf{T}_t$ to a scalar random variable by calculating the squared Mahalanobis distance for the characteristic $\mathbf{T}_t$:

$$D_t^2 = (\mathbf{T}_t - E_0(\mathbf{T}_t))'[Cov_0(\mathbf{T}_t)]^{-1}(\mathbf{T}_t - E_0(\mathbf{T}_t)). \qquad (3)$$

Then a univariate EWMA recursion is applied to this quantity by

$$Z_t = (1-r)Z_{t-1} + r\, D_t^2, \quad t \geq 1, \qquad Z_0 = E_0(D_t^2)\,, \tag{4}$$

where $r \in (0, 1]$ is a smoothing parameter. The charts gives a signal if Z_t is large enough.

4 Control Charts Based on Cumulative Sums

The MEWMA charts of Lowry et al. (1992) turn out to be directionally invariant for an independent random sample, i.e. their out-of-control distribution depends on the magnitude as well as on the direction of the mean vector of the underlying process. The univariate CUSUM approach is based on the sequential probability ratio test. In the multivariate framework, however, this attempt leads to a control chart which is no longer directionally invariant. The analysis of such schemes turns out to be quite difficult. For that reason we focus on the directionally invariant CUSUM control charts for the mean vector.

Let $\mathbf{T}_t$ be defined as in Section 3. Instead of exponential smoothing we now consider the sum of the deviations of $\mathbf{T}_t$ from its in-control expectation between the time periods $t = a$ and $t = b$. More precisely, $\mathbf{S}_{a,b} = \sum_{t=a+1}^{b}(\mathbf{T}_t - E_0(\mathbf{T}_t))$ for $a < b$. Next we adjust two CUSUM schemes of Pignatiello and Runger (1990) for our purposes.

4.1 MC1 Charts

The MC1 chart of Pignatiello and Runger is based on a multivariate CUSUM procedure. The CUSUM vector is reduced to a scalar by taking its Euclidean norm, given as $||\mathbf{S}_{a,b}|| = (\mathbf{S}'_{a,b}[Cov_0(\mathbf{S}_{a,b})]^{-1}\mathbf{S}_{a,b})^{1/2}$.

MC1 chart depends on the reference parameter K. It determines the quantity to be subtracted from the sum statistic. In the univariate case the optimal reference value should be equal to the half-distance between the in- and out-of-control expectations of the control statistic. Pignatiello and Runger (1990) propose to choose K as the distance between the expectation of the statistic in the in- and out-of-control states.

Here the MC1 chart control statistic is given by

$$MC1_t = \max\{\sqrt{n_t}||\mathbf{S}_{t-n_t,t}|| - n_t K/2, 0\}, \tag{5}$$

where n_t denotes the number of time periods since the last renewal, i.e.

$$n_t = \begin{cases} n_{t-1} + 1 & \text{for} \quad MC1_{t-1} > 0 \\ 1 & \text{for} \quad MC1_{t-1} = 0 \end{cases}.$$

A signal is given if the value of the $MC1_t$ statistic is sufficiently large.

4.2 MC2 Charts

MC2 is another multivariate CUSUM scheme of Pignatiello and Runger (1990). It is constructed by applying the CUSUM procedure to the univariate quantity obtained from the Mahalanonis transformation. Thus, this scheme can be seen as a CUSUM counterpart of the previously described procedure based on the univariate EWMA recursion.

The control statistic depends on the cumulative sum of the current squared distance D_t^2, which is defined in (3). This distance is normalized by subtracting a reference value of the form $K^2/2+k-1$. Note that the in-control expectation of D_t^2 is equal to $k-1$. The MC2 statistic is then given by

$$MC2_t = \max\{0, MC2_{t-1} + D_t^2 - K^2/2 - (k-1)\} \quad \text{for} \quad t \geq 1, \tag{6}$$

with $MC2_0 = 0$. A signal is given if the value of $MC2_t$ is large enough.

5 Comparison Study

The introduced control charts are compared within an extensive Monte Carlo simulation study. In the following, however, we restrict ourselves to the comparison of the charts based on the weight process $\{\hat{\mathbf{w}}_{t,n}\}$ and the approximation process $\{\mathbf{q}_t\}$. These processes seem to be the most attractive ones from the practical viewpoint. The required in-control covariance matrices are obtained from simulations as well.

The out-of-control ARL serves as a performance measure. Because no explicit formulas of the ARLs are available these quantities are estimated via a simulation study based on 10^6 repetitions. The control limits are determined under the condition that all charts have the same in-control ARL equal to 120. Since there are around 250 trading days per year at the stock exchange this in-control ARL roughly corresponds to six months of daily observations.

The introduced control schemes depend on further design parameters. We consider a multivariate EWMA recursion with equal smoothing values, i.e. $\mathbf{R} = r\mathbf{I}$. This means that all components of $\mathbf{T}_t$ are exponentially weighted in the same way. In our study we take the value $r \in \{0.1, 0.2, ..., 1.0\}$. For the CUSUM charts the reference value K is chosen from the set $\{0.0, 0.2, ..., 4.0\}$. The current study is restricted to a portfolio consisting of $k = 4$ risky assets. The in-control covariance matrix is given by $\mathbf{\Sigma} = \mathbf{\Theta}\mathbf{\Upsilon}_0\mathbf{\Theta}$ with

$$\mathbf{\Theta} = \begin{pmatrix} 0.135 & 0 & 0 & 0 \\ 0 & 0.15 & 0 & 0 \\ 0 & 0 & 0.17 & 0 \\ 0 & 0 & 0 & 0.19 \end{pmatrix} \quad \text{and} \quad \mathbf{\Upsilon}_0 = \begin{pmatrix} 1 & 0.4 & 0.35 & 0.3 \\ 0.4 & 1 & 0.45 & 0.4 \\ 0.35 & 0.45 & 1 & 0.5 \\ 0.3 & 0.4 & 0.5 & 1 \end{pmatrix}.$$

The matrix $\mathbf{\Theta}$ contains the annualized standard deviations. $\mathbf{\Upsilon}$ is a correlation matrix. These parameter values correspond to typical empirical characteristics of risky assets (cf. Michaud (1998)). Then the vector of the optimal GMVP weights is given by

$$\mathbf{w} = (0.477,\ 0.273,\ 0.147,\ 0.103)'.$$

The out-of-control state is modeled by changing the magnitude of the standard deviations of the first and the fourth asset returns. Thus, we define the out-of-control covariance matrix as

$$\boldsymbol{\Sigma}_1 = \boldsymbol{\Xi}\boldsymbol{\Sigma}\boldsymbol{\Xi}, \qquad \text{where} \quad \boldsymbol{\Xi} = \begin{pmatrix} a_1 & 0 & 0 & 0 \\ 0 & 1 & 0 & 0 \\ 0 & 0 & 1 & 0 \\ 0 & 0 & 0 & a_4 \end{pmatrix}.$$

The change parameters are chosen as $a_1 \in \{0.5, 1, 2, 3\}$ and $a_4 \in \{0.5, 1, 2, 3\}$.

The rolling estimator of the portfolio weight is based on the last n observations. The choice of n has a large impact on the chart performance of the $\{\hat{\mathbf{w}}_{t,n}\}$ process. In the in-control state n should be large to get a reliable estimator. However, in the out-of-control state n should be small to ensure a faster reaction on the changes. Here we present the results for $n \in \{25, 30, 40, 60\}$. The values $n \in \{10, 15\}$ lead to larger out-of-control ARLs and are not reported here.

The results of our simulation study are given in Tables 1 and 2, providing the results for the $\{\hat{\mathbf{w}}_{t,n}\}$ and $\{\mathbf{q}_t\}$ processes, respectively. Accordingly, the indices $\mathbf{w}$ and $\mathbf{q}$ are used to denote the underlying chart. EWMA denotes a univariate EWMA recursion, while MEWMA a multivariate one. Moreover, MCUSUM1 and MCUSUM2 denote the MC1 and the MC2 charts, respectively. The first value in a column gives the out-of-control ARL. The standard deviation of the run length, the optimal smoothing parameter and the reference values are given in parenthesis. Table 1 reports the optimal value of n for the weight process, while the control charts for $\{\mathbf{q}_t\}$ are independent of n. The out-of-control ARLs of the best control chart for a specific parameter constellation are given in bold.

The obtained evidence suggests that there is no overall best chart for all types of changes. In Table 1 for increasing variance, i.e. for $a_1 \geq 1$ and $a_4 \geq 1$, the results for both EWMA charts and the MC2 chart are similar, while the MC1 scheme shows a worse behavior. The best value of n is nearly always $n = 60$. If the variances increase the EWMA charts without memory, i.e. for $r = 1.0$, provide the smallest ARLs. If one of the variances decreases then the multivariate EWMA chart dominates the other schemes. It is difficult to give a unique recommendation about the choice of n, K, and r values. A change in the weights due to a variance decrease is detected faster with smaller r and n values, e.g. $r = 0.1$ and $n = 30$.

In Table 2 the multivariate EWMA chart turns out to be clearly the best for $a_1 \geq 1$ and $a_2 \geq 1$. The best choice of the smoothing parameter is $r = 0.1$. However, if both variances decrease then the MC1 chart outperforms the other ones. In this case the results for the other charts are very bad since the out-of-control ARLs are larger than the in-control value 120.

$a_1 \backslash a_4$	0.5	1.0	2.0	3.0
0.5	26.92(14.48, 2.0, 40) 26.10(13.55, 2.6, 40) 26.14(13.35, 0.5, 40) **25.48**(15.02, 0.1, 30)	26.74(16.57, 0.6, 40) 26.49(12.87, 2.8, 40) 26.52(12.71, 0.7, 40) **25.18**(14.26, 0.1, 25)	26.48(16.45, 0.6, 40) 26.05(12.40, 2.6, 40) 26.07(12.01, 0.4, 40) **24.80**(13.00, 0.1, 30)	27.19(15.87, 3.0, 60) 27.06(15.57, 2.4, 60) 27.04(14.93, 0.3, 60) **25.64**(16.57, 0.1, 60)
1.0	19.84(10.91, 0.8, 25) 19.17 (8.67, 3.0, 25) 19.08 (8.76, 1.0, 25) **19.07** (8.89, 0.7, 25)	$MCUSUM1^{(w)}$ $MCUSUM2^{(w)}$ $EWMA^{(w)}$ $MEWMA^{(w)}$	44.50(29.12, 0.2, 60) 47.32(35.59, 0.4, 60) 47.28(40.85, 0.1, 60) **44.21**(41.87, 0.1, 60)	**36.47**(21.91, 0.2, 60) 41.66(30.58, 0.0, 60) 43.36(39.43, 0.1, 60) 40.38(39.40, 0.1, 60)
2.0	13.18 (8.28, 3.0, 60) 12.54 (8.10, 2.6, 60) 12.53 (8.10, 1.0, 60) **12.51** (8.08, 1.0, 60)	15.47(10.74, 3.0, 60) **14.86**(10.65, 2.6, 60) 14.87(10.68, 1.0, 60) 14.87(10.69, 1.0, 60)	13.93 (9.49, 3.0, 60) 13.27 (9.38, 2.6, 60) **13.25** (9.38, 1.0, 60) 13.26 (9.39, 1.0, 60)	12.46 (8.57, 3.0, 60) 11.81 (8.51, 2.6, 60) **11.79** (8.51, 1.0, 60) 11.84 (8.52, 1.0, 60)
3.0	7.79 (5.10, 3.0, 60) 6.96 (4.82, 2.6, 60) 6.95 (4.82, 1.0, 60) **6.94** (4.82, 1.0, 60)	8.14 (5.55, 3.0, 60) 7.33 (5.30, 2.6, 60) **7.32** (5.31, 1.0, 60) 7.33 (5.32, 1.0, 60)	7.91 (5.46, 3.0, 60) 7.06 (5.12, 2.6, 60) 7.06 (5.12, 1.0, 60) **7.04** (5.09, 1.0, 60)	7.40 (5.06, 3.0, 60) 6.56 (4.74, 2.6, 60) 6.55 (4.74, 1.0, 60) **6.53** (4.72, 1.0, 60)

Table 1. Out-of-control ARLs of EWMA and CUSUM type charts based on the estimated weight process $\{\hat{\mathbf{w}}_{t,n}\}$.

$a_1 \backslash a_4$	0.5	1.0	2.0	3.0
0.5	**32.58** (12.81, 0.2) 369.39(369.10, 4.0) 347.62(330.28, 1.0) 72.97 (67.21, 0.1)	**33.02** (13.47, 0.2) 246.68(245.99, 4.0) 237.67(235.07, 1.0) 66.87 (63.08, 0.1)	10.20 (7.49, 0.8) 16.82(15.55, 1.8) 15.96(15.06, 0.1) **8.81** (8.03, 0.1)	4.90 (3.76, 1.2) 5.95 (5.21, 2.4) 5.69 (4.92, 0.1) **4.04** (3.57, 0.1)
1.0	**35.98** (15.97, 0.2) 155.53(155.01, 4.0) 152.73(152.21, 1.0) 64.94 (64.21, 0.1)	$MCUSUM1^{(q)}$ $MCUSUM2^{(q)}$ $EWMA^{(q)}$ $MEWMA^{(q)}$	10.38 (8.86, 1.0) 11.93(10.34, 1.6) 11.08 (9.93, 0.1) **8.32** (8.53, 0.1)	4.76 (3.91, 1.4) 5.12 (4.36, 2.4) 4.87 (4.07, 0.1) **3.81** (3.58, 0.1)
2.0	5.97 (4.48, 1.0) 7.47 (6.47, 2.0) 7.02 (6.10, 0.1) **4.77** (4.33, 0.1)	6.25 (4.96, 1.0) 7.06 (6.02, 2.0) 6.59 (5.63, 0.1) **4.86** (4.66, 0.1)	4.51 (3.77, 1.4) 4.49 (3.61, 2.2) 4.23 (3.35, 0.1) **3.32** (3.14, 0.1)	3.18 (2.56, 1.6) 3.13 (2.42, 2.6) 3.00 (2.26, 0.1) **2.42** (2.09, 0.1)
3.0	2.99 (2.21, 1.6) 3.25 (2.61, 3.0) 3.16 (2.47, 0.1) **2.46** (1.96, 0.1)	3.00 (2.25, 1.6) 3.17 (2.54, 3.0) 3.08 (2.38, 0.1) **2.44** (1.99, 0.1)	2.61 (1.98, 1.8) 2.64 (1.99, 3.0) 2.56 (1.86, 0.1) **2.08** (1.63, 0.1)	2.21 (1.61, 2.0) 2.20 (1.56, 3.2) 2.15 (1.48, 0.1) **1.79** (1.30, 0.1)

Table 2. Out-of-control ARLs of EWMA and CUSUM type control charts based on the approximated process $\{\mathbf{q}_t\}$.

A comparison of Tables 1 and 2 illustrates the advantages and disadvantages of both considered processes. For a variance increase the charts based on the process $\{\mathbf{q}_t\}$ perform better than the charts for the estimated portfolio weights. However, if both variances decrease then the schemes based on the weight process $\{\hat{\mathbf{w}}_{t,n}\}$ dominate the charts based on $\{\mathbf{q}_t\}$. In general, it can be recommended to combine the MEWMA chart for $\mathbf{q}_t$ and $\hat{\mathbf{w}}_{t,n}$. A signal from the $\mathbf{q}_t$ chart can indicate a change in the weights caused by a variance increase, while a signal for the $\hat{\mathbf{w}}_{t,n}$ chart points at a possible alteration in the weights due to a variance decrease. This study advocates the advantages of EWMA charts compared to CUSUM procedures.

6 Conclusions

Changes in the optimal portfolio weights should be detected sequentially. This paper introduces several control charts for the surveillance of the global minimum variance portfolio composition. These charts monitor the processes of the estimated optimal weights as well as their approximations. The changes in the weights are modeled via alterations in the variances of the asset returns. The average run length (ARL) criterion is applied for performance evaluation.

The considered schemes are compared to each other within a Monte Carlo simulation study. It turns out that no chart uniformly dominates the other ones. The performance of the charts depends on the fact whether the process variance increases or decreases. In case of a variance increase the MEWMA chart based on the approximated characteristic provides the smallest out-of-control ARLs. In case of a variance decrease the MEWMA chart based on the weights turns out to be the best scheme. Thus it is recommended to combine both procedures for practical applications. A simultaneous use of both MEWMA charts allows to interpret the obtained signals. The behavior of the schemes for a changing correlation structure requires further investigation.

References

1. Banerjee, A. and Urga, G. (2005). Modeling structural breaks, long memory and stock market volatility: an overview. Journal of Econometrics 129, 1-34.
2. Best, M. and Grauer, R. (1991). On the sensitivity of mean-variance-efficient portfolios to changes in asset means: some analytical and computational results. Review of Financial Studies 4, 315-342.
3. Bodnar, O. and Schmid, W. (2007). Surveillance of the mean behaviour of multivariate time series. Statistica Neerlandica 61, 383-406.
4. Engle, R. (1982). Autoregressive conditional heteroskedasticity with estimates of the variance of United Kingdom inflation. Econometrica 50, 987-1008.
5. Farrell, J. (1997). Portfolio Management. New-York:McGraw-Hill.
6. Golosnoy, V. and Schmid, W. (2007). EWMA control charts for optimal portfolio weights. Sequential Analysis 26, 195-224.
7. Golosnoy, V., Schmid, W., and Okhrin, I. (2007). Sequential monitoring of optimal portfolio weights, in Financial Surveillance, M. Frisen (Ed.), Wiley, New York, (2007), 179-210.
8. Golosnoy, V., Okhrin, I., and Schmid, W. (2008). Statistical methods for the surveillance of portfolio weights. To appear in: Statistics.
9. Härdle, W., Herwartz, H., and Spokoiny, V. (2003). Time inhomogeneous multiple volatility modelling. Journal of Financial Econometrics 1, 55-95.
10. Hsu, D., Miller, R., and Wichern, D. (1974). On the stable Paretian behaviour of stock market prices. Journal of American Statistical Association 69, 108-113.

11. Lowry, C. A., Woodall, W.H., Champ, C.W., and Rigdon, S.E. (1992). A multivariate exponentially weighted moving average control chart. Technometrics 34, 46-53.
12. Markowitz, H. (1952). Portfolio selection. Journal of Finance 7, 77-91.
13. Merton, R.C. (1980). On estimating the expected return on the market: an exploratory investigation, Journal of Financial Economics 8, 323-361.
14. Michaud, O. (1998). Efficient Asset Management, Boston, Massachusetts, Harvard Business School Press.
15. Montgomery, D.C. (2005). Introduction to Statistical Quality Control, Wiley.
16. Okhrin, Y. and Schmid, W. (2006). Distributional properties of optimal portfolio weights. Journal of Econometrics 134, 235-256.
17. Pignatiello, Jr., J. J. and Runger, G.C. (1990). Comparison of multivariate CUSUM charts. Journal of Quality and Technology 22, 173-186.
18. Rachev, S.T. and Mittnik, S. (2000). Stable Paretian Models in Finance, Wiley.

Part II

On-line Control

Surveillance Sampling and Sampling Plans

Principles for Multivariate Surveillance

Marianne Frisén

Statistical Research Unit, Department of Economics, Göteborg University

Summary. Multivariate surveillance is of interest in industrial production as it enables the monitoring of several components. Recently there has been an increased interest also in other areas such as detection of bioterrorism, spatial surveillance and transaction strategies in finance.

Several types of multivariate counterparts to the univariate Shewhart, EWMA and CUSUM methods have been proposed. Here a review of general approaches to multivariate surveillance is given with respect to how suggested methods relate to general statistical inference principles.

Suggestions are made on the special challenges of evaluating multivariate surveillance methods.

1 Introduction

The need is great for continuous observation of time series with the aim of detecting an important change in the underlying process as soon as possible after the change has occurred. The first versions of modern control charts (Shewhart (1931)) were made for industrial use. Multivariate surveillance is of interest in industrial production, for example in order to monitor several sources of variation in assembled products. Wärmefjord (2004) described the multivariate problem for the assembly process of the Saab automobile. Sahni et al. (2005) suggest that the raw material and different process variables in food industry should be analysed in order to assure the quality of the final product. Surveillance of several parameters (such as the mean and the variance) of a distribution is multivariate surveillance (see for example Knoth and Schmid (2002)).

In recent years, there has been an increased interest in statistical surveillance also in other areas than industrial production. The increased interest in surveillance methodology in the US following the 9/11 terrorist attack is notable. In the US and also in other countries several new types of data are now being collected. Since the collected data involve several related variables, this calls for multivariate surveillance techniques. Spatial surveillance is useful for the detection of a local change or a spread (for example of a disease or of a harmful agent). Spatial surveillance is multivariate since several locations are involved. Recently, there have

H.-J. Lenz et al. (eds.), *Frontiers in Statistical Quality Control 9*,
DOI 10.1007/978-3-7908-2380-6_9,

also been efforts to use multivariate surveillance for financial decision strategies (see for example Okhrin and Schmid (2007) and Golosnoy et al. (2007)) with respect to various assets and their relations and combinations.

We will focus on some principles for the construction of multivariate surveillance methods. These general approaches do not depend on the distributional properties of the process in focus, even though the implementation does. Note especially that both continuous and discrete data can be mixed. Reviews on multivariate surveillance methods can be found for example in Basseville and Nikiforov (1993), Lowry and Montgomery (1995), Ryan (2000), Woodall and Amiriparian (2002), Frisén (2003)and Sonesson and Frisén (2005).

In Section 2 the notations and specifications will be given. In Section 3 different approaches to the construction of multivariate surveillance methods are described and exemplified. In Section 4 we discuss the special challenges of evaluating multivariate surveillance methods. In Section 5 we give a simple example in order to illustrate approaches and evaluations. Concluding remarks are made in Section 6.

2 Notations

The multivariate process under surveillance is denoted by $\mathbf{Y} = \{\mathbf{Y}(t), \quad t = 1,2,...\}$. At each time point, t, a p-variate vector $\mathbf{Y}(t) = \begin{pmatrix} Y_1(t) & Y_2(t) & \cdots & Y_p(t) \end{pmatrix}^T$ of variables is observed. The components of the vector may be, for example, a measure of each of p different components of a produced item. When the process is in control and no change has occurred, $\mathbf{Y}(t)$ has a certain distribution for example with a certain mean vector $\boldsymbol{\mu}_0$ and a certain covariance matrix $\boldsymbol{\Sigma}_\mathbf{Y}$. The purpose of the surveillance method is to detect a deviation to a changed state as soon as possible in order to warn and to take corrective actions. We denote the current time point by s. At that time we want to determine whether or not a change in the distribution of $\mathbf{Y}$ has occurred up to now. Thus we want to discriminate between the events $\{\tau \le s\}$ and $\{\tau > s\}$, where τ denotes the time point of the change. In a multivariate setting, each component can change at different times τ_1, .. τ_p. A natural aim in many situations is to detect the first time that the process is no longer in control. Then it is natural to consider $\tau_{\min} = \min\{\tau_1,...\tau_p\}$. In order to detect the change, we can use all available observations of the process $\mathbf{Y}_s = \{\mathbf{Y}(t), \quad t \le s\}$ to form an alarm statistic denoted by $p(\mathbf{Y}_s)$. The surveillance method makes an alarm, indicating that the change has happened in the process at the first time point when $p(\mathbf{Y}_s)$ exceeds an alarm limit $G(s)$.

3 General approaches to multivariate surveillance

3.1. Reduction of dimension

It is useful to add any relevant structure to the problem in order to focus the detection ability. One way to reduce dimensionality is to consider the principal components instead of the original variables as proposed for example by Jackson (1985), Mastrangelo et al. (1996) and Kourti and MacGregor (1996). However, unless the principal components can be interpreted, a surveillance method based on them may be difficult to interpret. In Runger (1996) an alternative transformation, using so-called U^2 statistics, was introduced to allow the practitioner to choose the subspace of interest, and this is used for fault patterns in Runger et al. (2007). Another way to reduce the dimensionality is to use projection pursuit as done by Ngai and Zhang (2001) and Chan and Zhang (2001). Rosolowski and Schmid (2003) use the Mahalanobis distance to reduce the dimensionality of the statistic, thus expressing the distance from the target of the mean and the autocorrelation in a multivariate time series. After reducing the dimensionality, any of the approaches for multivariate surveillance described below can be used. There are also other reduction approaches which may be used after the initial reduction of dimensions. Vector accumulation was used by Scranton et al. (1996) after reduction to principal components and by Runger et al. (1999) after the U-transformation.

3.2 Reduction to scalar statistics

The reduction of the dimension can go as far as to summarise the components for each time point into one statistic. This is also a common way to handle multivariate surveillance problems. We start by transforming the vector from the current time point into a scalar statistic, which we then accumulate over time. In Sullivan and Jones (2002) this is referred to as "scalar accumulation". The relevant scalar statistics depend on the application. In spatial surveillance it is common to start by a purely spatial analysis for each time point as in Rogerson (1997). One natural reduction is to use the Hotelling T^2 statistic (Hotelling (1947)). This statistic is $T^2(t) = (\mathbf{Y}(t) - \boldsymbol{\mu}_0(t))^T \mathbf{S}_{\mathbf{Y}(t)}^{-1} (\mathbf{Y}(t) - \boldsymbol{\mu}_0(t))$, where the sample covariance matrix $\mathbf{S}_{\mathbf{Y}(t)}$ is used to estimate $\boldsymbol{\Sigma}_{\mathbf{Y}}$. When $\boldsymbol{\Sigma}_{\mathbf{Y}}$ is regarded as known and the statistic has a χ^2 distribution, it is referred to as the χ^2 statistic. Scalars based on regression and other linear weighting are suggested for example by Healy (1987), Kourti and MacGregor (1996) and Lu et al. (1998). The reduction to a univariate variable can be followed by univariate monitoring of any kind.

Originally, the Hotelling T^2 statistic was used in a Shewhart method, and this is often referred to as the Hotelling T^2 control chart. An alarm is triggered as soon as the statistic $T^2(t)$ is large enough. Note that over time, there is no accumulation of the observation vectors if the Shewhart method is used.

In order to achieve a more efficient method, all previous observations should be used in the alarm statistic. There are several suggestions of combinations where reduction to a scalar statistic is combined with different monitoring methods. Crosier (1988) suggested to first calculate the Hotelling T variable (the square root of $T^2(t)$) and then use this as the variable in a univariate CUSUM method, making it a scalar accumulation method.

A non-parametric scalar accumulation approach was used in Liu (1995), where the observation vector for a specific time point was reduced to a rank in order to remove the dependency on the distributional properties of the observation vector. Several methods were discussed for the surveillance step, including the CUSUM method. Yeh et al. (2003) suggested a transformation of multivariate data at each time to a distribution percentile, and the EWMA method was suggested for the detection of changes in the mean as well as in the covariance.

3.3 Parallel surveillance

This is a commonly used approach, where a univariate surveillance method is used for each of the individual components in parallel. This approach is referred to as combined univariate methods or parallel methods. One can then combine the univariate methods into a single surveillance procedure in several ways. The most common is to signal an alarm if any of the univariate methods signals. This is a use of the union-intersection principle for multiple inference problems. Sometimes the Bonferroni method is used to control a false alarm error, see Alt (1985). General references about parallel methods include Woodall and Ncube (1985), Hawkins (1991), Pignatiello and Runger (1990), Yashchin (1994) and Timm (1996).

Parallel methods suitable for different kinds of data have been suggested. In Skinner et al. (2003) a generalised linear model was used to model independent multivariate Poisson counts. Deviations from the model were monitored with parallel Shewhart methods. In Steiner et al. (1999) binary results were monitored using a parallel method of two individual CUSUM methods. However, to be able to detect also small simultaneous changes in both outcome variables, the method was complemented with a third alternative, which signals an alarm if both individual CUSUM statistics are above a lower alarm limit at the same time. The addition of the combined rule is in the same spirit as the vector accumulation methods presented below. Parallel CUSUM methods were used also by Marshall et al. (2004).

3.4 Vector accumulation

The accumulated information on each component is here utilised by a transformation of the vector of component-wise alarm statistics into a scalar alarm statistic. An alarm is triggered if this statistic exceeds a limit. This is naturally referred to as "vector accumulation".

Lowry et al. (1992) proposed a multivariate extension of the univariate EWMA method, which is referred to as MEWMA. This method uses a vector of univariate EWMA statistics $\mathbf{Z}(t)=\mathbf{\Lambda Y}(t)+(\mathbf{I}-\mathbf{\Lambda})\mathbf{Z}(t-1)$ where $\mathbf{Z}(0)=\mathbf{0}$ and $\mathbf{\Lambda}=\mathrm{diag}(\lambda_1,\lambda_2,...,\lambda_p)$. An alarm is triggered at $t_A=\min\{t;\ \mathbf{Z}(t)^T\mathbf{\Sigma}_{\mathbf{Z}(t)}^{-1}\mathbf{Z}(t)>L\}$ for the alarm limit, L. The MEWMA can be seen as the Hotelling T^2 control chart applied to EWMA statistics instead of the original data and is thus a vector accumulation method. Reynolds and Keunpyo (2005) studied sequential sampling design in connection with MEWMA.

One natural way to construct a multivariate version of the CUSUM method would be to proceed as for EWMA and construct the Hotelling T^2 control chart applied to univariate CUSUM statistics for the individual variables. One important feature of such a method is the lower barrier of each of the univariate CUSUM statistics (assuming we are interested in a positive change). This kind of multivariate CUSUM was suggested by Bodnar and Schmid (2004) and Sonesson and Frisén (2005). Other approaches to construct a multivariate CUSUM have also been suggested. Crosier (1988) suggested the MCUSUM method, and Pignatiello and Runger (1990) had another suggestion. Both these methods use a statistic consisting of univariate CUSUMs for each component and are thus vector accumulation methods. However, the way in which the components are used is different as compared with the MEWMA construction.

One important feature of these two methods is that the characteristic zero-return of the CUSUM technique is constructed in a way which is suitable when all the components change at the same time point. If all components change at the same time, a univariate reduction is optimal, however.

3.5 Joint solution

Stepwise constructions of methods for complicated problems are often useful. In this section, however, we discuss jointly optimal methods. such optimality is not guaranteed by the approaches described in the sections above, which start with a reduction in either time or space (or other multivariate setting). Sometimes a sufficient reduction will result in a separation of the spatial and the temporal components. However, such a sufficient separation is not always available. The use of the sufficient statistic implies that no information is lost. An example of this is the result by Wessman (1998) that when all the variables change at the same time, a sufficient reduction to univariate surveillance exists.

Healy (1987) derived the CUSUM method for the case of simultaneous change in a specified way for all the variables. The results are univariate CUSUMs for a function of the variables. The CUSUM method is minimax optimal. Thus, the multivariate methods by Healy (1987) are simultaneously minimax optimal for the specified direction when all variables change at the same time.

Another way to achieve a simultaneously optimal solution is by applying the full likelihood ratio method (Shiryaev (1963) and Frisén and de Maré (1991). This method can be used as soon as the event to be detected is specified. The full likelihood ratio method was used by Järpe (1999) in the case of clustering in a spatial log-linear model on a fixed lattice. The likelihood ratio method was used after a sufficient reduction of the spatio-temporal pattern. Another example is Järpe (2001), where the optimal way for detecting an increased radiation level was derived. In this application the shift process spread spatially with time.

4 Measures of the properties of multivariate surveillance

Optimality is hard to achieve and even hard to define for all multivariate problems, also in the surveillance case (seeFrisén (2003)). We have a spectrum of problems where one extreme is that there are hardly any relations between the multiple surveillance components. The other extreme is that we can reduce the problem to a univariate one by considering the relation between the components. Consider, for example, the case when we measure several components of an assembled item. If we restrict our attention to a general change in the factory, changes will be expected to occur at the same time. Then, the multivariate situation is easily reduced to a univariate one (Wessman (1998)) and we can easily derive optimal methods. For many applications, however, the specification of one general change is too restrictive.

The problem is how to determine which type of change to focus on and which not to. The method derived according to the specification of a general change will not be capable of detecting a change in only one of many components. On the other hand, if we focus on detecting all kinds of changes, the detection ability of the surveillance method for each specific type of change will be small. One way to focus the attention is to consider some type of dimension reduction transformation (Hawkins (1991) and Runger (1996)).

In hypothesis testing, the false rejection is considered most important. It is important to control the error in multiple testing since the rejection of a null hypothesis is considered as a proof that the null hypothesis is false. Important methods for controlling the risk of an erroneous rejection in multiple comparison procedures are described in the monograph by Hochberg and Tamhane (1987). When testing several drugs against the standard, the family-wise error rate is relevant. In other cases, such as when many aspects of a drug is tested, the False Discover Rate, FDR, suggested by Benjamini and Hochberg (1995) may be more relevant and does not reduce the power so much. This situation is more accepted as a screening than as hypothesis testing. In surveillance this is further stressed as all methods with a fair power to detect a change have a false alarm rate that tends to one (Bock (2007)). The problem with adopting FDR is that it uses a probability which is not constant in surveillance. Marshall et al. (2004) solve this problem as the monitoring is carried out over a fairly short period of time and they use only the properties of the early part of the run length distribution. FDR in surveillance has been advocated for example by Rolka et al. (2007). However, the question is whether control of FDR is necessary when surveillance is used as a screening instrument, which indicates that further examination should be made. For some such applications the ARL^0 of the combined procedure may be informative enough since it gives information about the expected time until (an unnecessary) screening. It will sometimes be easier to judge the practical burden with a too low alarm limit by the ARL^0 than by the FDR for that situation.

The timeliness in detection is of extreme interest in surveillance, and other measures than the ones traditionally used in hypothesis testing are important. To evaluate the timeliness, different measures such as the average run length, the conditional expected delay and the probability of successful detection (Frisén (1992) can be used with or without modification also in a multivariate setting.

The ARL^1 is the most commonly used measure of the detection ability also in the multivariate case. It is usually assumed that all variables change immediately. However, the results by Wessman (1998) are that univariate surveillance is always the best method in this setting.

It is not entirely satisfactory to restrict the evaluation of the methods to changes occurring at the same time as the surveillance starts, since the detection ability depends on when the change occurs. The conditional expected delay $CED(t) = \mathrm{E}\left[t_A - \tau \mid t_A \geq \tau = t\right]$ is a component in many measures which avoids the dependency on τ either by concentrating on just one value of τ (e.g. one, infinity or the worst value). Frisén (2003) advocated that the whole function of τ should be studied. This measure can be generalized by considering the delay from the first change

$$\tau_{\min} = \min\{\tau_1,\ldots\tau_p\}$$

$$CED(\tau_1,\ldots\tau_p) = E(t_A - \tau_{\min} \mid t_A \geq \tau_{\min}) .$$

This type of generalization was suggested by Andersson (2007).

The Probability of Successful Detection suggested by Frisén (1992) measures the probability of detection with a delay time shorter than d. In the multivariate case it can be defined as

$$PSD(d,\tau_1,\ldots\tau_p) = P(t_A - \tau_{\min} \leq d \mid t_A \geq \tau_{\min}) .$$

This measure is a function of both the times of the changes and the length of the interval in which the detection is defined as successful. Also when there is no absolute limit to the detection time it is often useful to describe the ability to detect the change within a certain time. In such cases it may be useful to calculate the PSD for different time limits d. This has been done for example by Marshall et al. (2004) in connection with use of the FDR. The ability to make a very quick detection (small d) is important in surveillance of sudden major changes, while the long term detection ability (large d) is more important in ongoing surveillance where smaller changes are expected.

Thorough evaluation, involving changes of different types occurring at different time points, is recommended. Several measures of evaluation may be useful.

5 Examples

In order to illustrate principles and measures we will compare one method using reduction to one scalar for each time (Method M1) with one using reduction to one scalar for each time parallel surveillance (Method M2). To keep it simple we base the methods on the Shewhart method. Our simple model is for two normally distributed variables which possibly have shifts at different time points

$$X(t) \sim \begin{cases} N(0,1) & t < \tau_X \\ N(1,1) & t \geq \tau_X \end{cases}$$
$$Y(t) \sim \begin{cases} N(0,1) & t < \tau_Y \\ N(2,1) & t \geq \tau_Y \end{cases}$$

Method M1 gives an alarm if the Shewhart method gives an alarm for the sum of the variables.

$$t_A = \min\{t; X(t) + Y(t) > G_{M1}\}$$

Method M2 gives an alarm if the Shewhart method gives an alarm for any of the variables.

$$t_A = \min\{t; X(t) > G_{M2} \cup Y(t) > G_{M2}\}$$

The limits were determined to G_{M1}=3,29 and G_{M2}=2,57 so that ARL^0=100 for both M1 and M2.

Several situations were examined. In the first one both variables shift at the same time. That is $\tau_X = \tau_Y$. For the Shewhart method the CED is constant in this case. The method M1 has the conditional expected delay CED=1,39 while M2 has (constant) CED=2,09. The probability to detect the out-of-control state immediately PSD(0,t) is also constant with respect to t. For the M1 method PSD=0,42 and for M2 we have PSD=0,32. Thus we see that if both methods shift at the same time it is best to use the univariate sum as alarm statistic. This is also in accordance with theory.

In the second situation one variable does not shift, while the other one does - but we do not know beforehand which one it might be. For the case when X in fact did not change ($\tau_X=\infty$) but Y did we have $\tau_{\min} = \tau_Y$. The method M1 has the conditional expected delay CED=4,53 and M2 has CED=2,49. For the M1 method PSD=0,18 and for M2 PSD=0,29. Thus, we see that if only one out of several processes changes the properties of M2 are much better.

In the third situation we know that only the distribution of Y can change. We can thus focus on Y only. If this had been the case the univariate Shewhart method would have had CED=1,69. The probability to detect the out-of-control state immediately would have been PSD(0,t)=0,37. Thus, the knowledge improves the detection ability (for the same ARL^0) considerably.

6 Conclusions

The construction of surveillance methods is challenging in many ways. It involves statistical theory, practical issues as to the collection of new types of data and computational ones such as the implementation of automated methods in large scale surveillance data bases. The data is sometimes highly dimensional and collected into huge data bases. Here the focus has been on the statistical inference aspects of the multivariate surveillance problem.

Methods can be characterised as scalar accumulating, parallel, vector accumulating or simultaneous. However, there is no sharp limit between some of these categories. What is regarded as a dimension reduction and what is regarded to be a scalar accumulation sometimes overlap. Many methods first reduce the dimension for example by principal components, and then one of the approaches for multivariate surveillance is used. Fuchs and Benjamini (1994) suggest Multivariate Profile Charts which demonstrate both the overall multivariate surveillance and individual surveillance in the same chart and thus combine two of the approaches.

The more specifically the aim is stated, the better the possibilities of the surveillance to meet this aim. Hauck et al. (1999) describe how a change may influence variables and also the relation between them. One way to focus the detection ability is by specifying a loss function with respect to the relative importance of changes in different directions. Mohebbi and Havre (1989) use weights from a linear loss function instead of the covariance for the reduction to a univariate statistic. Tsui and Woodall (1993) use a non-linear loss function and a vector accumulation method named MLEWMA. For some methods the detection ability depends only on one non-centrality parameter which measures the magnitude of the multidimensional change. Such methods are known as "directionally invariant". However, this is not necessarily a good property, since there often is an interest in detecting a certain type of change. Fricker (2007) stresses the importance of directionally sensitive methods for syndromic surveillance. Preferably, the specification should be governed by the application.

The question of which multivariate surveillance method is the best has no concise answer. Different methods are suitable for different problems. Some causes may lead to a simultaneous increase in several variables, and then one should use a reduction to a univariate surveillance method, as shown by Wessman (1998) and demonstrated here by the examples in Section 5. If the changes occur independently, one does not expect simultaneous changes and may instead prefer to use (for example) parallel methods. This may work better than the scalar reduction, as demonstrated by the examples in Section 5, where it is also shown that the best would be to utilize knowledge on which component to concentrate on.

One advantage with parallel methods is that the interpretation of alarms will be clear. The identification of why an alarm was raised is important. An example is the inability of the Hotelling T^2 control chart to distinguish between a change in the mean vector and a change in the covariance structure. Mason et al. (1995) provided a general approach by a decomposition of the T^2 statistic into independent components. Other suggestions include for example principal component analysis, see Pignatiello and Runger (1990), Kourti and MacGregor (1996) and Maravelakis et al. (2002). An example is a spatially restricted outbreak. The importance of knowledge about where to concentrate the effort after an alarm indicating a bioterrorist attack is discussed by Mostashari and Hartman (2003).

References

Alt, F. B. (1985) Multivariate quality control. In *Encyclopedia of Statistical Science*, Vol. 6 (Eds, Johnson, N. L. and Kotz, S.) Wiley, New York, pp. 110-122.

Andersson, E. (2007) Effect of dependency in systems for multivariate surveillance. Research Report 2007:1, Statistical Research Unit, University of Gothenburg.

Basseville, M. and Nikiforov, I. (1993) *Detection of abrupt changes- Theory and application,* Prentice Hall, Englewood Cliffs.

Benjamini, Y. and Hochberg, Y. (1995) Controlling the false discovery rate: a practical and powerful approach to multiple testing. *Journal of the Royal Statistical Society B,* **57,** 289-300.

Bock, D. (2007) Aspects on the control of false alarms in statistical surveillance and the impact on the return of financial decision systems. *Journal of Applied Statistics*, (in press).

Bodnar, O. and Schmid, W. (2004) CUSUM control schemes for multivariate time series. In *Frontiers in Statistical Quality Control. Intelligent Statistical Quality control*Warsaw.

Chan, L. K. and Zhang, J. (2001) Cumulative sum control charts for the covariance matrix. *Statistica Sinica,* **11,** 767-790.

Crosier, R. B. (1988) Multivariate generalizations of cumulative sum quality-control schemes. *Technometrics,* **30,** 291-303.

Fricker, R. D. (2007) Directionally Sensitive Multivariate Statistical Process Control Procedures with Application to Syndromic Surveillance *Advances in Disease Surveillance,* **3,** 1-17.

Frisén, M. (1992) Evaluations of Methods for Statistical Surveillance. *Statistics in Medicine,* **11,** 1489-1502.

Frisén, M. (2003) Statistical surveillance. Optimality and methods. *International Statistical Review,* **71,** 403-434.

Frisén, M. and de Maré, J. (1991) Optimal Surveillance. *Biometrika,* **78,** 271-80.

Fuchs, C. and Benjamini, Y. (1994) Multivariate profile charts for statistical process control. *Technometrics,* **36,** 182-195.

Golosnoy, V., Schmid, W. and Okhrin, I. (2007) Sequential Monitoring of Optimal PortfolioWeights. In *Financial surveillance* (Ed, Frisén, M.) Wiley, Chichester, pp. 179-210.

Hauck, D. J., Runger, G. C. and Montgomery, D. C. (1999) Multivariate statistical process monitoring and diagnosis with grouped regression-adjusted variables. *Communications in Statistics. Simulation and Computation,* **28,** 309-328.

Hawkins, D. M. (1991) Multivariate quality control based on regression-adjusted variables. *Technometrics,* **33,** 61-75.

Healy, J. D. (1987) A note on multivariate CUSUM procedures. *Technometrics,* **29,** 409-412.

Hochberg, Y. and Tamhane, A. C. (1987) *Multiple comparison procedures,* Wiley New York.

Hotelling, H. (1947) Multivariate quality control. In *Techniques of statistical analysis* (Eds, Eisenhart , C., Hastay, M. W. and Wallis, W. A.) McGraw-Hill, New York.

Jackson, J. E. (1985) Multivariate quality control. *Communications in Statistics. Theory and Methods,* **14,** 2657-2688.

Järpe, E. (1999) Surveillance of the interaction parameter of the Ising model. *Communications in Statistics. Theory and Methods,* **28,** 3009-3027.

Järpe, E. (2001) Surveillance, environmental. In *Encyclopedia of Environmetrics* (Eds, El-Shaarawi, A. and Piegorsh, W. W.) Wiley, Chichester.

Knoth, S. and Schmid, W. (2002) Monitoring the mean and the variance of a stationary process. *Statistica Neerlandica,* **56,** 77-100.

Kourti, T. and MacGregor, J. F. (1996) Multivariate SPC methods for process and product monitoring. *Journal of Quality Technology,* **28,** 409-428.

Liu, R. Y. (1995) Control charts for multivariate processes. *Journal of the American Statistical Association,* **90,** 1380-1387.

Lowry, C. A. and Montgomery, D. C. (1995) A review of multivariate control charts. *IIE Transactions,* **27,** 800-810.

Lowry, C. A., Woodall, W. H., Champ, C. W. and Rigdon, S. E. (1992) A multivariate exponentially weighted moving average control chart. *Technometrics,* **34,** 46-53.

Lu, X. S., Xie, M., Goh, T. N. and Lai, C. D. (1998) Control chart for multivariate attribute processes. *International Journal of Production Research,* **36,** 3477-3489.

Maravelakis, P. E., Bersimis, S., Panaretos, J. and Psarakis, S. (2002) Identifying the out of control variable in a multivariate control chart. *Communications in Statistics. Theory and Methods,* **31,** 2391-2408.

Marshall, C., Best, N., Bottle, A. and Aylin, P. (2004) Statistical issues in the prospective monitoring of health outcomes across multiple units. *Journal of the Royal Statistical Society A,* **167,** 541-559.

Mason, R. L., Tracy, N. D. and Young, J. C. (1995) Decomposition of T^2 for multivariate control chart interpretation. *Journal of Quality Technology,* **27,** 99-108.

Mastrangelo, C. M., Runger, G. C. and Montgomery, D. C. (1996) Statistical process monitoring with principal components. *Quality and Reliability Engineering International,* **12,** 203-210.

Mohebbi, C. and Havre, L. (1989) Multivariate control charts: A loss function approach. *Sequential Analysis,* **8,** 253-268.

Mostashari, F. and Hartman, J. (2003) Syndromic surveillance: a local perspective. *Journal of Urban Health,* **80,** I1-I7.

Ngai, H. M. and Zhang, J. (2001) Multivariate cumulative sum control charts based on projection pursuit. *Statistica Sinica,* **11,** 747-766.

Okhrin, Y. and Schmid, W. (2007) Surveillance of Univariate and Multivariate Nonlinear Time Series. In *Financial surveillance* (Ed, Frisén, M.) Wiley, Chichester, pp. 153-177.

Pignatiello, J. J. and Runger, G. C. (1990) Comparisons of multivariate CUSUM charts. *Journal of Quality Technology,* **22,** 173-186.

Reynolds, M. R. and Keunpyo, K. (2005) Multivariate Monitoring of the Process Mean Vector With Sequential Sampling. *Journal of Quality Technology,* **37,** 149-162.

Rogerson, P. A. (1997) Surveillance systems for monitoring the development of spatial patterns. *Statistics in Medicine,* **16,** 2081-2093.

Rolka, H., Burkom, H., Cooper, G. F., Kulldorff, M., Madigan, D. and Wong, W.-K. (2007) Issues in applied statistics for public health bioterrorism surveillance using multiple data streams: research needs. *Statistics in Medicine,* **26,** 1834-1856.

Rosolowski, M. and Schmid, W. (2003) EWMA charts for monitoring the mean and the autocovariances of stationary Gaussian processes. *Sequential Analysis,* **22,** 257-285.

Runger, G. C. (1996) Projections and the U^2 chart for multivariate statistical process control. *Journal of Quality Technology,* **28,** 313-319.

Runger, G. C., Barton, R. R., Del Castillo, E. and Woodall, W. H. (2007) Optimal Monitoring of Multivariate Data for Fault Detection". *Journal of Quality Technology,* **39,** 159-172.

Runger, G. C., Keats, J. B., Montgomery, D. C. and Scranton, R. D. (1999) Improving the performance of the multivariate exponentially weighted moving average control chart. *Quality and Reliability Engineering International,* **15,** 161-166.

Ryan, T. P. (2000) *Statistical methods for quality improvement,* Wiley, New York.

Sahni, N. S., Aastveit, A. H. and Naes, T. (2005) In-Line Process and Product COntrol Using Spectroscopy and Multivariate Calibration. *Journal of Quality Technology,* **37,** 1-20.

Scranton, R. D., Runger, G. C., Keats, J. B. and Montgomery, D. C. (1996) Efficient shift detection using multivariate exponentially-weighted moving average control charts and principal components. *Quality and Reliability Engineering International,* **12,** 165-171.

Shewhart, W. A. (1931) *Economic Control of Quality of Manufactured Product,* MacMillan and Co., London.

Shiryaev, A. N. (1963) On optimum methods in quickest detection problems. *Theory of Probability and its Applications,* **8,** 22-46.

Skinner, K. R., Montgomery, D. C. and Runger, G. C. (2003) Process monitoring for multiple count data using generalized linear model-based control charts. *International Journal of Production Research,* **41,** 1167-1180.

Sonesson, C. and Frisén, M. (2005) Multivariate surveillance. In *Spatial surveillance for public health* (Eds, Lawson, A. and Kleinman, K.) Wiley, New York, pp. 169-186.

Steiner, S. H., Cook, R. J. and Farewell, V. T. (1999) Monitoring paired binary surgical outcomes using cumulative sum charts. *Statistics in Medicine,* **18,** 69-86.

Sullivan, J. H. and Jones, L. A. (2002) A self-starting control chart for multivariate individual observations. *Technometrics,* **44,** 24-33.

Timm, N. H. (1996) Multivariate quality control using finite intersection tests. *Journal of Quality Technology,* **28,** 233-243.

Tsui, K. L. and Woodall, W. H. (1993) Multivariate control charts based on loss functions. *Sequential Analysis,* **12,** 79-92.

Wessman, P. (1998) Some Principles for surveillance adopted for multivariate processes with a common change point. *Communications in Statistics. Theory and Methods,* **27,** 1143-1161.

Woodall, W. H. and Amiriparian, S. (2002) On the economic design of multivariate control charts. *Communications in Statistics -Theory and Methods,* **31,** 1665-1673.

Woodall, W. H. and Ncube, M. M. (1985) Multivariate cusum quality control procedures. *Technometrics,* **27,** 285-292.

Wärmefjord, K. (2004) Multivariate quality control and Diagnosis of Sources of Variation in Assembled Products. Licentiat Thesis, Department of Mathematics, Göteborg University.

Yashchin, E. (1994) Monitoring Variance Components. *Technometrics,* **36,** 379-393.

Yeh, A. B., Lin, D. K. J., Zhou, H. H. and Venkataramani, C. (2003) A multivariate exponentially weighted moving average control chart for monitoring process variability. *Journal of Applied Statistics,* **30,** 507-536.

Research Issues and Ideas on Health-Related Surveillance

William H. Woodall[1], Olivia A. Grigg[2], and Howard S. Burkom[3],

[1]Department of Statistics, Virginia Tech, Blacksburg, VA 24061-0439, USA
bwoodall@vt.edu

[2]MRC Biostatistics Unit, Institute of Public Health, Cambridge, UK
Olivia.Grigg@cantab.net

[3]National Security Technology Department, The Johns Hopkins University, Applied Physics Laboratory, Laurel, MD 20723
Howard.Burkom@jhuapl.edu

Summary. In this overview paper, some of the surveillance methods and metrics used in health-related applications are described and contrasted with those used in industrial practice. Many of the aforesaid methods are based on the concepts and methods of statistical process control. Public health data often include spatial information as well as temporal information, and in this and other regards, public health applications could be considered more challenging than industrial applications. Avenues of research into various topics in health-related monitoring are suggested.

1 Introduction

The purpose of this paper is to give an introduction to health-related monitoring, particularly to those more familiar with research in industrial process monitoring. Some of the topics covered by Woodall (2006) are updated. A review of recently published papers, and some still in press, is given. Many of the methods currently used in public health surveillance are based on the concepts of statistical process control (SPC), commonly applied in industry and business. Indeed, Lawson and Kleinman (2005, p. 5) stated, "SPC has formed the basis for many disease surveillance systems."

There is a growing interest in health-related monitoring and improvement among industrial practitioners. There will be, for example, a special issue on this topic by *Quality Engineering*. There are many interesting and challenging research problems in public health surveillance and in health-related monitoring. Re-

H.-J. Lenz et al. (eds.), *Frontiers in Statistical Quality Control 9*,
DOI 10.1007/978-3-7908-2380-6_10, © Springer-Verlag Berlin Heidelberg 2010

searchers in industrial statistics are strongly encouraged to consider topics in these areas, a number of which are suggested in this paper.

There are several excellent review papers and books that can be very useful to anyone wanting to know more about health-related monitoring. Sonesson and Bock (2003) and Lawson and Kleinman (2005) provided very helpful reviews of public health surveillance. For more in-depth information, an online journal on public health surveillance topics, *Advances in Disease Surveillance*, can be found at http://www.isdsjournal.org/issue/current.

The discussion by Benneyan (2006) is recommended as a starting point for those interested in hospital applications. Thor et al. (2007) provided a review of the literature of healthcare applications of control charts. Control of infection rates is very important in hospitals, with a recent paper on this topic given by Sherlaw-Johnson et al. (2007). Tennant et al. (2007) reviewed applications of control charts for use with individual patients with chronic diseases such as diabetes and asthma.

The papers by Fienberg and Shmueli (2005), Shmueli and Burkom (2008), Buckeridge et al. (2005), Fricker and Rolka (2006), and Rolka et al. (2007) would be very useful sources for those interested in the very challenging area of syndromic surveillance. In syndromic surveillance data are drawn from a variety of sources, such as emergency room records, over-the-counter drug sales, absenteeism rates, and so forth, in an attempt to supplement traditional sentinel surveillance for natural disease outbreaks or bioterrorist attacks.

2 Contrasts between Industrial and Health-Related Monitoring

Some of the differences between industrial and health-related surveillance practice, performance metrics, and research literature are given in this section.

1. **Data:** Attribute data are of greater focus in the health setting than in the industrial setting. For example, in surveillance for outbreaks of disease the incidence rate, or number of new cases in the population of interest by unit of time, is typically monitored. More use is made of observed data, especially in the comparison of aspects of available methods, because of the evolving, non-stationary nature of the data streams and because processes in the health environment cannot be summarily corrected. For instance, rather than adopting a model for normal (non-outbreak) conditions, observed data may be used as testbeds on which to compare competing methods' ability in detecting simulated outbreak scenarios, despite the fact that effects of actual unknown outbreaks ay be present in these datasets. Indeed, there is somewhat of a cultural divide regarding the use of observed data for evaluation of methods in public health surveillance and the use of null models in industrial SPC, a divide discussed by Breiman (2001) in a more general context.
2. **Pattern of Outbreak:** Most industrial control chart methods are studied under the model of a constant baseline parameter, with a sustained step shift in

the parameter when the process goes "out-of-control". Outbreaks in health-related monitoring are more likely to be transitory, often with more gradual increases and decreases than with abrupt shifts. Health-related monitoring methods tend to be one-sided, to focus on the primary concern of increases in rates, whereas use of two-sided methods is typical in industrial practice. Where there is an intervention, such as a surgical or drug treatment, monitoring for improvement in health of an individual would be of interest.

3. **Resetting after a Signal:** Some public health surveillance methods are run without interruption or resetting when a signal indicates a potential outbreak. In industrial applications, on the other hand, a process adjustment or investigation is usually made following a signal and the chart then restarted.

4. **Performance Metrics:** In evaluating the statistical properties of charts, the focus in industrial monitoring has been on the time to the first signal. Since timely response to changes in the process is a key concern, this measure has been used to develop optimality criteria by which to compare charts. Discussions of the various metrics used in public health surveillance were given by Frisén (1992, 2003), Frisén and Wessman (1999), Frisén and Sonesson (2005), Buckeridge et al. (2005), and Fraker et al. (2008). Some of these metrics, based on medical test terminology such as "sensitivity" and "specificity", may be unfamiliar to industrial practitioners.

5. **Use of empirically derived *p*-values:** For health-related process monitoring, several thresholding approaches have been used in which *p*-values are calculated from the empirical distribution of historical or artificially generated values of test statistics. See, for example, Kulldorff (2001), Kleinman et al. (2004), Wallenstein and Naus (2004), and Spliid (2007). Associating chart statistics with a *p*-value measure (see Benjamini and Kling, 1999; Grigg and Spiegelhalter, 2008) can allow for use, at some level, of flexible stopping rules such as false discovery rate (FDR) control (see Benjamini and Hochberg, 1995; Storey et al, 2004). Use of *p*-values may also allow significance to be conveyed on a reference scale, without dictation of a formal stopping rule.

6. **Baseline:** In health-related applications the monitored process may be complex. To avoid alarms resulting from predictable factors, description of the in-control state often includes covariates such as patient risk factors and seasonal effects. A more detailed null model, or knowledge of unmodelled variability in the observed process, may make consideration of steady-state behaviour of a method more difficult. However, steady-state measures remain of interest as these allow the effects of the initialization of the chart, and any head-start features, to be removed.

7. **Updating Control Limits:** In some public health surveillance applications it is common to update parameter estimates and measures of uncertainty sur-

rounding those estimates at each new observation. Many health-related applications have accounted for the nonstationary nature of monitored data streams by updating parameters and control limits with a sliding baseline—a "moving Phase I" approach (as in the CDC EARS methods discussed by Fricker et al. (2007)). It has been recommended by Buckeridge et al. (2005), Fricker et al. (2007), and others that buffer intervals (guard bands) be used to exclude the most recent data values from this baseline to avoid having the early effects of an outbreak prevent the later, possibly more severe, effects from being detected A more traditional, calibrated SPC approach to updating control limits, illustrating the cultural divide, is the ACUSUM of Sparks (2000).

8. **Recurrence Interval:** In some applications of public health surveillance it is recommended that the "recurrence interval" be used to assess in-control performance, where the recurrence interval is the fixed number of time periods (not necessarily adjacent) across which the expected number of alarms under the null is one. Fraker et al. (2008) showed, somewhat surprisingly, that the time-to-signal performance can vary widely among charts with the same value of the recurrence interval.

9. **Literature:** The body of literature on health-related surveillance is smaller than that on industrial surveillance, and is somewhat less mathematical in nature. Editors and reviewers for health-related journals typically require, or very strongly request, an application using real data. Evaluation of performance of methods based on assumed models, or comparison with other methods, has not always been required. As Sonesson and Bock (2003) stated, "A notable feature in many of the methods suggested in the (public health) literature is the lack of evaluation by other means than in different case-studies." As well as opportunities for making application-specific comparisons, valuable research opportunities exist in making more general, model-based evaluations and comparisons of methods.

3 Using Bernoulli Data

One much-discussed problem in the monitoring of rates, such as those of congenital malformations, deals with detecting an increase in an incidence rate, where outcomes are assumed to be results from independent Bernoulli trials. One could imagine, for example, having information regarding each successive birth over time in a region of interest. A common objective is to be able to quickly detect sustained increases from a postulated, constant baseline rate. Various "sets-based" methods have been considered for this purpose, as discussed by Sego et al. (2008a). The basic sets method of Chen (1978) would signal an increased rate if the number of births between malformations (a set) is uniformly less than a threshold t for s successive sets. Sego et al. (2008a) showed that a Bernoulli-based cumulative sum (CUSUM) chart proposed by Reynolds and Stoumbos (1999) has better performance than the sets-based methods.

Ismail et al. (2003) and Naus and Wallenstein (2006) proposed a scan method that signals as soon as m out of the last n Bernoulli trials corresponded to incidences. Joner et al. (2008a) showed, however, that the Bernoulli CUSUM has better performance. The scan method has good performance, however, so some might prefer it to the CUSUM chart due to its simplicity. Spliid (2007) recently proposed a competing method based on the exponentially weighted moving average (EWMA). Comparisons of methods in the Bernoulli case are complicated by the fact that the in-control performance of competing methods cannot be matched exactly due to the discreteness of the data.

Much work remains to be done on detection of transitory rate increases. Also risk-adjustment of scan methods is important. Various SPC methods have been proposed for "high quality processes", usually based on an underlying geometric or negative binomial distribution (see, for example, Benneyan, 2001; Xie et al., 2002). These methods are applicable in monitoring rare health events, where one would like to be able to detect any increase in incidence, and thus merit further study in that context.

4 Risk adjustment

A key characteristic of health-related monitoring is that the human element of the monitored process is likely to be great. In evaluating hospital and physician care, it should be taken into account that arriving patients may vary (from patient to patient, from hospital to hospital, or from physician to physician) with regard to their general state of health. Grigg and Farewell (2004a) gave a review of risk-adjusted monitoring. Steiner et al. (2000) proposed a risk-adjusted CUSUM method. Grigg and Farewell (2004b) demonstrated the use of a graphical risk-adjusted sets method and Grigg and Spiegelhalter (2007) recently proposed a risk-adjusted EWMA method.

In risk-adjusted monitoring one usually monitors a measure, such as a mortality rate, conditional on the value of covariates such as patient risk factors. The conditioning is carried out through use of a risk model, for example a logistic regression model, where risk is predicted given observed values for the covariates. The in-control incidence rate then varies from person to person and the focus moves to monitoring of an odds ratio (odds of a case under changed conditions to odds of a case under baseline conditions). In monitoring the performance of heart surgeons, patient 30-day mortality following surgery is a typical measure to consider. Changes in the (risk-adjusted) 30-day mortality rate might be ascribed to changes in surgical technique. Sego et al. (2008b) recently proposed a way to use more of the information in the data by treating the survival time as a censored random variable, instead of converting it to a dichotomous variable.

Perfect adjustment for patient mix is difficult, as explained by Winkel and Zhang (2007), so any inferences made following adjustment for risk should perhaps be made with care. The adequacy of adjustments made might be checked by considering the fit of the risk model to training or Phase I data. Lack of fit might be ascribed to common cause overdispersion in the training data, a description of which might be incorporated into the risk model, as discussed by Marshall et al. (2004) and by Grigg, Spiegelhalter, and Jones (2008).

Risk-adjusted charting methodology might be applicable to a variety of healthcare settings. Porter and Teisberg (2006) considered risk-adjusted comparisons of healthcare providers to be an essential component of improving healthcare in the U.S. Recently Axelrod et al. (2006) gave an application involving the monitoring of liver and kidney transplant success rate performance, while Novick et al. (2006) considered the monitoring of heart surgery outcomes.

There is potential for comparisons relating to the statistical performance and other aspects of risk-adjusted charting methods. These comparisons might be made for specific patient populations. The effects of null model misspecification and error in estimation of the measures of interest warrant more study. Robust specification of the null model and control limits, in the style of Phase I type methods, is an area that might be investigated further in the health setting. In addition, other outcomes, such as quality of life measures, could be considered.

5 Spatiotemporal Monitoring Opportunities

It is important to detect geographic clusters corresponding to increased incidence rates in the public health setting. One common situation is to collect and screen, at regular intervals, count data from several subregions in a region of interest. For example, monthly counts from counties within a U.S. state might be collected and their spatial distribution examined. The in-control distribution must be available using direct or indirect standardisation from census-type tables or from data history, or by modelling sub-regional streams. It is assumed that any null model is correct in terms of form and in terms of correct covariate selection; whereas it is assumed with standardization that the effective size of the population for which a rate is calculated is accurate.

Expected counts calculated might vary across subregions and possibly over time. When there is no increase which is of particular concern, the counts for the subregions at a given time might be mutually independent, but there could be correlation between sub-regional counts at any given time. Counts are traditionally assumed to be Poisson random variables, although experience has shown that allowance for overdispersion is often indicated. Various methods have been proposed for this spatiotemporal problem, and the scan statistics approach of Kulldorff (1997, 2001, 2005) has been widely popular because it identifies clusters of variable location and size, includes some measure of control for multiple testing, and is available in a freely downloadable software package.

For the monitoring of distributed data streams, with or without spatial information, multivariate control charts have also been proposed by Joner et al. (2008b), Fricker (2007), Fricker et al. (2008) and others. Directionally invariant multivariate control charts monitor for general shifts in multivariate space and thus react indiscriminately towards increases and decreases in rates. These charts have been adapted to react only to increases in rates. As the number of streams increases, dimensionality renders the anomaly identification more complex, and efficient strategies for combining multivariate and univariate charts is increasingly important, as suggested in Hawkins (1993).

Monitoring incidence rates of chronic diseases may entail less processing of information than monitoring rates of infectious diseases. Traditional SPC applications designed to monitor for persistent mean shifts are more naturally suited to detect sustained increases in chronic disease rates caused by elevated risk factors. Background behavior of infectious or volatile diseases is commonly influenced by cyclic or seasonal effects. Just as in industrial applications, time series models have been used to remove such effects; see, for example, Cowling et al. (2006) and Burr et al. (2006). In the health surveillance context, regular cyclic effects, such as weekly patterns in daily emergency rooms visit rates, are more prominent than in the industrial context.

There are many opportunities for research into spatiotemporal monitoring, a topic that has not been considered in the industrial monitoring literature. There is a need for thorough comparisons of possible methods. Comparisons are complicated by the many factors that can vary: the number of subregions, the locations and sizes of any clusters, the correlation structure of the data, and so forth. Identifying (significant) clusters of cases would be of most value if the cause or causes of the clusters can be discovered.

Public health surveillance applications are often characterized by a large number of multiple streams of data. For example, there could be data available over time on a number of different subregions, hospitals or physicians. See, e.g., Grigg and Spiegelhalter (2006). There has been research on multiple stream processes in industrial SPC, but the assumptions may be too restrictive for these methods to be applied directly to health-related monitoring. When many patient sub-groupings or several healthcare indicators are simultaneously considered, it can be a challenge to control the rate of false alarms and yet retain power to detect meaningful outbreaks. Marshall et al. (2004) and Grigg, Spiegelhalter and Jones (2008) applied FDR principles to obtain manageable alert rates when applying simultaneous CUSUM charts to data streams from a large collection of health districts.

As discussed by Woodall et al. (2008), little study has been done on the time-to-signal performance of the spatiotemporal scan methods. Evaluations under relatively simple models of the data could give some indication of the expected performance under the often complex data environment of public health applications. If one considers the incidence rate as a function over the region of interest, then the spatiotemporal application can be considered to be an example of profile monitoring, as discussed by Woodall et al. (2004). A very recent CUSUM approach to spatiotemporal monitoring was proposed by Sonesson (2007), who provided considerable insight on Kulldorff's (2001) scan methods.

6 Conclusions

Researchers in industrial process monitoring are strongly encouraged to investigate topics in health-related monitoring, where there is an abundance of promising research opportunities. The information and sources given here are intended to aid those interested in making this transition.

The study of the following areas seems especially promising: prospective scan-based methods, health-related applications of control charts for "high-quality" processes, risk-adjusted control charting, spatiotemporal surveillance, and (one-sided) multivariate control charts. System-level strategies for combining the various methods to achieve distributed, multivariate sensitivity at manageable false alarm rates is a difficult but important challenge.

References

Axelrod DA, Guidinger MK, Metzger RA, Wiesner RH, Webb RL, Merion RM (2006) Transplant Center Quality Assessment Using a Continuously Updatable, Risk-Adjusted Technique (CUSUM). American Journal of Transplantation 6: 313—323.

Benjamini Y, Hochberg Y (1995) Controlling the False Discovery Rate: a Practical and Powerful Approach to Multiple Testing. Journal of the Royal Statistical Society, Series B, 57(1): 289—300.

Benjamini Y, Kling Y (1999) A Look at Statistical Process Control Through the *p*-values. Technical Report RP-SOR-99-08. Tel Aviv University, Israel. `http://www.math.tau.ac.il/~ybenja/KlingW.html` Accessed on 10/04/05.

Benneyan JC (2001) Number-Between g-Type Statistical Control Charts for Monitoring Adverse Events. Health Care Management Science 4: 305—318.

Benneyan JC (2006) Discussion of "Use of Control Charts in Health Care Monitoring and Public Health Surveillance" by WH Woodall. Journal of Quality Technology 38: 113—123.

Breiman L (2001) Statistical Modelling: The Two Cultures (with discussion). Statistical Science 16: 199—231.

Buckeridge DL, Burkom H, Campbell M, Hogan WR, Moore AW (2005) Algorithms for Rapid Outbreak Detection: A Research Synthesis. Journal of Biomedical Informatics 38: 99—113.

Burr T, Graves T, Klamann R, Michalak S, Picard R, Hengartner N (2006) Accounting for Seasonal Patterns in Syndromic Surveillance Data for Outbreak Detection. BMC Medical Informatics and Decision Making 6:40.

Chen R (1978) A Surveillance System for Congenital Malformations. Journal of the American Statistical Association 73: 323—327.

Cowling BJ, Wong IOL, Ho LM, Riley S, Leung GM (2006) Methods for Monitoring Influenza Surveillance Data. International Journal of Epidemiology 35: 1314—1321.

Fienberg SE, Shmueli G (2005) Statistical Issues and Challenges Associated with the Rapid Detection of Terrorist Outbreaks. Statistics in Medicine 24: 513—529.

Fraker SE, Woodall WH, Mousavi S (2008) Performance Metrics for Surveillance Schemes, To appear in Quality Engineering.

Fricker RD Jr (2007) Directionally Sensitive Multivariate Statistical Process Control Procedures with Application to Syndromic Surveillance. Advances in Disease Surveillance 3: 1—17.

Fricker RD Jr., HR Rolka (2006) Protecting Against Biological Terrorism: Statistical Issues in Electronic Biosurveillance, *Chance* 19: 4-13.

Fricker RD Jr, Hegler, BL, Dunfee DA (2007). Comparing Syndromic Surveillance Detection Methods: EARS' versus a CUSUM-based Methodology. Statistics in Medicine 27: 3407-3429.

Fricker RD Jr, Knitt M C, Hu CX (2008). Comparing Directionally Sensitive MCUSUM and MEWMA Procedures with Application to Biosurveillance. To appear in Quality Engineering.

Frisén M (1992) Evaluation of Methods for Statistical Surveillance. Statistics in Medicine 11: 1489—1502.

Frisén M (2003) Statistical Surveillance, Optimality and Methods. International Statistical Review 71:403—434.

Frisén M, Sonesson C (2005) Optimal Surveillance. In: Lawson AB, Kleinman K (eds) Spatial & Syndromic Surveillance John Wiley & Sons, Ltd., New York, NY, pp. 31—52.

Frisén M, Wessman P (1999) Evaluation of Likelihood Ratio Methods for Surveillance. Communications in Statistics – Simulation and Computation 28: 597—622.

Grigg O, Farewell V (2004a) An Overview of Risk-Adjusted Charts. Journal of the Royal Statistical Society A 167: 523—539.

Grigg O, Farewell V (2004b) A Risk-adjusted Sets Method for Monitoring Adverse Medical Outcomes. Statistics in Medicine 23: 1593-1602.

Grigg O, Spiegelhalter D (2006) Discussion of "Use of Control Charts in Health Care Monitoring and Public Health Surveillance" by WH Woodall. Journal of Quality Technology 38: 124—126.

Grigg O, Spiegelhalter D (2007) A Simple Risk-Adjusted Exponentially Weighted Moving Average. Journal of the American Statistical Association 102: 140—152.

Grigg O, Spiegelhalter D (2008) The Null Steady-State Distribution of the CUSUM Statistic. To appear in Technometrics.

Grigg O, Spiegelhalter D, Jones H (2008) Local and Marginal Control Charts Applied to MRSA Bacteraemia Reports in UK Acute NHS Trusts. To appear in the Journal of the Royal Statistical Society, Series A.

Hawkins DM (1993) Regression Adjustment for Variables in Multivariate Quality Control. Journal of Quality Technology 25: 170-81.

Ismail NA, Pettitt AN, Webster RA (2003) 'Online' Monitoring and Retrospective Analysis of Hospital Outcomes Based on a Scan Statistic. Statistics in Medicine 22: 2861—2876.

Joner MD Jr, Woodall WH, Reynolds MR, Jr (2008a) On Detecting a Rate Increase Using a Bernoulli-based Scan Statistic. Statistics in Medicine 27: 2555-2575.

Joner MD Jr, Woodall WH, Reynolds MR Jr, Fricker RD Jr (2008b) A One-sided MEWMA Chart for Health Surveillance. in *Quality and Reliability Engineering International* 24:503-518.

Kleinman K, Lazarus R, Platt, R (2004) A Generalized Linear Models Approach for Detecting Incident Clusters of Disease in Small Areas, with an Application to Biological Terrorism (with discussion). American Journal of Epidemiology 159: 217—228.

Kulldorff M (1997) A Spatial Scan Statistic. Communications in Statistics – Theory and Methods 26: 1481—1496.

Kulldorff M (2001) Prospective Time Periodic Geographical Disease Surveillance Using a Scan Statistic. Journal of the Royal Statistical Society A 164: 61—72.

Kulldorff, M (2005) *SaTScan*™: Software for the spatial, temporal, and space-time scan statistics, version 5.1 [computer program]. Information Management Services 2005; Available: http://www.satscan.org/

Lawson AB, Kleinman K (2005) *Spatial & Syndromic Surveillance for Public Health*, John Wiley & Sons, Inc., New York.

Marshall C, Best N, Bottle A, Aylin P (2004) Statistical Issues in the Prospective Monitoring of Health Outcomes Across Multiple Units. Journal of the Royal Statistical Society, Series A, 167(3): 541—559.

Naus J, Wallenstein S (2006) Temporal Surveillance Using Scan Statistics. Statistics in Medicine 25: 311—324.

Novick RJ, Fox SA, Stitt LW, Forbes TL, Steiner S (2006) Direct Comparison of Risk-Adjusted and Non-Risk-Adjusted CUSUM Analyses of Coronary Artery Bypass Surgery Outcomes. The Journal of Thoracic and Cardiovascular Surgery 132: 386—391.

Porter, ME, Teisberg EO (2006) Redefining Health Care, Creating Value-Based Competition Based on Results. Harvard Business School Press, Boston MA.

Reynolds MR Jr, Stoumbos ZG (1999) A CUSUM Chart for Monitoring a Proportion When Inspecting Continuously. Journal of Quality Technology 31: 87—108.

Rolka H, Burkom H, Cooper GF, Kulldorff M, Madigan D, Wong WK (2007) Issues in Applied Statistics for Public Health Bioterrorism Surveillance Using Multiple Data Streams: Some Research Needs. Statistics in Medicine 26: 1834—1856.

Sego LH, Woodall WH, Reynolds MR Jr (2008a) A Comparison of Surveillance Methods for Small Incidence Rates. Statistics in Medicine 27: 1225-1247.

Sego LH, Reynolds MR Jr, Woodall WH (2008b) Risk-Adjusted Monitoring of Survival Times, provisionally accepted by Statistics in Medicine.

Sherlaw-Johnson C, Wilson P, Gallivan S (2007) The Development and Use of Tools for Monitoring the Occurrence of Surgical Wound Infections. Journal of the Operational Research Society 58: 228—234.

Shmueli G, Burkom HS (2008) Statistical Challenges in Modern Biosurveillance, to appear in Technometrics.

Sonesson C (2007) A CUSUM Framework for Detection of Space-Time Disease Clusters Using Scan Statistics. Statistics in Medicine 26(26): 4770-4789.

Sonesson C, Bock D (2003) A Review and Discussion of Prospective Statistical Surveillance in Public Health. Journal of the Royal Statistical Society A 166: 5—21.

Sparks R (2000) CUSUM Charts for Signaling Varying Location Shifts. Journal of Quality Technology 32: 157-71.

Spliid H (2007) Monitoring Medical Procedures by Exponential Smoothing. Statistics in Medicine 26: 124—138.

Steiner SH, Cook RJ, Farewell VT, Treasure T (2000) Monitoring Surgical Performance Using Risk-Adjusted Cumulative Sum Charts. Biostatistics 1: 441—452.

Storey JD, Taylor JE, Siegmund D (2004) Strong Control, Conservative Point Estimation and Simultaneous Conservative Consistency of False Discovery Rates: a Unified Approach. Journal of the Royal Statistical Society, Series B, 66: 187—205.

Tennant R, Mohammed MA, Coleman JJ, Martin U (2007) Monitoring Patients Using Control Charts: A Systematic Review. To appear in the International Journal for Quality in Health Care.

Thor, J., Lundberg, J., Ask, J., Olsson, J., Carli, C., Härenstam, K. P., and Brommels, M. (2007), "Application of Statistical Process Control in Healthcare Improvement: Systematic Review", *Quality and Safety in Health Care*, 16, pp. 387-399.

Wallenstein S, Naus J (2004) Scan Statistics for Temporal Surveillance for Biologic Terrorism. Morbidity and Mortality Weekly Report 53 (Suppl): 74—78.

Winkel P, Zhang NF (2007) *Statistical Development of Quality in Medicine.* John Wiley & Sons Ltd: Chichester, pp. 173—181.

Woodall WH (2006) Use of Control Charts in Health Care Monitoring and Public Health Surveillance (with discussion). Journal of Quality Technology 38: 89—104 (available at http://www.asq.org/pub/jqt).

Woodall WH, Marshall JB, Joner MD Jr, Fraker SE, Abdel-Salam AG (2008) On the Use and Evaluation of Prospective Scan Methods in Health-Related Surveillance. Journal of the Royal Statistical Society A 171: 223-237.

Woodall WH, Spitzner DJ, Montgomery DC, Gupta S (2004) Using Control Charts to Monitor Process and Product Quality Profiles. Journal of Quality Technology 36: 309—320.

Xie M, Goh TN, Kuralmani V (2002) Statistical Models and Control Charts for High Quality Processes, Kluwer Academic Publishers, Norwell, MA.

Surveillance Sampling Schemes

David H Baillie

Chesham, UK
david.baillie@which.net

Summary. A new type of sampling scheme for process fraction nonconforming is proposed for the simultaneous acceptance inspection of any number of large lots of similar size, which provides the advantage of scale without losing the ability to sentence lots individually. The methodology is described and illustrated by a flow chart. Attributes and variables sampling schemes are developed. For beta prior probability density functions for the process fractions nonconforming, formulae are derived for the expected proportion of lots accepted and the expected number of items inspected per lot. The advantages of the method are illustrated by some examples.

1 Surveillance sampling by attributes

1.1 General

The sampling schemes presented in this paper are referred to as surveillance sampling schemes, as the motivating application was the need for economic surveillance of the quality of munitions kept in storage.

A surveillance sampling scheme by attributes for k lots is defined by the sample sizes n, n_0 and n_1 together with the integer acceptance numbers $\boldsymbol{c}_k = (c_{k1}, c_{k2}, \ldots, c_{kk})$ where $0 \le c_{ki} < i$ for $i = 1, 2, \ldots, k$ and $c_{k1} \le c_{k2} \le \ldots \le c_{kk}$. An initial random sample of size n is taken from each lot. All lots for which the initial sample yields 2 or more nonconforming items are immediately non-accepted. The number of remaining lots is denoted by m. The total number of nonconforming items in the initial samples from these m lots is denoted by x. If $x \le c_{km}$, all m remaining lots are accepted; otherwise, if $x > c_{km}$, a second sample of size n_0 or n_1 is taken from the lots for which the initial sample yielded 0 or 1 nonconforming item respectively. Lots for which the second sample yields no nonconforming items are accepted, while lots for which the second sample yields one or more nonconforming item are non-accepted.

Figure 1 is a flow chart of the procedure for applying surveillance sampling by attributes.

H.-J. Lenz et al. (eds.), *Frontiers in Statistical Quality Control 9*,
DOI 10.1007/978-3-7908-2380-6_11, © Springer-Verlag Berlin Heidelberg 2010

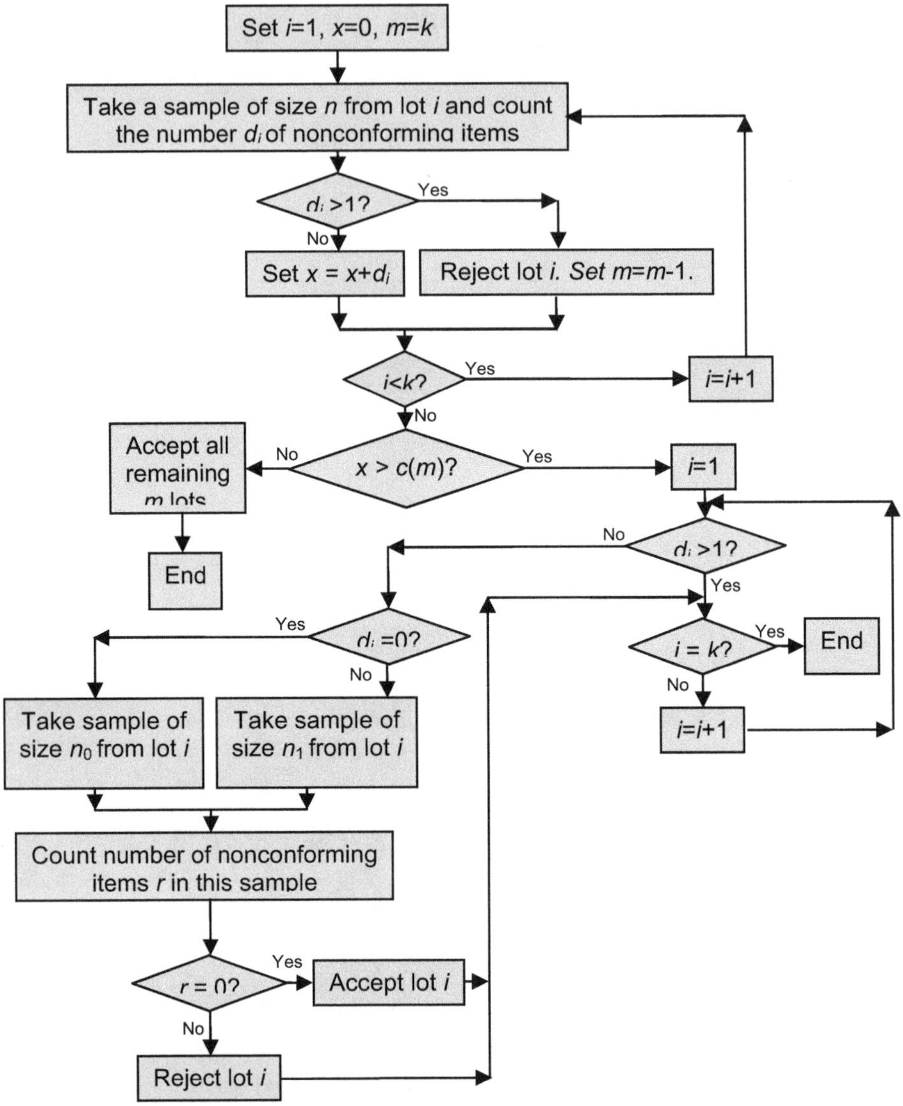

We suppose throughout that the fraction nonconforming varies from lot to lot in accordance with a beta distribution with prior probability density function

$$f(p) = p^{r-1}(1-p)^{s-1} / B(r,s) \text{ for } 0 \le p \le 1,$$

where $r > 0$, $s > 0$ and $B(r,s) = \Gamma(r)\Gamma(s)/\Gamma(r+s)$, with $\Gamma(.)$ representing the gamma function. Thus, sampling is being used to distinguish between a prior beta distribution with mean $r_0/(r_0+s_0)$ and coefficient of variation $\sqrt{s_0/\{r_0(r_0+s_0+1\}}$ and a prior beta distribution with mean $r_1/(r_1+s_1)$ and coefficient of variation $\sqrt{s_1/\{r_1(r_1+s_1+1\}}$. Note that the coefficient of variation of a beta distribution with parameters (r,s) decreases as $r+s$ increases subject to the ratio s/r remaining constant.

In practice, a minimum expected proportion of lots accepted will be specified at an acceptable beta prior probability density with parameters (r_0, s_0) and a maximum expected proportion of lots accepted will be specified at an unacceptable beta prior probability density with parameters (r_1, s_1).

In this section, for each value of k considered, the values of the $k + 3$ parameters (i.e. the k integers in array $\boldsymbol{c}_k$ together with the 3 sample sizes) are determined that minimise the maximum expected number of items inspected per lot with respect to the quality level, subject to these two constraints on the operating characteristic curve.

1.2 The expected number of items inspected per lot

The probability generating function for the posterior probability p_{kmx} that m of the k first samples contain no more than one nonconforming item *and* that x of these m samples have precisely one nonconforming item is

$$
\begin{aligned}
G_1(t,v;k) &= \sum_{m=0}^{k}\sum_{x=0}^{m} p_{kmx} t^m v^x \\
&= \prod_{i=1}^{k}\left[E\left\{1-(1-p_i)^n - np_i(1-p_i)^{n-1}\right\} + E\left\{(1-p_i)^n + np_i(1-p_i)^{n-1}v\right\}t\right] \\
&= \left[E\left\{b_{\geq 2}(n,p)\right\} + E\left\{b_0(n,p)\right\}t + E\left\{b_1(n,p)\right\}vt\right]^k \\
&= \left[\alpha_1 + \beta_0 t + \beta_1 vt\right]^k \qquad (1)
\end{aligned}
$$

where

$$
\begin{aligned}
\beta_0 &= E\left\{b_0(n,p)\right\} = E(1-p)^n = B(r,s+n)/B(r,s) \\
\beta_1 &= E\left\{b_1(n,p)\right\} = En(1-p)^{n-1} = nB(r,s+n-1)/B(r,s) \\
\beta_2 &= E\left\{b_{\geq 2}(n,p)\right\} = 1-\beta_0-\beta_1;
\end{aligned}
$$

$G_1(t,v;k)$ can be expanded as

$$G_1(t,v;k)=\sum_{m=0}^{k}\binom{k}{m}\beta_2^{k-m}(\beta_0+\beta_1 v)^m t^m$$
$$=\sum_{m=0}^{k}\binom{k}{m}\beta_2^{k-m}\sum_{x=0}^{m}\binom{m}{x}\beta_0^{m-x}\beta_1^x t^m v^x.$$

Hence,

$$p_{kmx}=\binom{k}{m}\binom{m}{x}\beta_2^{k-m}\beta_0^{m-x}\beta_1^x \text{ for } m=0,1,\ldots,k; x=0,1,\ldots,m. \quad (2)$$

If $x \le c_{km}$, the total number of items inspected is kn, whereas if $x > c_{km}$, the total number of items inspected is $kn+(m-x)n_0+xn_1$. The expected number of items inspected per lot is therefore

$$A=n+\frac{1}{k}\sum_{m=0}^{k}\sum_{x=c_{km}+1}^{m}p_{kmx}\{(m-x)n_0+xn_1\}$$
$$=n+\frac{1}{k}\sum_{m=1}^{k}\sum_{x=c_{km}+1}^{m}p_{kmx}\{(m-x)n_0+xn_1\}. \quad (3)$$

1.3 The proportion of lots accepted

The probability generating function for y lots having no nonconforming items in their second samples is given by

$$G_2(t,v,w;k)=\sum_{m=0}^{k}\sum_{x=0}^{m}\sum_{y=0}^{m}p_{kmxy}t^m v^x w^y$$

$$=E\prod_{i=1}^{k}[b_{\ge 2}(n,p_i)+b_0(n,p_i)\{b_{\ge 1}(n_0,p_i)+b_0(n_0,p_i)w\}t$$
$$+b_1(n,p_i)\{b_{\ge 1}(n_1,p_i)+b_0(n_1,p_i)w\}vt]$$

This can be written

$$G_2(t,v,w;k)=[\alpha_1+\alpha_2 t+\alpha_3 wt+\alpha_4 vt+\alpha_5 vwt]^k \quad (4)$$

where

$$\alpha_1=\beta_2;$$
$$\alpha_2=\{B(r,s+n)-B(r,s+n+n_0)\}/B(r,s)=\beta_0-\alpha_3;$$
$$\alpha_3=B(r,s+n+n_0)/B(r,s);$$
$$\alpha_4=n\{B(r+1,s+n-1)-B(r+1,s+n+n_1-1)\}/B(r,s)=\beta_1-\alpha_5;$$
$$\alpha_5=nB(r+1,s+n+n_1-1)/B(r,s).$$

$G_2(t,v,w;k)$ can be expanded as

$$G_2(t,v,w;k)=\sum_{m=0}^{k}\sum_{x=0}^{m}\sum_{y=0}^{m}p_{kmxy}t^m v^x w^y$$

$$=\{(\alpha_1+t(\alpha_2+\alpha_4 v+\alpha_3 w+\alpha_5 vw)\}^k$$

$$=\sum_{m=0}^{k}\binom{k}{m}\alpha_1^{k-m}\{(\alpha_2+\alpha_3 w)+v(\alpha_4+\alpha_5 w)\}^m t^m$$

$$=\sum_{m=0}^{k}\binom{k}{m}\alpha_1^{k-m}\sum_{x=0}^{m}\binom{m}{x}(\alpha_2+\alpha_3 w)^{m-x}(\alpha_4+\alpha_5 w)^x v^x t^m$$

$$=\sum_{m=0}^{k}\binom{k}{m}\alpha_1^{k-m}\sum_{x=0}^{m}\binom{m}{x}\sum_{y=0}^{m}\sum_{j=\max(0,y-x)}^{\min(m-x,y)}\binom{x}{y-j}\binom{m-x}{j}\alpha_2^{m-x-j}\alpha_3^j\alpha_4^{x-y+j}\alpha_5^{y-j}t^m v^x w^y,$$

so

$$p_{kmxy}=\binom{k}{m}\binom{m}{x}\alpha_1^{k-m}\sum_{j=\max(0,y-x)}^{\min(m-x,y)}\binom{x}{y-j}\binom{m-x}{j}\alpha_2^{m-x-j}\alpha_3^j\alpha_4^{x-y+j}\alpha_5^{y-j}. \qquad (5)$$

for $m=0,1,\ldots,k;\ x=0,1,\ldots,m;\ y=0,1,\ldots,m.$

Note that $\alpha_2=0$ when $n_0=0$ and that $\alpha_4=0$ when $n_1=0$. It follows that:

$$p_{kmxy}=\begin{cases}\binom{k}{m}\binom{m}{x}\binom{x}{m-y}\alpha_1^{k-m}\alpha_3^{m-x}\alpha_4^{m-y}\alpha_5^{y-m+x} \\ \quad \text{when } n_0=0,\ n_1>0 \text{ and } y=m-x,m-x+1,\ldots,m; \\ \binom{k}{m}\binom{m}{x}\binom{m-x}{y-x}\alpha_1^{k-m}\alpha_2^{m-y}\alpha_3^{y-x}\alpha_5^{x} \\ \quad \text{when } n_0>0,\ n_1=0 \text{ and } y=x,x+1,\ldots,m; \\ \binom{k}{m}\binom{m}{x}\alpha_1^{k-m}\alpha_3^{m-x}\alpha_5^{x} \\ \quad \text{when } n_0=0,\ n_1=0 \text{ and } y=m,\end{cases} \qquad (6)$$

for $m=0,1,\ldots,k;\ x=0,1,\ldots,m.$

Taking into account the restrictions imposed by the array $\boldsymbol{c}_k$, the expected proportion of lots accepted is

$$P_a = \frac{1}{k}\left\{\sum_{m=0}^{k}\sum_{x=0}^{c_{km}}\sum_{y=0}^{m} m p_{kmxy} + \sum_{m=0}^{k}\sum_{x=c_{km}+1}^{m}\sum_{y=0}^{m} y p_{kmxy}\right\}$$

$$= \frac{1}{k}\left\{\sum_{m=1}^{k} m \sum_{x=0}^{c_{km}} p_{kmx} + \sum_{m=1}^{k}\sum_{x=c_{km}+1}^{m}\sum_{y=1}^{m} y p_{kmxy}\right\}. \qquad (7)$$

1.4 Examples

We denote the expected proportion of lots accepted as $P_a(0)$ when $(r,s)=(r_0,s_0)$ and as $P_a(1)$ when $(r,s)=(r_1,s_1)$, and the expected number of items inspected per lot for given beta prior, sample sizes and c_k array by $A(k,r,s,n,n_0,n_1,\boldsymbol{c}_k)$.

For all three examples of sampling by attributes given below, the prior beta distributions have been chosen to have means $r_0/(r_0+s_0)=0.005=0.5\%$ and $r_1/(r_1+s_1)=0.05=5\%$, with $r_0+s_0=r_1+s_1$. For examples 1 to 3, $r_0+s_0=r_1+s_1$ has been chosen to be 200, 400 and 1000 respectively, to provide a decreasing coefficient of variation.

The selection of surveillance sampling schemes by attributes given below were determined so as to provide a producer's risk of at most $\boldsymbol{\alpha}=0.05$ (i.e. $P_a(0)\geq 95\%$) at a prior density defined by $(r,s)=(r_0,s_0)$, and a consumer's risk of at most $\boldsymbol{\beta}=0.10$ (i.e. $P_a(1)\leq 10\%$) at a prior density defined by $(r,s)=(r_1,s_1)$, subject to minimising the maximum value $A_{\max}$ of

$$A(k,\gamma r_0+(1-\gamma)r_1,\gamma s_0+(1-\gamma)s_1,n,n_0,n_1,\boldsymbol{c}_k)$$

over the range $0\leq\gamma<r_1/(r_1-r_0)$. The expected number of items inspected per lot at $(r,s)=(r_0,s_0)$ and at $(r,s)=(r_1,s_1)$ are denoted by A_0 and A_1 respectively.

Example 1: Sampling by attributes with
$\alpha = 0.05$, $\beta = 0.10$, $(r_0, s_0) = (1,199)$, $(r_1, s_1) = (10,190)$

k	n	n_0	n_1	c_k	$P_a(0)\%$	$P_a(1)\%$	A_0	A_{max}	A_1
1	*** No sampling scheme of this nature was found for this combination of k, α and β ***								
2	*** No sampling scheme of this nature was found for this combination of k, α and β ***								
3	51	46	64	0,0,2	95.0998	9.9989	52.6038	69.8531	64.8751
4	51	40	52	0,0,1,3	95.6306	9.9836	51.5065	66.7614	62.8590
5	43	42	53	0,0,0,1,4	96.5606	9.9776	43.6809	64.0345	59.5857
6	34	46	54	0,0,0,0,1,3	97.5241	9.9983	34.8082	61.4618	57.4774
7	32	47	53	0,0,0,0,0,2,4	98.0013	9.9974	32.2709	59.5017	56.9654
8	30	45	49	0,0,0,0,0,0,2,5	98.1825	9.9925	30.3183	58.0433	55.3793
9	30	45	47	0,0,0,0,0,0,1,3,7	98.2790	9.9910	30.1126	57.1823	54.9028
10	29	42	48	0,0,0,0,0,0,0,1,4,6	98.3869	9.9989	29.0945	56.5558	54.4954
k	n	n_0	n_1	c_k	$P_a(0)\%$	$P_a(1)\%$	A_0	A_{max}	A_1

Example 2: Sampling by attributes with
$\alpha = 0.05$, $\beta = 0.10$, $(r_0, s_0) = (2,398)$, $(r_1, s_1) = (20,380)$

k	n	n_0	n_1	c_k	$P_a(0)\%$	$P_a(1)\%$	A_0	A_{max}	A_1
1	*** No sampling scheme of this nature was found for this combination of k, α and β ***								
2	*** No sampling scheme of this nature was found for this combination of k, α and β ***								
3	51	26	47	0,0,2	95.9183	9.9996	52.1061	64.8489	60.7534
4	44	39	49	0,0,0,3	96.6189	9.9955	45.4171	62.8735	57.9856
5	41	38	47	0,0,0,1,4	97.3564	9.9840	41.5035	60.3282	56.3836
6	34	40	49	0,0,0,0,1,4	98.0277	9.9958	34.5455	58.0194	54.7524
7	32	43	46	0,0,0,0,0,2,5	98.3764	9.9957	32.1942	56.3443	53.9013
8	28	43	45	0,0,0,0,0,0,2,4	98.6937	9.9592	28.2302	55.2437	52.9301
9	29	41	44	0,0,0,0,0,0,1,3,8	98.6870	9.9952	29.0755	54.4606	52.5420
10	28	42	42	0,0,0,0,0,0,0,1,4,6	98.7770	9.9953	28.0580	53.9118	52.0259
k	n	n_0	n_1	c_k	$P_a(0)\%$	$P_a(1)\%$	A_0	A_{max}	A_1

Example 3: Sampling by attributes with
$\alpha = 0.05$, $\beta = 0.10$, $(r_0, s_0) = (5,995)$, $(r_1, s_1) = (50,950)$

k	n	n_0	n_1	c_k	$P_a(0)\%$	$P_a(1)\%$	A_0	A_{max}	A_1
1	*** No sampling scheme of this nature was found for this combination of k, α and β ***								
2	56	0	36	0,1	95.4955	9.9307	57.7465	65.2388	61.7588
3	50	25	41	0,0,2	96.5886	9.9932	50.9346	62.5602	58.6629
4	42	37	48	0,0,0,3	97.3320	9.9978	43.1164	60.6337	56.2812
5	39	37	46	0,0,0,1,4	97.9402	9.9837	39.3894	58.2082	54.7872
6	33	42	44	0,0,0,0,1,4	98.4226	9.9823	33.4405	56.1376	53.0102
7	31	39	44	0,0,0,0,0,2,4	98.7030	9.9934	31.1630	54.5560	52.2062
8	29	36	44	0,0,0,0,0,0,2,5	98.8421	9.9899	29.1768	53.5974	51.3696
9	29	38	41	0,0,0,0,0,0,1,3,7	98.8966	9.9981	29.0605	52.8880	50.8941
10	28	38	42	0,0,0,0,0,0,0,2,4,6	98.9860	9.9938	28.0182	52.3982	50.9212
k	n	n_0	n_1	c_k	$P_a(0)\%$	$P_a(1)\%$	A_0	A_{max}	A_1

2 Surveillance sampling by variables

2.1 General

Two types of sampling scheme for surveillance sampling by variables are developed below. Both can be used, with suitable changes to the parameter values, for the case of known or unknown process variability. It is initially assumed that the quality characteristic is normally distributed with a single specification limit.

Some further notation is necessary. Suppose that a sample of size n from a normally distributed process with fraction nonconforming p yields a minimum variance unbiased estimate $\hat{p}$ of p. Then the probability that $\hat{p} \le p^*$ will be denoted by $d(n, p, p^*)$.

When the numerical value of the process standard deviation, σ, is known, then

$$d(n,p,p^*) = \Phi\left(\sqrt{n}K_p - \sqrt{n-1}K_{p^*}\right) \text{ for } n \ge 2,$$

where K_p denotes the upper p-fractile of the standard normal distribution and $\Phi(.)$ denotes the standard normal distribution function, thus $K_p = \Phi^{-1}(1-p)$.

When the numerical value of the process standard deviation is unknown, then

$$d(n,p,p^*) = 1 - F_{n-1,\sqrt{n}K_p}\left((n-1)(1-2\beta_{(n-2)/2,p^*})\right) \text{ for } n \ge 3,$$

where $F_{\nu,\lambda}(t)$ denotes the distribution function of the non-central t-distribution with degrees of freedom ν and non-centrality parameter λ, and $\beta_{(n-2)/2,p^*}$ denotes the p^*-fractile of the symmetric beta distribution with both parameters equal to $(n-2)/2$.

2.2 Type A surveillance sampling schemes

2.2.1 Definition

We define a Type A surveillance sampling scheme by variables for k lots by $k+7$ parameters, namely the sample sizes n, n_0 and n_1 together with the acceptability constants p_a^*, p_r^*, p_0^*, p_1^* and $c_k = c_{k1}, c_{k2}, \ldots, c_{kk}$, where $0 \le p_a^* \le p_r^* < 1$, $0 \le p_0^*, p_1^* < 1$ and $0 \le c_{k1} \le c_{k2} \le \ldots \le c_{kk} < k$. From each lot an initial random sample of size n is taken and the minimum variance unbiased estimators (MVUEs) $\hat{p}_1$, $\hat{p}_2$, ..., $\hat{p}_k$ of their process fractions nonconforming are calculated. The ith lot is immediately non-accepted if $\hat{p}_i \ge p_r^*$, for $i = 1, 2, \ldots, k$. The number of remaining lots is denoted by m. A second sample of size n_0 is taken from each of the remaining lots for which $\hat{p}_i \le p_a^*$ and a sample of size n_1 is taken from each of the remaining lots for which $p_{\mathrm{a}}^{*} < \hat{p}_i \le p_{\mathrm{r}}^{*}$. For each new sample, i, the MVUE, $\hat{p}_i^{(2)}$, of the process fraction nonconforming is calculated. Lots for which the sample of size n_j yields $\hat{p}_i^{(2)} \le p_j^{*}$ for $j = 0$ or 1 are accepted, otherwise they are non-accepted.

2.2.2 Expected number of items inspected per lot

There is a close correspondence between the parameters of this type of variables sampling scheme and the surveillance sampling scheme by attributes described earlier. Thus, replacing

$b_0(n, p_i)$ by $d(n, p_i, p_{\mathrm{a}}^{*})$;

$b_1(n, p_i)$ by $d(n, p_i, p_r^{*}) - d(n, p_i, p_{\mathrm{a}}^{*})$; and

$b_{\ge 2}(n, p_i)$ by $1 - d(n, p_i, p_r^{*})$

in (1), it is found that the probability generating function for m and x under Type A surveillance sampling by variables is given by

$$G_1^{(\mathrm{A})}(t, v; k) = \sum_{m=0}^{k} \sum_{x=0}^{m} p_{kmx} t^m v^x$$

$$= \prod_{i=1}^{k} \left\{ E\left(1 - d(n, p_i, p_{\mathrm{r}}^{*})\right) + E\left\{ d(n, p_i, p_{\mathrm{a}}^{*}) \right\} t + E\left(d(n, p_i, p_{\mathrm{r}}^{*}) - d(n, p_i, p_{\mathrm{a}}^{*}) \right) vt \right\}. \quad (8)$$

But p_{kmx} is given by (2) with

$$\begin{aligned} \beta_0 &= E(d(n, p, p_{\mathrm{a}}^{*})); \\ \beta_1 &= E(d(n, p, p_{\mathrm{r}}^{*}) - d(n, p, p_{\mathrm{a}}^{*})); \\ \beta_2 &= 1 - \beta_0 - \beta_1. \end{aligned} \quad (9)$$

Hence the expected number of items inspected per lot is given by substituting for p_{kmx} from (2) and (9) into (3).

2.2.3 The expected proportion of lots accepted

Similarly, it is found that the Type A probability generating function for m, x and y is

$$G_2^{(\mathrm{A})}(t, v, w; k) = \left(\alpha_1 + \alpha_2 t + \alpha_3 wt + \alpha_4 vt + \alpha_5 vwt\right)^k$$

where

$$\begin{aligned} \alpha_1 &= \beta_2; \\ \alpha_2 &= E\left\{ d(n, p, p_{\mathrm{a}}^{*})\left(1 - d(n_0, p, p_0^{*})\right)\right\} = \beta_0 - \alpha_3; \end{aligned} \quad (10)$$

$$\alpha_3 = E\left\{d(n,p,p_a^*)d(n_0,p,p_0^*)\right\};$$

The value of p_{kmxy} may be determined by substituting from (10) into (5) and (6). Thus, taking into account the restrictions imposed by the array $c_{k1}, c_{k2}, \ldots, c_{kk}$, the expected proportion of lots accepted is found by substituting the values of p_{kmxy} so obtained into (7).

2.2.4 Examples of Type A sampling schemes

Examples 4 and 5 show some Type A variables sampling schemes with the same constraints and prior distributions as for Example 1. Example 4 is for unknown process standard deviation and Example 5 is for known process standard deviation.

Example 4: Type A sampling by variables, unknown σ, with $\alpha = 0.05$, $\beta = 0.10$, $(r_0, s_0) = (1{,}199)$, $(r_1, s_1) = (10{,}190)$

k	n	n_0	n_1	p_a^*	p_r^*	p_0^*	p_1^*	c_k	$P_a(0)$ %	$P_a(1)$ %	A_0	A_{max}	A_1
1	39	--	24	.01116	.02486	--	.01903	0	95.0133	9.9922	41.0103	45.0103	42.1624
2	34	3	3	.02847	.10249	.00000	.00000	0,0	95.1930	9.8453	36.9999	36.9999	36.7956
3	22	10	3	.03042	.13604	.00022	.00000	0,0,0	95.0038	9.9888	31.7924	31.9084	27.4765
4	19	9	3	.01972	.12669	.00034	.00000	0,0,0,1	98.2364	9.8636	27.5789	27.8188	23.4124
5	16	10	3	.02594	.14983	.00001	.00000	0,0,0,0,1	98.5144	9.3520	25.6526	25.8454	21.4614
6	14	11	3	.02448	.15138	.00029	.00000	0,0,0,0,0,1	97.4462	6.6409	24.5384	24.7928	19.8782
7	13	11	3	.03165	.15385	.00103	.00000	0,0,0,0,0,0,1	98.4328	9.5582	23.6753	23.8503	19.5255
8	12	10	3	.03577	.16328	.00011	.00000	0,0,0,0,0,0,0,1	98.3809	9.0445	21.7551	21.8832	18.4061
9	9	9	3	.02716	.15162	.00001	.00000	0,0,0,0,0,0,0,0,1	95.0333	7.9468	17.6443	17.8819	14.6865
10	7	9	3	.03129	.16736	.00001	.00000	0,0,0,0,0,0,0,0,0,1	95.0001	9.8562	15.6742	15.8766	13.1300
k	n	n_0	n_1	p_a^*	p_r^*	p_0^*	p_1^*	c_k	$P_a(0)$ %	$P_a(1)$ %	A_0	A_{max}	A_1

Example 5: Type A sampling by variables, known σ, with $\alpha = 0.05$, $\beta = 0.10$, $(r_0, s_0) = (1{,}199)$, $(r_1, s_1) = (10{,}190)$

k	n	n_0	n_1	p_a^*	p_r^*	p_0^*	p_1^*	c_k	$P_a(0)$ %	$P_a(1)$ %	A_0	A_{max}	A_1
1	11	--	10	.01101	.02457	--	.01585	0	95.0014	9.9996	11.9360	14.3044	13.0200
2	7	--	3	.00675	.03185	--	.00033	0,1	95.0001	9.9850	7.5398	8.4965	8.1532
3	5	--	2	.00353	.03497	--	.00001	0,0,2	95.0005	9.9898	5.5383	6.1625	5.9567
4	3	2	3	.00255	.18543	.00001	.00002	0,0,0,2	95.0306	9.9742	5.2558	5.6816	5.6749
5	3	--	2	.00053	.04361	--	.00001	0,0,0,2,4	95.0007	9.9058	3.8868	4.4088	4.2544
6	3	--	2	.00072	.03168	--	.00001	0,0,1,1,3,5	95.0011	9.9398	3.7806	4.2500	4.0788
7	2	2	2	.00205	.10598	.00001	.00001	0,0,0,0,0,0;3	95.0002	9.9824	3.9809	3.9903	3.7198
8	2	2	2	.00596	.07548	.00001	.00001	0,0,0,0,0,0,1,2	95.0001	9.9975	3.9702	3.9855	3.6297
9	2	2	2	.00081	.02865	.00000	.00001	0,0,0,0,1,2,3,4,5	95.0009	9.9295	3.9199	3.9636	3.3410
10	2	2	2	.00143	.02550	.00001	.00001	0,0,0,0,0,1,2,3,4,5	95.0002	9.9356	3.9117	3.9601	3.3049
k	n	n_0	n_1	p_a^*	p_r^*	p_0^*	p_1^*	c_k	$P_a(0)$ %	$P_a(1)$ %	A_0	A_{max}	A_1

2.3 Type B surveillance sampling schemes

2.3.1 Definition

It may seem perverse not to immediately accept a lot for which the initial sample yields an estimate of zero for the process fraction nonconforming, which can happen with non-zero probability for the case of unknown process standard deviation. To accommodate this viewpoint, Type A sampling schemes for unknown process standard deviation can be modified so that, after the initial sample has been drawn from the ith lot, that lot is not only immediately non-accepted if $\hat{p}_i \ge p_{\text{r}}^*$, but also immediately *accepted* if $\hat{p}_i = 0$, for $i = 1,2,...,k$. Such sampling schemes are developed below, and will be referred to as Type B surveillance sampling schemes.

2.3.2 Expected number of items inspected per lot

The probability generating function for the probability p_{kmx} that m of the k initial samples have $0 < \hat{p} \le p_{\text{r}}^*$ *and* that x of these m samples have $\hat{p} > p_{\text{a}}^*$ is

$$G_1^{(B)}(t,v;k) = \sum_{m=0}^{k}\sum_{x=0}^{m} p_{kmx} t^m v^x$$

$$= E\left[\prod_{i=1}^{k}\left\{1 - d\left(n,p_i,p_{\text{r}}^*\right) + d\left(n,p_i,0\right) + \left(d\left(n,p_i,p_{\text{a}}^*\right) - d\left(n,p_i,0\right)\right)t + \left(d\left(n,p_i,p_{\text{r}}^*\right) - d\left(n,p_i,p_{\text{a}}^*\right)\right)vt\right\}\right]$$

Thus p_{kmx} is given by (9) with

$$\begin{aligned}
\beta_0 &= E\left(d\left(n,p_i,p_{\text{a}}^*\right) - d\left(n,p_i,0\right)\right);\\
\beta_1 &= E\left(d\left(n,p_i,p_{\text{r}}^*\right) - d\left(n,p_i,p_{\text{a}}^*\right)\right); \qquad (11)\\
\beta_2 &= 1 - \beta_0 - \beta_1.
\end{aligned}$$

Hence the expected number of items inspected per lot is given by substituting for p_{kmx} from (2) and (11) into (3).

2.3.3 The expected proportion of lots accepted

Similarly,

$$G_2^{(B)}(t,v,w;k) = \left(\alpha_1 + \alpha_0 w + \alpha_2 t + \alpha_3 wt + \alpha_4 vt + \alpha_5 vwt\right)^k \qquad (12)$$

Where

$$\begin{aligned}
\boldsymbol{\alpha}_0 &= E\left\{d(n,p,0)\right\};\\
\boldsymbol{\alpha}_1 &= E\left\{1-d\left(n,p,p_{\mathrm{r}}^*\right)\right\};\\
\boldsymbol{\alpha}_2 &= E\left\{\left(d\left(n,p,p_{\mathrm{a}}^*\right)-d(n,p,0)\right)\left(1-d\left(n_0,p,p_0^*\right)\right)\right\}=\boldsymbol{\beta}_0-\boldsymbol{\alpha}_3;\\
\boldsymbol{\alpha}_3 &= E\left\{\left(d\left(n,p,p_{\mathrm{a}}^*\right)-d(n,p,0)\right)d\left(n_0,p,p_0^*\right)\right\};\\
\boldsymbol{\alpha}_4 &= E\left\{\left(d\left(n,p,p_{\mathrm{r}}^*\right)-d\left(n,p,p_{\mathrm{a}}^*\right)\right)\left(1-d\left(n_1,p,p_1^*\right)\right)\right\}=\boldsymbol{\beta}_1-\boldsymbol{\alpha}_5;\\
\boldsymbol{\alpha}_5 &= E\left\{\left(d\left(n,p,p_{\mathrm{r}}^*\right)-d\left(n,p,p_{\mathrm{a}}^*\right)\right)d\left(n_1,p,p_1^*\right)\right\}.
\end{aligned} \tag{13}$$

Expanding (12),

$$G_2^{(B)}(t,v,w;k)=\sum_{m=0}^{k}\sum_{x=0}^{m}\sum_{y=0}^{m}p_{kmxy}t^m v^x w^y$$

$$=\left(\boldsymbol{\alpha}_1+\boldsymbol{\alpha}_0 w+\boldsymbol{\alpha}_2 t+\boldsymbol{\alpha}_3 wt+\boldsymbol{\alpha}_4 vt+\boldsymbol{\alpha}_5 vwt\right)^k$$

$$=\sum_{m=0}^{k}\binom{k}{m}\left\{\left(\boldsymbol{\alpha}_2+\boldsymbol{\alpha}_3 w\right)+v\left(\boldsymbol{\alpha}_4+\boldsymbol{\alpha}_5 w\right)\right\}^m t^m\left(\boldsymbol{\alpha}_1+\boldsymbol{\alpha}_0 w\right)^{k-m}$$

$$=\sum_{m=0}^{k}\binom{k}{m}\sum_{x=0}^{m}\binom{m}{x}\sum_{l=0}^{m}\sum_{j=\max(0,l-x)}^{\min(m-x,l)}\binom{x}{l-j}\binom{m-x}{j}$$

$$\times\boldsymbol{\alpha}_2^{m-x-j}\boldsymbol{\alpha}_3^{j}\boldsymbol{\alpha}_4^{x-l+j}\boldsymbol{\alpha}_5^{l-j}t^m v^x w^l\sum_{i=0}^{k-m}\binom{k-m}{i}\boldsymbol{\alpha}_1^{k-m-i}\left(\boldsymbol{\alpha}_0 w\right)^i$$

$$=\sum_{m=0}^{k}\binom{k}{m}\sum_{x=0}^{m}\binom{m}{x}\sum_{y=0}^{k}\sum_{i=\max(0,y-m)}^{\min(y,k-m)}\binom{k-m}{i}\sum_{j=\max(0,y-i-x)}^{\min(m-x,y-i)}\binom{x}{y-i-j}$$

$$\times\binom{m-x}{j}\boldsymbol{\alpha}_1^{k-m-i}\boldsymbol{\alpha}_0^{i}\boldsymbol{\alpha}_2^{m-x-j}\boldsymbol{\alpha}_3^{j}\boldsymbol{\alpha}_4^{x-y+i+j}\boldsymbol{\alpha}_5^{y-i-j}t^m v^x w^y$$

from which it is seen that

$$p_{kmxy}=\binom{k}{m}\binom{m}{x}\sum_{i=\max(0,y-m)}^{\min(y,k-m)}\binom{k-m}{i}\sum_{j=\max(0,y-i-x)}^{\min(m-x,y-i)}\binom{x}{y-i-j}\binom{m-x}{j}$$

$$\times\boldsymbol{\alpha}_0^{i}\boldsymbol{\alpha}_1^{k-m-i}\boldsymbol{\alpha}_2^{m-x-j}\boldsymbol{\alpha}_3^{j}\boldsymbol{\alpha}_4^{x-y+i+j}\boldsymbol{\alpha}_5^{y-i-j}$$

$$=\sum_{i=\max(0,y-m)}^{\min(y,k-m)}\sum_{j=\max(0,y-i-x)}^{\min(m-x,y-i)}\frac{k!\boldsymbol{\alpha}_0^{i}\boldsymbol{\alpha}_1^{k-m-i}\boldsymbol{\alpha}_2^{m-x-j}\boldsymbol{\alpha}_3^{j}\boldsymbol{\alpha}_4^{x-y+i+j}\boldsymbol{\alpha}_5^{y-i-j}}{i!(k-m-i)!(m-x-j)!(x-y+i+j)!(y-i-j)!}. \tag{14}$$

for $m=0,1,\dots,k;\ x=0,1,\dots,m;\ y=0,1,\dots,k.$

Note again that $\alpha_2 = 0$ when $n_0 = 0$ and that $\alpha_4 = 0$ when $n_1 = 0$. It follows that

$$p_{kmxy} = \begin{cases} \displaystyle\sum_{i=\max(0,y-m)}^{\min(y,k-m,y-m+x)} \frac{k!\alpha_0^i \alpha_1^{k-m-i} \alpha_3^{m-x} \alpha_4^{m-y+i} \alpha_5^{y-i-m+x}}{i!(k-m-i)!(m-x)!(m-y+i)!(y-m+x-i)!} \\ \qquad \text{when } n_0 = 0,\ n_1 > 0 \text{ and } y = m-x, m-x+1, \ldots, k; \\ \displaystyle\sum_{i=\max(0,y-m)}^{\min(y,k-m,y-x)} \frac{k!\alpha_0^i \alpha_1^{k-m-i} \alpha_2^{m-y+i} \alpha_3^{y-x-i} \alpha_5^{x}}{i!(k-m-i)!(m-y+i)!(y-x-i)!x!} \\ \qquad \text{when } n_0 > 0,\ n_1 = 0 \text{ and } y = x, x+1, \ldots, k; \\ \displaystyle\frac{k!\alpha_0^{y-m} \alpha_1^{k-y} \alpha_3^{m-x} \alpha_5^{x}}{(y-m)!(k-y)!(m-x)!x!} \\ \qquad \text{when } n_0 = 0,\ n_1 = 0 \text{ and } y = m, m+1, \ldots, k, \end{cases}$$

for $m = 0,1,\ldots,k;\ x = 0,1,\ldots,m.$

The values of p_{kmx} and p_{kmxy} for given n_0 and n_1 can be found by substituting the values of α_0, α_1, α_2, α_3, α_4, α_5, β_0 and β_1 from (11) and (13) into (2), (14) and (15). Taking into account the restrictions imposed by the array $c_1, c_2, \ldots, c_k$, the expected proportion of lots accepted is given by

$$P_a = \alpha_0 + \frac{1}{k}\left\{ \sum_{m=0}^{k} \sum_{x=0}^{c_{km}} \sum_{y=0}^{k} m p_{kmxy} + \sum_{m=0}^{k} \sum_{x=c_{km}+1}^{m} \sum_{y=0}^{k} y p_{kmxy} \right\}$$

$$= \alpha_0 + \frac{1}{k}\left\{ \sum_{m=1}^{k} m \sum_{x=0}^{c_{km}} p_{kmx} + \sum_{m=1}^{k} \sum_{x=c_{km}+1}^{m} \sum_{y=1}^{k} y p_{kmxy} \right\}.$$

2.4 Examples of Type B sampling schemes

Example 6 shows some Type B variables sampling schemes for unknown process standard deviations, with the same constraints and prior distributions as in Examples 1, 4 and 5.

Comparing Table 6 with Table 4, it is seen that Type B yields a dramatic reduction in average sampling effort for two or more lots, suggesting that Type B sampling schemes may be superior to Type A schemes.

3 Concluding remarks

It is clear from the examples that the methodology proposed in this paper provides worthwhile reductions in the average amount of inspection needed to satisfy the

Example 6: Type B sampling by variables, unknown σ, with

$\alpha = 0.05$, $\beta = 0.10$, $(r_0, s_0) = (1,199)$, $(r_1, s_1) = (10,190)$

k	n	n_0	n_1	p_a^*	p_r^*	p_0^*	p_1^*	c_k	$P_a(0)$ %	$P_a(1)$ %	A_0	A_{max}	A_1
1	39	--	24	.01219	.02346	--	.01985	0	95.0133	9.9922	41.0103	45.0103	42.1624
2	26	--	3	.00920	.03052	--	.00000	0,1	95.0011	9.9883	26.4317	27.1995	26.7999
3	24	--	3	.00918	.02628	--	.00000	0,0,2	95.0008	9.9938	24.3932	24.9929	24.6385
4	16	--	3	.00335	.03244	--	.00000	0,0,1,3	95.0001	9.9932	16.7752	17.4494	17.0536
5	13	--	3	.00041	.03599	--	.00000	0,0,0,3,4	95.0000	9.9655	14.2264	14.8010	14.3101
6	8	3	3	.00351	.05637	.00000	.00000	0,0,0,0,1,3	95.0007	9.9903	10.9317	10.9664	9.9838
7	7	3	3	.00059	.05158	.00000	.00000	0,0,0,0,0,2,4	95.0014	9.9889	9.9105	9.9560	8.9493
8	6	3	3	.00261	.07757	.00000	.00000	0,0,0,0,0,0,1,3	95.0005	9.9927	8.9459	8.9720	8.2278
9	7	0	4	.00034	.05579	.00000	.00000	0,0,0,0,0,0,1,2,6	95.0006	9.9948	8.9096	8.9545	8.0340
10	5	3	3	.00090	.09535	.00000	.00000	0,0,0,0,0,0,0,0,1,4	95.0002	9.9987	7.9519	7.9745	7.3471
k	n	n_0	n_1	p_a^*	p_r^*	p_0^*	p_1^*	c_k	$P_a(0)$ %	$P_a(1)$ %	A_0	A_{max}	A_1

producer's and consumer's risks. Examples 1, 2 and 3 for sampling by attributes show a relatively modest reduction of approaching 20% between the maximum average effort required to inspect each of three lots and the maximum average effort required for each of ten lots. Compensation is the decrease in the producer's risk as the number of lots increases and as the coefficients of variation of the prior beta distributions decrease.

There is a much more substantial reduction for sampling by variables. For the prior beta distributions considered here, the Type B sampling schemes for unknown process variation were found to be superior to the Type A sampling schemes and showed a reduction of more than 80% in the average amount of sampling per lot between inspecting one lot and inspecting ten lots. As expected, the sampling schemes for known process variation required the smallest sample sizes, and showed more than 70% improvement between one and ten lots. For each number of lots under type B, the optimum sampling scheme had producer's risks and consumer's risks close enough to their nominal values to be considered, for all practical purposes, to be equal to their nominal values.

By comparing the three attributes examples, it is seen that increased discrimination between the prior densities (in the sense of larger values of r and s but keeping the same ratios) leads to a reduced sampling requirement, although perhaps not so much of a reduction as might have been expected.

The sampling schemes for sampling by variables were developed for a single quality characteristic with a single specification limit. However, it was shown by Baillie [1] that when, as in this paper, the process fraction nonconforming is estimated using the minimum variance unbiased estimator, the band of operating characteristics is quite narrow for double specification limits and also for multivariate sampling, particularly for unknown process variance(s). The formulae developed in this paper may therefore be used to determine surveillance sampling schemes by variables for the multivariate case and for double specification limits.

A potential improvement to the Type B sampling schemes would be to replace the acceptance criterion $\hat{p} = 0$ at the initial sample by $\hat{p} \le p_z^*$, say, where $p_z^* \ge 0$. Another potential improvement after second samples had been randomly selected would be to use the MVUE based on the initial and second samples combined. The benefits of these modifications have yet to be investigated.

Acknowledgement

My thanks go to an anonymous referee for perceptive comments on an earlier draft that led to a number of improvements in the presentation of this paper.

Reference

1. Baillie, D.H. (1987). Multivariate acceptance sampling. In: *Frontiers in Statistical Quality Control 3*, Lenz, H.-J. et al., Eds., Physica-Verlag, Heidelberg, 83-115.

Selective Assembly for Maximizing Profit in the Presence and Absence of Measurement Error

Shun Matsuura[1] and Nobuo Shinozaki[2]

[1] Research Fellow of the Japan Society for the Promotion of Science, Graduate School of Science and Technology, Keio University, 3-14-1, Hiyoshi, Kohoku, Yokohama 223-8522, Japan
matsuura@ae.keio.ac.jp
[2] Faculty of Science and Technology, Keio University, 3-14-1, Hiyoshi, Kohoku, Yokohama 223-8522, Japan
nshino@ae.keio.ac.jp

Summary. Selective assembly is an effective approach for improving the quality of a product assembled from two components when the quality characteristic is the clearance between the mating components. A component is rejected if its dimension is outside specified limits of the dimensional distribution. Acceptable components are sorted into several classes by their dimensions, and the product is assembled from randomly selected mating components from the corresponding classes. We assume that the two component dimensions are normally distributed with equal variance, and that measurement error, if any, is also normally distributed. Taking into account the quality loss of a sold product, the selling price of an assembled product, the component manufacturing cost, and the income from a rejected component, we discuss the optimal partitioning of the dimensional distribution to maximize expected profit, including the optimal choice of the distribution limits or truncation points. Equations for a set of optimal partition limits are given and its uniqueness is established in the presence and absence of measurement error. It is shown that the expected profit based on the optimal partition decreases with increasing variance of the measurement error. In addition, some numerical results are presented to compare the optimal partitions for the cases when the truncation points are and are not fixed.

1 Introduction

We considered a product assembled from two components in which the quality characteristic is the difference of the relevant dimensions of the mating components (i.e., clearance) or their sum. Some variation is inevitable in any

H.-J. Lenz et al. (eds.), *Frontiers in Statistical Quality Control 9*,
DOI 10.1007/978-3-7908-2380-6_12,

production process, and the quality of the assembled product is dependent on the dimensional variability of the component parts. Random assembly of the mating components may lead to an unacceptably large variance in product quality. In such a situation, selective assembly should be an effective means for reducing the variance. We considered the cases in which a measurement error of each component dimension is either present or absent. Any component with its dimension outside specified limits of the dimensional distribution is rejected, and the remaining components are sorted into several classes by their dimensions. The product is then assembled by randomly selecting mating components from the corresponding classes. This approach enables the assembly of high-precision products from relatively low-precision components.

For example, consider the assembly of a piston and a cylinder in which quality is measured as the clearance between the mating components. Pistons and cylinders are sorted into several classes according to their diameters, and the product is assembled by selecting a piston and a cylinder from the corresponding classes. The smaller pistons are matched with the smaller cylinders, and the bigger pistons with the bigger cylinders to reduce the variation in the clearance. Other applications of selective assembly include pin and bushing assembly, hole and shaft matching, and valve body and spool assembly.

A great deal of research and development effort has been devoted to the subject of selective assembly. Pugh [14] studied equal width partitioning, assuming that the two component dimensions were normally distributed. Fang and Zhang [3] presented an algorithm to decrease surplus components that result from the difference between the numbers of components in the corresponding classes. Kannan and Jayabalan [5] proposed some partitioning schemes to reduce surplus components when the two component dimensions were normally distributed with unequal variances. Pugh [15] demonstrated with a simulation study that equal area partitioning is superior to equal width partitioning in terms of the acceptance probability. Pugh [16] also studied equal area partitioning in the case of dissimilar variances, and proposed to truncate the dimensional distribution with the larger variance to make the two variances equal.

In the case in which the two component dimensions have dissimilar variances, Kannan, Jayabalan, and Ganesan [4] and Kannan and Jayabalan [7] proposed a method of manufacturing the component with the smaller variance at three different means. Kannan and Jayabalan [6] studied selective assembly for three components to minimize surplus components. Kannan and Jayabalan [8] and Kannan, Asha, and Jayabalan [9] proposed a method using a genetic algorithm to reduce the assembly variation and surplus components.

Kwon, Kim, and Chandra [10] were the first to study optimal partitioning to minimize expected squared error loss, assuming that the two component dimensions had the same normal distribution. Mease, Nair, and Sudjianto [12] extensively studied some optimal partitioning strategies for several loss functions and distributions. Importantly, they developed optimal partitioning for minimizing the expected squared error loss for the case in which the two component dimensions are identically distributed. They gave equations for optimal partition limits, and showed that the solution to them is unique when the density function of the component dimension is strongly unimodal. They also showed that the optimal partition is much better than the two heuristic strategies of equal width and equal area partitioning. They pointed out that this optimization problem is related to the optimal partition of one distribution, which arises in many applications such as stratified sampling (Dalenius [1]), piecewise constant approximation of a function (Eubank [2]), and principal points (Tarpey and Flury [18]). Matsuura and Shinozaki [11] studied optimal partitioning under squared error loss when measurement error is present.

Mease and Nair [13] studied optimal partitions under general convex loss functions in "one-sided" selective assembly, in which only one of the two components is sorted into several classes. One-sided selective assembly is useful in the situation when one of the two components has a much smaller variance than the other.

These papers have fixed acceptable limits or truncation points of the dimensional distributions, at the mean $\pm$ 3 standard deviations, for example. In this paper, we choose optimal truncation points as well, taking into account the quality loss of a sold product, the selling price of the assembled product, the manufacturing cost of the components, and the income from rejected components. This paper extends the results of Mease et al. [12] and Matsuura and Shinozaki [11]. Assuming normality, we discuss the optimal partitioning strategy to maximize expected profit in the presence and absence of measurement error. We present equations for a set of optimal partition limits and establish its uniqueness. We also show that the expected profit based on the optimal partition decreases with increasing variance of measurement error. As well, we present some numerical comparisons of the optimal partitions in the presence and absence of measurement error, and also compare the optimal partitions when the truncation points are and are not fixed.

2 Optimal partitioning in the absence of measurement error

2.1 Models, notation, and assumptions

Let the two respective component dimensions, X and Y, be independently normally distributed with equal variance σ^2. Note that although we use the clearance as the assembly dimension of interest, our discussion is equally valid in the case in which it is the sum of the dimensions of the mating components. Suppose that the expected values of X and Y can be adjusted so that their difference is equal to a given target clearance. We can let the target be zero and the two expected values be the same without any loss of generality. Then we denote the common dimensional distribution of X and Y as $N(\mu, \sigma^2)$. In this section, we consider the case in which no measurement error occurs. Since the true dimension of a component is observed without error, the distributions of X and Y are partitioned into several classes. Let n denote the number of classes which is not subject to optimization but predetermined. The two partition limits for X and Y are the same so that we will not have a surplus of either component. Then for n partition classes,

$$(x_0, x_1, x_2, \ldots, x_{n-1}, x_n)$$

will be the common partition limits for X and Y. Since components with X and Y values in the intervals $(-\infty, x_0]$ and (x_n, ∞) are rejected, we call x_0 and x_n the truncation points. Components with X and Y values in the interval $(x_{i-1}, x_i]$ are sorted into the ith class of X and Y, respectively. Then, the product is assembled by randomly selecting mating components from the corresponding classes as shown in Figure 1. $P(x_{i-1} < X \leq x_i)$ is denoted by

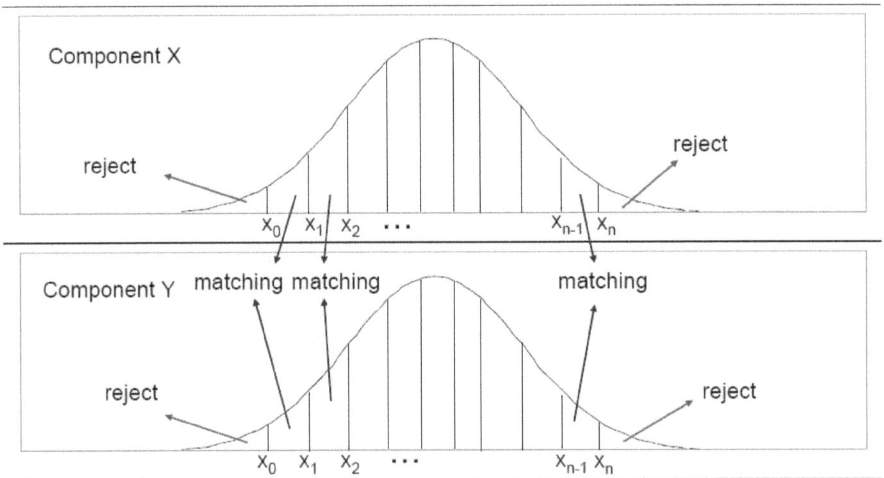

Fig. 1. Selective assembly

p_i, and $P(X \leq x_0)$ and $P(X > x_n)$ are denoted by p_0 and p_{n+1}, respectively. Then $p_0 + p_{n+1}$ is the probability of component rejection. Let X_i and Y_i denote the truncated random variables of X and Y defined on $(x_{i-1}, x_i]$, respectively.

Let C_X (C_Y) denote the manufacturing cost for each component X (Y). Note that the next subsection shows that the values of C_X and C_Y do not affect our optimization problem. Rejected components may be sold at a low price or scrapped for reprocessing. We let D_X (D_Y) denote the income from each rejected component X (Y). Let V denote the selling price for each assembled product. Throughout this paper, we assume that $V > D_X + D_Y$. We note that if this inequality is violated, then rejecting all components maximizes the profit. Let the quality loss of a sold product be proportional to the square of the deviation from the target clearance, as Taguchi [17, page 15] proposed for instance. We denote the constant of this proportion as k. For a specific value of k, we may follow Taguchi [17, page 15], who suggested that k is $(V - D_X - D_Y)/\Delta^2$ where Δ denotes a given tolerance specification of clearance and $V - D_X - D_Y$ is the rejection cost. Now our problem is to find the set of partition limits $(x_0, x_1, x_2, \ldots, x_{n-1}, x_n)$ that maximize the expected profit

$$I_0(x_0, x_1, \ldots, x_n) = V \sum_{i=1}^{n} p_i - k \sum_{i=1}^{n} E[(X_i - Y_i)^2] p_i \\ -(C_X + C_Y) + (D_X + D_Y)(p_0 + p_{n+1}). \qquad (1)$$

2.2 Equations for optimal partition limits

Since X_i and Y_i are independently and identically distributed, we see that

$$\sum_{i=1}^{n} E[(X_i - Y_i)^2] p_i = 2 \sum_{i=1}^{n} E[X_i^2] p_i - 2 \sum_{i=1}^{n} (E[X_i])^2 p_i.$$

Then the expected profit (1) is rewritten as

$$I_0(x_0, x_1, \ldots, x_n) = (V - C_X - C_Y) - 2k \sum_{i=1}^{n} E[X_i^2] p_i + 2k \sum_{i=1}^{n} (E[X_i])^2 p_i \\ -(V - D_X - D_Y)(p_0 + p_{n+1}).$$

Thus, maximizing (1) is equivalent to minimizing

$$\sum_{i=1}^{n} E[X_i^2] p_i - \sum_{i=1}^{n} (E[X_i])^2 p_i + C(p_0 + p_{n+1}), \qquad (2)$$

where $C = (V - D_X - D_Y)/(2k)$. Note that if we choose $k = (V - D_X - D_Y)/\Delta^2$, then $C = \Delta^2/2$. We only require the values of μ, σ, and C to solve this problem. Let f denote the density function of $N(\mu, \sigma^2)$. Since (2) can be rewritten as

$$\int_{x_0}^{x_n} x^2 f(x) dx - \sum_{i=1}^{n} \frac{\left(\int_{x_{i-1}}^{x_i} x f(x) dx\right)^2}{\int_{x_{i-1}}^{x_i} f(x) dx} + C\left(\int_{-\infty}^{x_0} f(x) dx + \int_{x_n}^{\infty} f(x) dx\right),$$

then its partial derivative with respect to x_i for $0 \le i \le n$ is given as

$$f(x_0)(E[X_1] - x_0 + \sqrt{C})(x_0 - E[X_1] + \sqrt{C}),$$
$$f(x_i)(E[X_{i+1}] - E[X_i])(2x_i - E[X_{i+1}] - E[X_i]),\ i = 1, 2, \ldots, n-1,$$
$$f(x_n)(x_n - E[X_n] + \sqrt{C})(x_n - E[X_n] - \sqrt{C}).$$

Since we are assuming that $V > D_X + D_Y$, then $C > 0$. Thus we see that the set of optimal partition limits satisfies these equations:

$$x_0 = E[X_1] - \sqrt{C}, \tag{3}$$

$$x_i = \frac{E[X_i] + E[X_{i+1}]}{2}, \quad i = 1, 2, \ldots, n-1, \tag{4}$$

$$x_n = E[X_n] + \sqrt{C}. \tag{5}$$

We show in the next subsection that the solution to (3)-(5) is unique. Thus we can obtain the set of optimal partition limits using the following numerical algorithm:

1. Select an initial set of partition limits $(x_0^0, x_1^0, \ldots, x_n^0)$.
2. Compute

$$x_0^1 = E[X|x_0^0 < X < x_1^0] - \sqrt{C},$$
$$x_i^1 = \frac{E[X|x_{i-1}^0 < X < x_i^0] + E[X|x_i^0 < X < x_{i+1}^0]}{2}, \quad i = 1, 2, \ldots, n-1,$$
$$x_n^1 = E[X|x_{n-1}^0 < X < x_n^0] + \sqrt{C}.$$

3. Go back to step 2 using $(x_0^1, x_1^1, \ldots, x_n^1)$ in place of $(x_0^0, x_1^0, \ldots, x_n^0)$ and iterate until convergence.

Note that Mease et al. [12] showed that for fixed x_0 and x_n, the set of partition limits minimizing $\sum_{i=1}^{n} E[(X_i - Y_i)^2]p_i$ satisfies the equations (4), and proposed an algorithm for solving them. Our results in this section extend those results to the case in which we choose truncation points as well.

2.3 Uniqueness of optimal partition limits

In this section, we prove the uniqueness of the optimal partition limits. Note that a proof of the uniqueness of solution to (4) for fixed x_0 and x_n has already been given by Mease et al. [12].

We show that if two different solutions satisfy (3)-(5), then we have a contradiction. Let $(x_0^A, x_1^A, \ldots, x_n^A)$ and $(x_0^B, x_1^B, \ldots, x_n^B)$ denote the two solutions. We see that

$$x_0^A = E[X|x_0^A < X < x_1^A] - \sqrt{C}, \tag{6}$$

$$x_i^A = \frac{E[X|x_{i-1}^A < X < x_i^A] + E[X|x_i^A < X < x_{i+1}^A]}{2}, \quad 1 \le i \le n-1, \tag{7}$$

$$x_n^A = E[X|x_{n-1}^A < X < x_n^A] + \sqrt{C}, \tag{8}$$

$$x_0^B = E[X|x_0^B < X < x_1^B] - \sqrt{C}, \tag{9}$$

$$x_i^B = \frac{E[X|x_{i-1}^B < X < x_i^B] + E[X|x_i^B < X < x_{i+1}^B]}{2}, \quad 1 \le i \le n-1, \tag{10}$$

$$x_n^B = E[X|x_{n-1}^B < X < x_n^B] + \sqrt{C}. \tag{11}$$

Let a denote the largest i such that $x_i^A \neq x_i^B$. Without any loss of generality, we assume that $x_a^A < x_a^B$, and equivalently $D_a > 0$ where $D_i = x_i^B - x_i^A$, $i = 0, 1, \ldots, n$. We note that since X is normally distributed, we have

$$\text{Condition (A)} \ : \quad E[X|t < X < t+u] - t = \frac{\int_t^{t+u} xf(x)dx}{\int_t^{t+u} f(x)dx} - t$$
$$\text{is decreasing in } t \text{ for all } u > 0,$$

whose proof is very similar to the one given by Mease et al. [12] (Lemma 1) and is omitted here. We first discuss the case in which $a = n$. From (8) and (11), using $D_n > 0$ and Condition (A), we see that

$$\begin{aligned} E[X|x_{n-1}^B < X < x_n^B] - x_n^B &= -\sqrt{C} = E[X|x_{n-1}^A < X < x_n^A] - x_n^A \\ &> E[X|x_{n-1}^A + D_n < X < x_n^B] - x_n^B. \end{aligned}$$

This means that $x_{n-1}^B > x_{n-1}^A + D_n$, which is equivalent to $D_{n-1} > D_n$. From $D_{n-1} > D_n > 0$ and Condition (A), we have

$$\begin{aligned} E[X|x_{n-1}^A < X < x_n^A] - x_{n-1}^A &> E[X|x_{n-1}^B < X < x_n^A + D_{n-1}] - x_{n-1}^B \\ &> E[X|x_{n-1}^B < X < x_n^B] - x_{n-1}^B. \end{aligned} \tag{12}$$

Now we require the following lemma whose proof is given in the appendix.

Lemma 1. *If for some* $1 \leq i \leq n-1$,

$$x_i^A < x_i^B,$$

$$E[X|x_i^A < X < x_{i+1}^A] - x_i^A > E[X|x_i^B < X < x_{i+1}^B] - x_i^B,$$

and (7) and (10) hold, then

$$x_{i-1}^A < x_{i-1}^B$$

and

$$E[X|x_{i-1}^A < X < x_i^A] - x_{i-1}^A > E[X|x_{i-1}^B < X < x_i^B] - x_{i-1}^B.$$

Let us now return to the proof of uniqueness. From $D_{n-1} > 0$ and (12), applying Lemma 1 repeatedly for $i = n-1, n-2, \ldots, 1$, we obtain $x_0^A < x_0^B$ and

$$E[X|x_0^A < X < x_1^A] - x_0^A > E[X|x_0^B < X < x_1^B] - x_0^B. \tag{13}$$

However, this gives a contradiction since (6) and (9) imply that both sides are the same.

Next we discuss the case in which $1 \leq a \leq n-1$. From (7), (10), $D_{a+1} = 0$, $D_a > 0$, and Condition (A), it follows that

$$\begin{aligned}
x_a^A - E[X|x_{a-1}^A < X < x_a^A] &= E[X|x_a^A < X < x_{a+1}^A] - x_a^A \\
&> E[X|x_a^B < X < x_{a+1}^A + D_a] - x_a^B \\
&> E[X|x_a^B < X < x_{a+1}^B] - x_a^B \\
&= x_a^B - E[X|x_{a-1}^B < X < x_a^B],
\end{aligned}$$

which is equivalent to

$$E[X|x_{a-1}^B < X < x_a^B] - x_a^B > E[X|x_{a-1}^A < X < x_a^A] - x_a^A. \tag{14}$$

From (14), $D_a > 0$, and Condition (A), we see that

$$\begin{aligned}
E[X|x_{a-1}^B < X < x_a^B] - x_a^B &> E[X|x_{a-1}^A < X < x_a^A] - x_a^A \\
&> E[X|x_{a-1}^A + D_a < X < x_a^B] - x_a^B.
\end{aligned}$$

This means that $x_{a-1}^B > x_{a-1}^A + D_a$, which is equivalent to $D_{a-1} > D_a$. Since $D_{a-1} > D_a > 0$, using Condition (A), we have

$$\begin{aligned}
E[X|x_{a-1}^A < X < x_a^A] - x_{a-1}^A &> E[X|x_{a-1}^B < X < x_a^A + D_{a-1}] - x_{a-1}^B \\
&> E[X|x_{a-1}^B < X < x_a^B] - x_{a-1}^B. \qquad (15)
\end{aligned}$$

From $D_{a-1} > 0$ and (15), applying Lemma 1 repeatedly for $i = a-1, a-2, \ldots, 1$, we obtain $x_0^A < x_0^B$ and (13), which gives a contradiction.

Finally, we discuss the case in which $a = 0$. From $D_0 > 0$, $D_1 = 0$, and Condition (A), we have

$$\begin{aligned}
E[X|x_0^A < X < x_1^A] - x_0^A &> E[X|x_0^B < X < x_1^A + D_0] - x_0^B \\
&> E[X|x_0^B < X < x_1^A] - x_0^B,
\end{aligned}$$

which also gives a contradiction. This completes the proof of uniqueness.

3 Optimal partitioning in the presence of measurement error

3.1 Models, notation, and assumptions

In this section, we study the optimal partitioning strategy when measurement error is present. The two measurement errors of the respective components, denoted by W^X and W^Y, are independent of X and Y, and follow the same normal distribution $N(0, \tau^2)$ independently. We note that if τ is 0, then this model reduces to the one in which measurement error is absent. Let Z^X and Z^Y denote the observations of the two respective components, i.e., $Z^X = X + W^X$ and $Z^Y = Y + W^Y$. Since X and Y are distributed as $N(\mu, \sigma^2)$, then Z^X and Z^Y follow the same normal distribution $N(\mu, \sigma^2 + \tau^2)$. In selective assembly when measurement error is present, it is the two distributions of

Z^X and Z^Y, not those of X and Y that are partitioned into several classes because we only observe Z^X and Z^Y. We let

$$(z_0, z_1, z_2, \dots, z_{n-1}, z_n)$$

be the common partition limits for Z^X and Z^Y. Here we use $p_i = P(z_{i-1} < Z^X \le z_i)$ for $i = 1, 2, \dots, n-1$, $p_0 = P(Z^X \le z_0)$, and $p_n = P(Z^X > z_n)$. Let Z_i^X and Z_i^Y denote the truncated random variables of Z^X and Z^Y defined on $(z_{i-1}, z_i]$, respectively. We further let X_i and Y_i denote the true dimensions of the two components X and Y conditioned so that the corresponding observations Z^X and Z^Y are on the interval $(z_{i-1}, z_i]$, respectively. We note that although Z_i^X and Z_i^Y take values in the interval $(z_{i-1}, z_i]$, X_i and Y_i may take values outside that interval. Then the problem is to find the set of partition limits $(z_0, z_1, z_2, \dots, z_{n-1}, z_n)$ that maximize the expected profit

$$\begin{aligned} I_{\tau^2}(z_0, z_1, \dots, z_n) = V \sum_{i=1}^{n} p_i - k \sum_{i=1}^{n} E[(X_i - Y_i)^2] p_i \\ -(C_X + C_Y) + (D_X + D_Y)(p_0 + p_{n+1}). \end{aligned} \quad (16)$$

We show in the next subsection that this problem reduces to the one discussed in Section 2 if we replace X with Z^X, and $C = (V - D_X - D_Y)/(2k)$ with C' given later.

3.2 Optimal partition and its uniqueness

We first give the following lemma which is easily shown using the fact that X given the condition $X + W^X = z$ is distributed as $N\left((\sigma^2 z + \tau^2 \mu)/(\sigma^2 + \tau^2), \sigma^2\tau^2/(\sigma^2 + \tau^2)\right)$.

Lemma 2. *For any $1 \le i \le n$, expected values of X_i and X_i^2 are expressed as follows:*

$$E[X_i] = \frac{\sigma^2}{\sigma^2 + \tau^2} E[Z_i^X] + \frac{\tau^2}{\sigma^2 + \tau^2} \mu,$$

$$E[X_i^2] = \frac{\sigma^4}{(\sigma^2 + \tau^2)^2} E[(Z_i^X)^2] + \frac{2\sigma^2\tau^2}{(\sigma^2 + \tau^2)^2} \mu E[Z_i^X] + \frac{\tau^4}{(\sigma^2 + \tau^2)^2} \mu^2 + \frac{\sigma^2\tau^2}{\sigma^2 + \tau^2}.$$

Note that Y_i has the same distribution as X_i; so using Lemma 2, we rewrite $\sum_{i=1}^{n} E[(X_i - Y_i)^2] p_i$ as

$$\begin{aligned} & 2 \sum_{i=1}^{n} E[X_i^2] p_i - 2 \sum_{i=1}^{n} (E[X_i])^2 p_i \\ &= \frac{2\sigma^4}{(\sigma^2 + \tau^2)^2} \sum_{i=1}^{n} \{E[(Z_i^X)^2] p_i - (E[Z_i^X])^2 p_i\} + \frac{2\sigma^2\tau^2}{\sigma^2 + \tau^2} \sum_{i=1}^{n} p_i. \end{aligned}$$

Then, the expected profit (16) is rewritten as

$$I_{\tau^2}(z_0, z_1, \dots, z_n) = \left(V - \frac{2k\sigma^2\tau^2}{\sigma^2+\tau^2} - C_X - C_Y\right)$$
$$-\frac{2k\sigma^4}{(\sigma^2+\tau^2)^2}\sum_{i=1}^{n}\{E[(Z_i^X)^2]p_i - (E[Z_i^X])^2 p_i\}$$
$$-\left(V - \frac{2k\sigma^2\tau^2}{\sigma^2+\tau^2} - D_X - D_Y\right)(p_0 + p_{n+1}).$$

Thus, maximizing (16) is equivalent to minimizing

$$\sum_{i=1}^{n} E[(Z_i^X)^2]p_i - \sum_{i=1}^{n}(E[Z_i^X])^2 p_i + C'(p_0 + p_{n+1}), \tag{17}$$

where

$$C' = \frac{(\sigma^2+\tau^2)^2}{\sigma^4}\left(C - \frac{\sigma^2\tau^2}{\sigma^2+\tau^2}\right).$$

We assume $C' > 0$ in this subsection, and we discuss the case in which $C' \leq 0$ in the next subsection. Note that (17) is of the same form as (2) if we replace X and C by Z^X and C', respectively. Therefore, we see that the set of optimal partition limits satisfies the equations,

$$z_0 = E[Z_1^X] - \sqrt{C'}, \tag{18}$$
$$z_i = \frac{E[Z_i^X] + E[Z_{i+1}^X]}{2}, \ i = 1, 2, \dots, n-1, \tag{19}$$
$$z_n = E[Z_n^X] + \sqrt{C'}. \tag{20}$$

From the uniqueness result given in Subsection 2.3, we see that the solution to (18)-(20) is also unique.

Note that the results in this section extend the work of Matsuura and Shinozaki [11] who have studied optimal partitioning in the presence of measurement error when the two truncation points z_0 and z_n are fixed. The main results of Matsuura and Shinozaki [11] are as follows. In the case in which the two truncation points z_0 and z_n are fixed, since the set of optimal partition limits satisfies the equations (19), which depend only on the distribution of the observation Z^X, we can clearly obtain the optimal partition limits without worrying about whether measurement error is present or absent. In other words, we need not specify the values of the variances σ^2, τ^2 separately and we only need the values of μ and $\sigma^2 + \tau^2$ to obtain the optimal partition limits. However, we see that when the two truncation points z_0 and z_n are not fixed, we do need the individual values of σ^2 and τ^2 to determine the optimal partition limits.

3.3 Some properties of the optimal partitioning strategy

We let $(x_0^*, x_1^*, \dots, x_n^*)$ denote the set of optimal partition limits in the absence of measurement error, and we specify its expected profit $I_0(x_0^*, x_1^*, \dots, x_n^*)$ by

I_0^*. We also let $(z_0^*, z_1^*, \ldots, z_n^*)$ represent the set of optimal partition limits in the presence of measurement error, and we denote its expected profit $I_{\tau^2}(z_0^*, z_1^*, \ldots, z_n^*)$ by $I_{\tau^2}^*$. We note that $(x_0^*, x_1^*, \ldots, x_n^*)$ and $(z_0^*, z_1^*, \ldots, z_n^*)$ are symmetrical about μ since a normal distribution is symmetrical about its expected value and the set of optimal partition limits is unique.

We give the following proposition whose proof is given in the appendix. From (i) we see that the expected profit based on the optimal partition increases when the number of classes increases. This implies that a cost optimal choice of n will be possible by balancing out the cost of partitioning with one more class against the gain in expected profit by an additional class. From (ii) and (iii) we can see that the expected profit based on the optimal partition decreases with increasing variance of measurement error.

Proposition 1. *Assume that* $C' > 0$ $(C - \sigma^2\tau^2/(\sigma^2+\tau^2) > 0)$. *Then:*
(i) I_0^* *and* $I_{\tau^2}^*$ *are increasing in* n.
(ii) For any $\tau^2 > 0$, $I_0^* > I_{\tau^2}^*$ *holds.*
(iii) $I_{\tau^2}^*$ *are decreasing in* τ^2.

Next, we discuss the case in which $C' \leq 0$.

Proposition 2. *In the case in which measurement error is present, if* $C' \leq 0$ $(C - \sigma^2\tau^2/(\sigma^2+\tau^2) \leq 0)$, *then rejecting all components* $(z_0^* = z_1^* = \cdots = z_n^*)$ *maximizes the expected profit.*

Although this result may seem strange, one can show it as follows.

Consider two components X and Y whose observations are z^x and z^y, respectively. If the two components are assembled, then the profit is $V - k(x-y)^2 - (C_X + C_Y)$ where we let x and y denote the respective true dimensions. Note that the values of x and y are unknown because we only know the observations z^x and z^y, and we see that
$x \sim N\left((\sigma^2 z^x + \tau^2\mu)/(\sigma^2+\tau^2), \sigma^2\tau^2/(\sigma^2+\tau^2)\right)$ and
$y \sim N\left((\sigma^2 z^y + \tau^2\mu)/(\sigma^2+\tau^2), \sigma^2\tau^2/(\sigma^2+\tau^2)\right)$. Since we see that
$x - y \sim N\left(\sigma^2(z^x - z^y)/(\sigma^2+\tau^2), 2\sigma^2\tau^2/(\sigma^2+\tau^2)\right)$, the conditional expected profit when the two components are assembled is given as

$$V - \frac{k\sigma^4}{(\sigma^2+\tau^2)^2}(z^x - z^y)^2 - \frac{2k\sigma^2\tau^2}{\sigma^2+\tau^2} - (C_X + C_Y).$$

However, if the two components are rejected, then the profit is $-(C_X+C_Y)+(D_X+D_Y)$. Since $C - \sigma^2\tau^2/(\sigma^2+\tau^2) \leq 0$, we see that for any z^x and z^y,

$$-(C_X+C_Y)+(D_X+D_Y) \geq V - \frac{k\sigma^4}{(\sigma^2+\tau^2)^2}(z^x - z^y)^2 - \frac{2k\sigma^2\tau^2}{\sigma^2+\tau^2} - (C_X+C_Y).$$

Thus we see that rejecting all components maximizes the expected profit.

4 Numerical results

In this section, we provide some numerical results to compare the optimal partitions when the truncation points are and are not fixed. We also compare the optimal partitions in the presence and absence of measurement error, and evaluate the effect of the measurement error on the expected profit.

We let X and Y be distributed as $N(0,1)$, and set $k = 1$, $C_X = C_Y = 0.1$, and $D_X = D_Y = 0.05$. We give numerical results for the following cases:

Case 1: $V = 0.9$ $(C = (V - D_X - D_Y)/(2k) = 0.4)$,
Case 2: $V = 2.1$ $(C = (V - D_X - D_Y)/(2k) = 1)$,
Case 3: $V = 4.1$ $(C = (V - D_X - D_Y)/(2k) = 2)$.

Note that the set of optimal partition limits depends only on the value of C (or C' in the case in which measurement error is present) since it is determined by (3)-(5) (or (18)-(20)). Table 1 gives the results of the optimal partition for Cases 1-3 when the truncation points are fixed at ± 3 and no measurement error exists. Tables 2-4 give the results for Cases 1-3, respectively, when the truncation points are not fixed and no measurement error exists. These numerical results are given for $n = 1, 2, 3, 4, 5, 6$. Tables 5-8 give the results when measurement error is present for $n = 6$ and $\tau^2 = 0, 0.2, 0.4, 0.6, 0.8$. Table 5 gives the results of the optimal partition for Cases 1-3 when the truncation points are fixed at $\pm 3\sqrt{1+\tau^2}$. Tables 6-8 give the results for Cases 1-3, respectively, when the truncation points are not fixed.

Comparing Table 1 with Tables 2-4, and Table 5 with Tables 6-8, we see that the optimal partitioning strategy when the truncation points are not fixed is superior to the one when the truncation points are fixed. This provides considerable improvement for the case in which the number of classes is small. From Tables 2-4 or Tables 6-8, we see that the optimal truncation points (the value of $x_n^* = -x_0^*$ or $z_n^* = -z_0^*$) get larger with value of $C = (V - D_X - D_Y)/(2k)$. This partially explains why the optimal partition studied in this paper represents a significant improvement over the one with the truncation points fixed at ± 3 (or $\pm 3\sqrt{1+\tau^2}$) when the value of C is small. A smaller value of C implies that the rejection cost $V - D_X - D_Y$ is small relative to the quality loss, and in such a case it is better to set $|x_0|$ and $|x_n|$ smaller and reject components with a larger probability.

From Tables 6-8, we see that the expected profit based on the optimal partition decreases with increasing variance of measurement error, as is shown analytically in Proposition 1. In Case 1 with measurement error $N(0, 0.8)$ (Table 6), we see that rejecting all components is the optimal strategy because $C - \frac{\sigma^2\tau^2}{\sigma^2+\tau^2} = 0.4 - \frac{0.8}{1.8} < 0$ (see Proposition 2).

Table 1. Truncation points are fixed at ±3 and measurement error is absent.

n	optimal partition limits	I_0 (Case 1)	I_0 (Case 2)	I_0 (Case 3)
1	±3	-1.2436	-0.0468	1.9478
2	0 ±3	0.0049	1.2017	3.1963
3	±0.604 ±3	0.3410	1.5377	3.5323
4	0 ±0.964 ±3	0.4805	1.6773	3.6719
5	±0.375 ±1.215 ±3	0.5518	1.7486	3.7432
6	0 ±0.643 ±1.405 ±3	0.5932	1.7899	3.7845

Table 2. Truncation points are not fixed and measurement error is absent (Case 1).

n	optimal partition limits	I_0^* (Case 1)
1	±0.632	0.1588
2	0 ±1.146	0.3449
3	±0.450 ±1.532	0.4601
4	0 ±0.782 ±1.826	0.5305
5	±0.328 ±1.040 ±2.056	0.5751
6	0 ±0.582 ±1.246 ±2.244	0.6046

Table 3. Truncation points are not fixed and measurement error is absent (Case 2).

n	optimal partition limits	I_0^* (Case 2)
1	±1	0.8679
2	0 ±1.661	1.3565
3	±0.544 ±2.088	1.5777
4	0 ±0.905 ±2.388	1.6892
5	±0.363 ±1.171 ±2.615	1.7521
6	0 ±0.633 ±1.378 ±2.796	1.7907

Table 4. Truncation points are not fixed and measurement error is absent (Case 3).

n	optimal partition limits	I_0^* (Case 3)
1	±1.414	2.4156
2	0 ±2.157	3.2440
3	±0.588 ±2.591	3.5387
4	0 ±0.957 ±2.886	3.6722
5	±0.377 ±1.223 ±3.106	3.7434
6	0 ±0.652 ±1.427 ±3.280	3.7858

Table 5. Truncation points are fixed at $\pm 3\sqrt{1+\tau^2}$ and measurement error is present.

τ^2	optimal partition limits	I_{τ^2} (Case 1)	I_{τ^2} (Case 2)	I_{τ^2} (Case 3)
0	0 ±0.643 ±1.405 ±3	0.5932	1.7899	3.7845
0.2	0 ±0.705 ±1.539 ±3.286	0.2782	1.4749	3.4695
0.4	0 ±0.761 ±1.662 ±3.550	0.0532	1.2499	3.2445
0.6	0 ±0.814 ±1.777 ±3.795	-0.1156	1.0812	3.0758
0.8	0 ±0.863 ±1.885 ±4.025	-0.2468	0.9500	2.9446

Table 6. Truncation points are not fixed and measurement error is present (Case 1).

τ^2	optimal partition limits	$I^*_{\tau^2}$ (Case 1)
0	0 ±0.582 ±1.246 ±2.244	0.6046
0.2	0 ±0.606 ±1.287 ±2.237	0.2935
0.4	0 ±0.590 ±1.236 ±2.056	0.0758
0.6	0 ±0.424 ±0.864 ±1.344	-0.0736
0.8	0 0 0 0 0 0 0	-0.1000

Table 7. Truncation points are not fixed and measurement error is present (Case 2).

τ^2	optimal partition limits	$I^*_{\tau^2}$ (Case 2)
0	0 ±0.633 ±1.378 ±2.796	1.7907
0.2	0 ±0.694 ±1.510 ±3.062	1.4756
0.4	0 ±0.749 ±1.631 ±3.308	1.2505
0.6	0 ±0.801 ±1.743 ±3.536	1.0817
0.8	0 ±0.850 ±1.849 ±3.751	0.9504

Table 8. Truncation points are not fixed and measurement error is present (Case 3).

τ^2	optimal partition limits	$I^*_{\tau^2}$ (Case 3)
0	0 ±0.652 ±1.427 ±3.280	3.7858
0.2	0 ±0.716 ±1.568 ±3.675	3.4712
0.4	0 ±0.774 ±1.697 ±4.052	3.2465
0.6	0 ±0.829 ±1.817 ±4.417	3.0781
0.8	0 ±0.880 ±1.930 ±4.770	2.9471

5 Conclusion

We studied the optimal partitioning of dimensional distributions to maximize the expected profit in selective assembly when truncation points are chosen in conjunction, in the presence and absence of measurement error. We gave equations for a set of optimal partition limits and established its uniqueness. We showed that the expected profit based on the optimal partition decreases with increasing variance of measurement error. We presented some numerical results to show that the optimal partitioning strategy studied in this paper is much superior in some cases to the one with the truncation points fixed.

6 Appendices

Proof of Lemma 1. Using (7), (10), and the condition

$$E[X|x_i^A < X < x_{i+1}^A] - x_i^A > E[X|x_i^B < X < x_{i+1}^B] - x_i^B,$$

we see that

$$\begin{aligned} E[X|x_{i-1}^B < X < x_i^B] - x_{i-1}^B &= E[X|x_{i-1}^B < X < x_i^B] - x_i^B + (x_i^B - x_{i-1}^B) \\ &> E[X|x_{i-1}^A < X < x_i^A] - x_i^A + (x_i^B - x_{i-1}^B) \\ &= E[X|x_{i-1}^A < X < x_i^A] - (x_{i-1}^B - D_i). \end{aligned}$$

Using Condition (A) and this inequality, we obtain

$$\begin{aligned} &E[X|x_{i-1}^B - D_i < X < x_i^A] - (x_{i-1}^B - D_i) \\ &\geq E[X|x_{i-1}^B < X < x_i^B] - x_{i-1}^B \\ &> E[X|x_{i-1}^A < X < x_i^A] - (x_{i-1}^B - D_i). \end{aligned}$$

From this, we see that $x_{i-1}^B - D_i > x_{i-1}^A$*, and we have*

$$D_{i-1} = x_{i-1}^B - x_{i-1}^A > D_i > 0. \tag{21}$$

Using Condition (A) and (21), we have

$$\begin{aligned} E[X|x_{i-1}^B < X < x_i^B] - x_{i-1}^B &\leq E[X|x_{i-1}^A < X < x_i^B - D_{i-1}] - x_{i-1}^A \\ &< E[X|x_{i-1}^A < X < x_i^A] - x_{i-1}^A. \end{aligned}$$

This completes the proof.

Proof of Proposition 1. (i) is easily shown although we omit its proof. We show (ii) in the following. Without any loss of generality, we assume $\mu = 0$*. Now recall that*

$$\begin{aligned} I_{\tau^2}(z_0, z_1, \ldots, z_n) = V\sum_{i=1}^{n} p_i - k\sum_{i=1}^{n} E[(X_i - Y_i)^2]p_i \\ -(C_X + C_Y) + (D_X + D_Y)(p_0 + p_{n+1}). \end{aligned}$$

From Lemma 2, we see that

$$E[X_i] = \frac{\sigma^2}{\sigma^2 + \tau^2} E[Z_i^X],\ i = 1, 2, \ldots, n$$

and

$$E[X_i^2] = \frac{\sigma^4}{(\sigma^2+\tau^2)^2} E[(Z_i^X)^2] + \frac{\sigma^2\tau^2}{\sigma^2+\tau^2}, \ i = 1, 2, \ldots, n.$$

Thus, noting that X *has the same distribution as* $\sigma Z^X / \sqrt{\sigma^2+\tau^2}$ *and* $C = (V - D_X - D_Y)/(2k)$, *we see that*

$$\begin{aligned}
I_{\tau^2}^* = & \, I_0(x_0^\dagger, x_1^\dagger, \ldots, x_n^\dagger) - \frac{2k\sigma^2\tau^2}{\sigma^2+\tau^2} \\
& + \frac{2k\tau^2}{\sigma^2+\tau^2} \sum_{i=1}^n E[X^2 | x_{i-1}^\dagger < X \le x_i^\dagger] Pr(x_{i-1}^\dagger < X \le x_i^\dagger) \\
& - \frac{2k\tau^2}{\sigma^2+\tau^2} \sum_{i=1}^n (E[X | x_{i-1}^\dagger < X \le x_i^\dagger])^2 Pr(x_{i-1}^\dagger < X \le x_i^\dagger) \\
& + \frac{2k\sigma^2\tau^2}{\sigma^2+\tau^2} \{Pr(X \le x_0^\dagger) + Pr(x_n^\dagger < X)\},
\end{aligned}$$

where $x_i^\dagger = \sigma z_i^* / \sqrt{\sigma^2+\tau^2}$, $i = 0, 1, \ldots, n$. *From* $I_0(x_0^\dagger, x_1^\dagger, \ldots, x_n^\dagger) \le I_0^*$, *we have*

$$\begin{aligned}
I_{\tau^2}^* \le & \, I_0^* - \frac{2k\sigma^2\tau^2}{\sigma^2+\tau^2} Pr(x_0^\dagger < X \le x_n^\dagger) \\
& + \frac{2k\tau^2}{\sigma^2+\tau^2} \sum_{i=1}^n E[X^2 | x_{i-1}^\dagger < X \le x_i^\dagger] Pr(x_{i-1}^\dagger < X \le x_i^\dagger) \\
& - \frac{2k\tau^2}{\sigma^2+\tau^2} \sum_{i=1}^n (E[X | x_{i-1}^\dagger < X \le x_i^\dagger])^2 Pr(x_{i-1}^\dagger < X \le x_i^\dagger) \\
= & \, I_0^* - \frac{2k\tau^2}{\sigma^2+\tau^2} \Bigg\{ \sigma^2 Pr(x_0^\dagger < X \le x_n^\dagger) \\
& - \sum_{i=1}^n E[X^2 | x_{i-1}^\dagger < X \le x_i^\dagger] Pr(x_{i-1}^\dagger < X \le x_i^\dagger) \Bigg\} \\
& - \frac{2k\tau^2}{\sigma^2+\tau^2} \sum_{i=1}^n (E[X | x_{i-1}^\dagger < X \le x_i^\dagger])^2 Pr(x_{i-1}^\dagger < X \le x_i^\dagger).
\end{aligned}$$

Noting that $x_n^\dagger = -x_0^\dagger$ *holds since* $(z_0^*, z_1^*, \ldots, z_n^*)$ *is symmetrical about* $\mu = 0$, *we have*

$$\begin{aligned}
\frac{\sum_{i=1}^n E[X^2 | x_{i-1}^\dagger < X \le x_i^\dagger] Pr(x_{i-1}^\dagger < X \le x_i^\dagger)}{Pr(x_0^\dagger < X \le x_n^\dagger)} & = E[X^2 | x_0^\dagger < X \le x_n^\dagger] \\
& < E[X^2] = \sigma^2.
\end{aligned}$$

Thus we see that

$$I^*_{\tau^2} < I^*_0 - \frac{2k\tau^2}{\sigma^2+\tau^2}\sum_{i=1}^{n}(E[X|x^\dagger_{i-1} < X \le x^\dagger_i])^2 Pr(x^\dagger_{i-1} < X \le x^\dagger_i)$$
$$< I^*_0,$$

which completes the proof of (ii). Similarly, we can show (iii), but we omit its proof.

Acknowledgements

This research was supported by Grant-in-Aid for JSPS Fellows, 20·381.

References

1. Dalenius, T. (1950): "The problem of optimal stratification". *Skandinavisk Aktuarietidskrift*, 33, 203-212.
2. Eubank, R.L. (1988): "Optimal grouping, spacing, stratification, and piecewise constant approximation". *SIAM Review*, 30, 404-420.
3. Fang, X.D., Zhang, Y. (1995): "A new algorithm for minimizing the surplus parts in selective assembly". *Computers and Industrial Engineering*, 28, 341-350.
4. Kannan, S.M., Jayabalan, V., Ganesan, S. (1997): "Process design to control the mismatch in selective assembly by shifting the process mean". *Proceedings of International Conference on Quality Engineering and Management*, P.S.G. College of Technology, Coimbatore, South India, 85-91.
5. Kannan, S.M., Jayabalan, V. (2001): "A new grouping method for minimizing the surplus parts in selective assembly". *Quality Engineering*, 14, 67-75.
6. Kannan, S.M., Jayabalan, V. (2001): "A new grouping method to minimize surplus parts in selective assembly for complex assemblies". *International Journal of Production Research*, 39, 1851-1864.
7. Kannan, S.M., Jayabalan, V. (2002): "Manufacturing mean design for selective assembly to minimize surplus parts". *Proceedings of International Conference on Quality and Reliability*, I.C.Q.R. RMIT University, Melbourne, Australia, 259-264.
8. Kannan, S.M., Jayabalan, V. (2003): "Genetic algorithm for minimizing assembly variation in selective assembly". *International Journal of Production Research*, 41, 3301-3313.
9. Kannan, S.M., Asha, A., Jayabalan, V. (2005): "A new method in selective assembly to minimize clearance variation for a radical assembly using genetic algorithm". *Quality Engineering*, 17, 595-607.
10. Kwon, H.M., Kim, K.J., Chandra, M.J. (1999): "An economic selective assembly procedure for two mating components with equal variance". *Naval Research Logistics*, 46, 809-821.
11. Matsuura, S., Shinozaki, N. (2007): "Optimal binning strategies under squared error loss in selective assembly with measurement error". *Communications in Statistics–Theory and Methods*, 36(16), 2863-2876.

12. Mease, D., Nair, V.N., Sudjianto, A. (2004): "Selective assembly in manufacturing: Statistical issues and optimal binning strategies". *Technometrics*, 46, 165-175.
13. Mease, D., and Nair, V.N. (2006): "Unique optimal partitions of distributions and connections to hazard rates and stochastic ordering". *Statistica Sinica*, 16, 1299-1312.
14. Pugh, G.A. (1986): "Partitioning for selective assembly". *Computers and Industrial Engineering*, 11, 175-179.
15. Pugh, G.A. (1986): "Group formation in selective assembly". *SME Ultratech Conference Proceedings*, 2117-2123.
16. Pugh, G.A. (1992): "Selective assembly with components of dissimilar variance". *Computers and Industrial Engineering*, 23, 487-491.
17. Taguchi, G. (1986): "*Introduction to Quality Engineering: Designing Quality into Products and Processes*". *Asian Productivity Organization.*
18. Tarpey, T., Flury, B. (1996): "Self-consistency: A fundamental concept in statistics". *Statistical Science*, 11, 229-243.

A New Approach to Bayesian Sampling Plans

Peter-Th. Wilrich

Institut für Statistik und Ökonometrie, Freie Universität Berlin, Garystrasse 21, D-14195 Berlin, Germany, wilrich@wiwiss.fu-berlin.de

Summary. A large number of papers exists that deal with Bayesian sampling plans. Hald defines Bayesian sampling plans as "plans obtained by minimizing average costs, consisting of inspection, acceptance and rejection costs". In order to obtain such a plan one starts with an *a priori* distribution of the fraction of nonconforming items in the lots, i.e. an assumption about the process curve, and calculates the sampling plan that minimizes the Bayesian risk or cost (if cost parameters are given). However, once these plans have been obtained they are applied in the classical manner just by making acceptance/rejection decisions for the inspected lots. For a Bayesian, the calculation of the *a posteriori* distribution of the fraction nonconforming in the lot is the essential step of the Bayesian analysis because for him the complete information combining prior knowledge and sample information is incorporated in the *a posteriori* distribution. Hence, in this paper the lot acceptance decision is directly based on the *a posteriori* distribution of the fraction nonconforming in the lot and especially the *a posteriori* estimate of the probability of the fraction of nonconforming items in the lot being larger than the acceptance quality limit p_{AQL}. This Bayesian method is applied to sampling by attributes based on a beta-binomial model.

1 Introduction

ISO 2859–1 [3] specifies an acceptance sampling system for inspection by attributes to be used for a continuing series of lots stemming from one and the same production process. "Its purpose is

(1) to induce a supplier through the economic and psychological pressure of lot non-acceptance to maintain a process average at least as good as the specified acceptance quality limit, while at the same time

(2) providing an upper limit for the risk of the consumer of accepting the occasional poor lot" ([3], p.1).

In order to achieve purpose (2) sampling plans (n, c) are provided: an incoming lot is accepted if the number of nonconforming items in the sample

H.-J. Lenz et al. (eds.), *Frontiers in Statistical Quality Control 9*,
DOI 10.1007/978-3-7908-2380-6_13,

of size n chosen randomly from the lot is not larger than the acceptance constant c, otherwise it is rejected.

In order to achieve purpose (1) rules for switching between three inspection stages (normal inspection, reduced inspection, tightened inspection) with different sampling plans (n, c) and a fourth inspection stage "discontinuation of inspection" have to be applied.

Users of ISO 2859–1 very often complain about two deficiencies of this acceptance sampling system:

1. Regardless of whether a lot is rejected or accepted they would like to get a probability of the lot being actually unacceptable, i.e. having a fraction of nonconforming items larger than the acceptance quality limit p_{AQL}. Instead, they only know the probability of the lot being accepted or rejected in relation to its fraction of nonconforming items, i.e. they get probabilities describing the quality of the acceptance sampling method and not the quality of the inspected lot.
2. They are aware of switching between the inspection stages but they would like to know what that means in terms of the distribution of the fraction of nonconforming items in the lots, called the process curve. Especially they are interested in estimates of the process average and of the process standard deviation of the process curve and in estimates of the fraction of unacceptable lots produced by the production process.

In this paper we offer a Bayesian approach that answers both questions.

A large number of papers exists that deal with Bayesian sampling plans. Hald [1] defines Bayesian sampling plans as "plans obtained by minimizing average costs, consisting of inspection, acceptance and rejection costs". In order to obtain such a plan one starts with an *a priori* distribution of the fraction of nonconforming items in the lots, i.e. an assumption about the process curve, and calculates the sampling plan that minimizes the Bayesian risk or cost (if cost parameters are given). However, once these plans have been obtained they are applied in the classical manner just by making acceptance/rejection decisions for the inspected lots. Lenz and Rendtel [2] discuss some existing and introduce some new sampling inspection schemes that include an adaptation of the sampling plan. There schemes consist of monitoring procedures for the parameters of the process curve and a switch to a new sampling plan in case a change of these parameters has been signalled.

For a Bayesian, the calculation of the *a posteriori* distribution of the fraction nonconforming in the lot is the essential step of the Bayesian analysis because for him the complete information combining prior knowledge and sample information is incorporated in the *a posteriori* distribution. Hence, in this paper the lot acceptance decision is directly based on the *a posteriori* distribution of the fraction nonconforming in the lot and especially the *a posteriori* estimate of the probability of the fraction of nonconforming items in the lot being larger than the acceptance quality limit p_{AQL}. This Bayesian method is applied to sampling by attributes based on a beta-binomial model.

2 The model

We use a hierarchical Bayes model with two levels of random variation; at the first level we have randomly distributed fractions of nonconforming items in the lots, and at the second level we have randomly distributed numbers of nonconforming items in the samples drawn randomly from the lots.

Level 1:

We assume that after production of N consecutively produced items a lot t ($t = 1, 2, \dots$) is formed and that the lot size N is large ($N > 1000$). Under the latter assumption we can model the fraction of nonconforming items in lot t as a continuous random variable P_t that takes values p_t between 0 and 1. As the distribution of P_t, called the process curve, we choose a beta distribution with density

$$f_0(p_t) = \frac{p_t^{\alpha_t - 1}(1 - p_t)^{\beta_t - 1}}{B(\alpha_t, \beta_t)} \tag{1}$$

where $\alpha_t > 0$ and $\beta_t > 0$ are two parameters; $B(\alpha_t, \beta_t)$ is the complete beta function. This distribution is a flexible distribution model that includes unimodal distributions for $(\alpha_t > 1, \beta_t > 1)$, the rectangular distribution for $(\alpha_t = 1, \beta_t = 1)$, J-shaped distributions for $(\alpha_t \leq 1 < \beta_t)$ or $(\alpha_t > 1 \geq \beta_t)$, and U-shaped distributions for $(\alpha_t < 1, \beta_t < 1)$. The expectation and the variance of P_t are

$$E(P_t) = \frac{\alpha_t}{\alpha_t + \beta_t}, \tag{2}$$

$$V(P_t) = \frac{\alpha_t \beta_t}{(\alpha_t + \beta_t)^2(\alpha_t + \beta_t + 1)}. \tag{3}$$

Given p_{AQL} as the borderline between acceptable and unacceptable lots the probability of a lot being acceptable is

$$PA_t = \int_0^{p_{AQL}} f_0(p_t) dp_t. \tag{4}$$

However, since we do not know $f_0(p_t)$ we do not know $E(P_t)$, $V(P_t)$ and PA_t.

Level 2:

Under the condition that lot t has a fixed fraction p_t of nonconforming items the number of nonconforming items in a sample of size n drawn randomly from the lot of size N, follows the hypergeometric distribution with parameters N , n and p_t. Under the assumption that the sample size n is not larger than 10% of the lot size N, $n \leq 0.1N$, the hypergeometric distribution can be approximated by the binomial distribution with the probability function

$$f_t(x) = P(X = x) = \binom{n}{x} p_t^x (1 - p_t)^{n-x} \; ; \quad x = 0, 1, \dots, n. \tag{5}$$

The expectation and the variance of X are

$$E_{p_t}(X) = np_t, \tag{6}$$

$$V_{p_t}(X) = np_t(1 - p_t), \tag{7}$$

respectively.

Level 1 and level 2 combined:

The number X of nonconforming items in a sample of size n, randomly chosen from lot t that has been randomly chosen from the beta process curve, is beta-binomial (or Polya-) distributed with probability mass function

$$f_t(x) = P(X = x) = \binom{n}{x} \frac{B(x + \alpha, n - x + \beta)}{B(\alpha, \beta)}. \tag{8}$$

The expectation and the variance of X are

$$\mu_t = E_t(X) = \frac{n\alpha_t}{\alpha_t + \beta_t}, \tag{9}$$

$$\sigma_t^2 = V_t(X) = \frac{n\alpha_t\beta_t(n + \alpha_t + \beta_t)}{(\alpha_t + \beta_t)^2(1 + \alpha_t + \beta_t)}, \tag{10}$$

respectively. σ_t^2 is bounded:

$$\frac{\mu_t(n - \mu_t)}{n} \leq \sigma_t^2 \leq \mu_t(n - \mu_t); \tag{11}$$

the lower bound is the variance of the binomial distribution.

3 The estimation of the parameters of the process curve and the lot acceptance decision

Having reached lot k we estimate the parameters of the process curve directly (using an empirical Bayes approach) from the observed numbers $x_1, x_2, \ldots, x_k$ of nonconforming items in the samples taken from the lots $t = 1, 2, \ldots, k$.

Under the assumption that the parameters of the process curve do not change in time, $\mu_t = \mu$ and $\sigma_t^2 = \sigma^2$ for $t = 1, 2, \ldots, k$, the simplest estimates of μ and σ^2 would be

$$\hat{\mu}_k = \bar{x}_k = \frac{1}{k} \sum_{t=1}^{k} x_t \tag{12}$$

and

$$\hat{\sigma}_k^2 = s_k^2 = \frac{1}{k} \sum_{t=1}^{k} (x_t - \hat{\mu}_k)^2, \tag{13}$$

respectively. However, the assumption of a constant process curve is contradictory to the purpose of the sampling procedure to detect a possible alteration of the process average. Hence, we have to estimate the parameters of the process curve more based on the most recent observations (by giving them a larger weight) than on the more elderly observations. To do this, we use exponentially weighted moving averages as estimators,

$$\begin{aligned} \hat{\mu}_k &= (1-\gamma)\hat{\mu}_{k-1} + \gamma x_k \\ \hat{\sigma}_k^2 &= (1-\gamma)\hat{\sigma}_{k-1}^2 + \gamma(x_k - \hat{\mu}_k)^2 \end{aligned} \qquad k = 2, 3, \ldots \quad (14)$$

and start the estimation with

$$\hat{\mu}_1 = \max(x_1, 1) \quad (15)$$

and

$$\hat{\sigma}_1^2 = \frac{\hat{\mu}_1(n-\hat{\mu}_1)}{n} \cdot \frac{n+2}{3}; \quad (16)$$

$\hat{\mu}_1$ is adjusted so that it cannot be zero because then the updated process curve would be a one-point distribution at $p = 0$; $\hat{\sigma}_1^2$ is the largest possible variance of a unimodal ($\alpha \geq 1; \beta \geq 1$) beta distribution with $\mu = \hat{\mu}_1$. γ can be chosen arbitrarily between 0 and 1. For $\gamma = 1$ the estimate $\hat{\mu}_k$ is equal to the observed number of nonconforming items in the present sample, x_k. As smaller γ, as more $\hat{\mu}_k$ is based on the x's observed in the past. We choose $\gamma = 0.1$.

An assumption underlying the application of sampling systems is a production process with an acceptable process curve, i.e. one that causes very few lot rejections. Hence, we base the updating procedure for the estimation of the parameters of the process curve only on accepted lots. Lots being rejected are interpreted as outlier lots with an abnormal large fraction of nonconforming items that are not generated by the process curve but by an irregular situation. On the other hand, frequently occurring rejections of lots might signalize a deterioration of the process curve. Hence, if more than two consecutive lots are rejected, the updating procedure starts to include all consecutively following rejected lots. However, in practice one would rather stop the application of the sampling system in such a situation and investigate the reason for the deterioration of the process curve.

With the updated estimates $\hat{\mu}_k$ and $\hat{\sigma}_k^2$ the estimates of the parameters α_k and β_k of the beta-binomial distribution are

$$\hat{\alpha}_k = \hat{\mu}_k \frac{\hat{\mu}_k(n-\hat{\mu}_k) - \hat{\sigma}_k^2}{n\hat{\sigma}_k^2 - \hat{\mu}_k(n-\hat{\mu}_k)}, \quad (17)$$

$$\hat{\beta}_k = (n-\hat{\mu}_k) \frac{\hat{\mu}_k(n-\hat{\mu}_k) - \hat{\sigma}_k^2}{n\hat{\sigma}_k^2 - \hat{\mu}_k(n-\hat{\mu}_k)}. \quad (18)$$

Within the beta-binomial distribution, the estimates of expectation and variance of the beta distribution generating the fractions p of nonconforming items in the lots are

$$\hat{\mu}_k(P) = \frac{\hat{\mu}_k}{n} = \frac{\hat{\alpha}_k}{\hat{\alpha}_k + \hat{\beta}_k}, \tag{19}$$

$$\hat{\sigma}_k^2(P) = \frac{\hat{\sigma}_k^2}{n(n + \hat{\alpha}_k + \hat{\beta}_k)} = \frac{\hat{\alpha}_k \hat{\beta}_k}{(\hat{\alpha}_k + \hat{\beta}_k)^2 (1 + \hat{\alpha}_k + \hat{\beta}_k)}. \tag{20}$$

This beta distribution is the estimate of the process curve being obtained from lots $t = 1, 2, \ldots, k$, and it is used as the *a priori* distribution $f(p_{k+1}; \hat{\alpha}_k, \hat{\beta}_k)$ for the Bayesian analysis of the observed number x_{k+1} of nonconforming items in the sample drawn from lot $k+1$. The *a posteriori* distribution of the fraction p_{k+1} of nonconforming items in lot $k+1$ is

$$g(p_{k+1}) \sim l(p_{k+1}; n, x_{k+1}) \cdot f(p_{k+1}; \hat{\alpha}_k, \hat{\beta}_k) \tag{21}$$

where

$$l(p_{k+1}; n, x_{k+1}) = \binom{n}{x_{k+1}} x_{k+1}^{p_{k+1}} (n - x_{k+1})^{1 - p_{k+1}} \tag{22}$$

is the likelihood function of the binomial distribution. Since the beta distribution and the binomial distribution are conjugated (see [4]), the *a posteriori* distribution $g(p_{k+1})$ is also a beta distribution and hence, we only need to determine its parameters as

$$\alpha_{post} = \hat{\alpha}_k + x_{k+1}, \tag{23}$$

$$\beta_{post} = \hat{\beta}_k + n - x_{k+1}, \tag{24}$$

see [4], p.175 ff.

The probability of lot $k+1$ having a fraction of nonconforming items smaller than p_{AQL} is *a priori*

$$PA_{k+1,priori} = \int_0^{p_{AQL}} f(p_{k+1}; \hat{\alpha}_k, \hat{\beta}_k) dp_{k+1} \tag{25}$$

and *a posteriori*

$$PA_{k+1,posteriori} = \int_0^{p_{AQL}} g(p_{k+1}; \alpha_{post}, \beta_{post}) dp_{k+1}. \tag{26}$$

The latter one is used for the acceptance decision. In the classical approach to sampling inspection we accept lot $k+1$ if the number x_{k+1} of nonconforming items in the sample drawn randomly from lot $k+1$ is not larger than the acceptance constant c of the sampling plan. In our Bayes approach we accept lot $k+1$ if the *a posteriori* probability $PA_{k+1,posteriori}$ for lot $k+1$ having a fraction p_{k+1} of nonconforming items smaller than p_{AQL}, is not smaller than a predetermined probability limit L , $PA_{k+1,posteriori} \geq L$, and reject it otherwise. The sampling plan now consists of the sample size n and the probability limit L. Since we assume that the sampling procedure is intended to be applied to a production process of lots having a process average below p_{AQL} we choose $L = 0.5$.

This decision procedure works for all lots except for the first one because in the beginning we do not have an estimate of the process curve and hence no *a priori* distribution of the fraction of nonconforming items in that lot. Therefore, we start with a "non-informative" *a priori* distribution and choose the uniform distribution in the range between $p = 0$ and $p = 1$ as "non-informative". I do not intend to start a discussion whether this is non-informative in the strict sense. However, it says that each fraction nonconforming in the lot is equally likely. There is the argument that this is not a realistic assumption: If we have agreed upon $p_{AQL} = 10\%$, say, we do not equally likely expect lots with 10% and with 90% nonconforming items! Although this is true, it does not meet the point. The fundamental Bayes equation multiplies the *a priori* density with the likelihood, and the result of a multiplication is (almost) zero if one of the factors is (almost) zero. Hence, the *a priori* density plays a role only in the range of p for which the likelihood is not approaching zero. And this is the range in which the fraction nonconforming is likely to be expected on the basis of the sample result. If we had started with a uniform *a priori* distribution of p in this range (instead of the range between 0 and 1), the result of the Bayesian analysis would be practically identical.
In other words, to start with a non-informative uniform *a priori* distribution of p in the lots is essentially equivalent to the following assumption: In the range of fractions nonconforming in the inspected lot that we likely expect on the basis of the sample result, we expect each of the fractions nonconforming to be equally likely!

Quantitatively, if we start with the uniform *a priori* distribution, i.e. the beta distribution with $\alpha = 1$ and $\beta = 1$, and we observe x_1 nonconforming items in the sample the *a posteriori* distribution would be a beta distribution with

$$\alpha_{post} = \alpha + x_1 = x_1 + 1, \tag{27}$$

$$\beta_{post} = \beta + n - x_1 = n - x_1 + 1; \tag{28}$$

its expectation is

$$E_{post}(p) = \frac{\alpha_{post}}{\alpha_{post} + \beta_{post}} = \frac{x_1 + 1}{n + 2}; \tag{29}$$

instead of the classical estimate x_1/n of the fraction p_1 of nonconforming items in lot 1 we have $E_{post}(p)$, i.e. the "non-informative" *a priori* distribution changes the classical estimate as if one more nonconforming item had found in a sample of a size two items larger.

The decision rule $PA_{k+1,posteriori} \geq L$ contains an intrinsic adaptive procedure: if the numbers of nonconforming items in consecutive samples are large the estimated process curve shifts to the right and hence, the *a posteriori* probabilities of the fractions of nonconforming items below p_{AQL}, $PA_{k+1,posteriori}$, decrease so that rejections become more likely, and vice versa. In the classical sampling inspection system of ISO 2859-1 the same effect is achieved by switching to tightened inspection under which the sample size remains unchanged but the acceptance constant becomes $c-1$ or $c-2$. This switch is unnecessary in the Bayesian approach.

4 Some simulated scenarios

The quality requirement for a series of lots of equal size $N = 1000$ is expressed by an acceptance quality limit $p_{AQL} = 4.0\%$, i.e. lots are acceptable if they come from a process with a process average not larger than 4.0%. For $N = 1000$ and $p_{AQL} = 4.0\%$, ISO 2859-1, Tables 2-A and 2-B, give the sample size $n = 50$, the acceptance constant $c_1 = 5$ for normal inspection and $c_0 = 3$ for tightened inspection. The switching rules of the Standard are applied, except that switching to reduced inspection is not installed. The Bayesian approach is used in parallel, with $n = 50$, $p_{AQL} = 4.0\%$ and $L = 0.5$.

Scenario 1: Process curve constant in time

Fractions p_t of nonconforming items of a series of 100 lots are generated by drawing from a beta process curve with process average $\mu(P) = 0.02$ and process standard deviation $\sigma(P) = 0.01$. A sample of size $n = 50$ is drawn randomly from each lot, and the number x of nonconforming items in the sample is determined.

Figure 1 shows the process average as blue bold line and the 95% range (between the 0.025-quantile and the 0.975-quantile as blue lines). The double points indicate the fractions of nonconforming items p_t of lots $t = 1, \ldots, 100$. The larger circle of the double point is green/red if the lot is accepted/rejected by the Bayesian procedure, the smaller circle is green/red if the lot is accepted/rejected by the ISO 2859 method. All lots have been accepted by ISO, 98 of 100 by Bayes. In addition, symbols "N", "T" and "S" on the horizontal line p_{AQL} indicate switches to normal, tightened and discontinued inspection, respectively. There are not any switches in this scenario.

Figure 2 shows the estimated process curve: the bold black line is the estimated process average, the other two black lines indicate the estimated 95% range. The corresponding blue lines for the true process curve of Figure 1 are repeated.

Figure 3 shows, for each lot, the *a posteriori* probability of its fraction of nonconforming items being larger than p_{AQL}. The horizontal line for $L = 0.5$ separates the accepted lots (points below p_{AQL}) from the rejected lots (points above p_{AQL}).

The Bayesian analysis for lot 29 is depicted in Figure 4. The graph in the middle is the likelihood function, based on the sample result of $x = 5$ noncon-

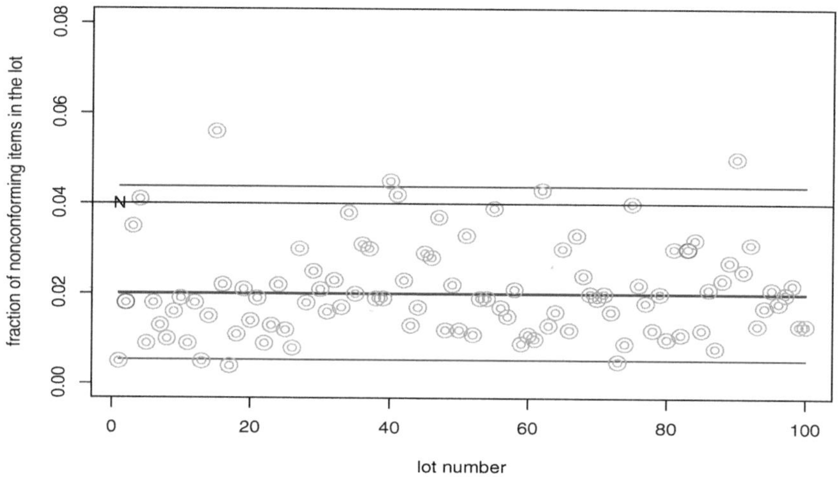

Figure 1. Fractions of nonconforming items in the lots and acceptance decisions of scenario 1; green indicates acceptance, red rejection. Larger circles indicate the Bayes decision, smaller circles the ISO 2859 decision

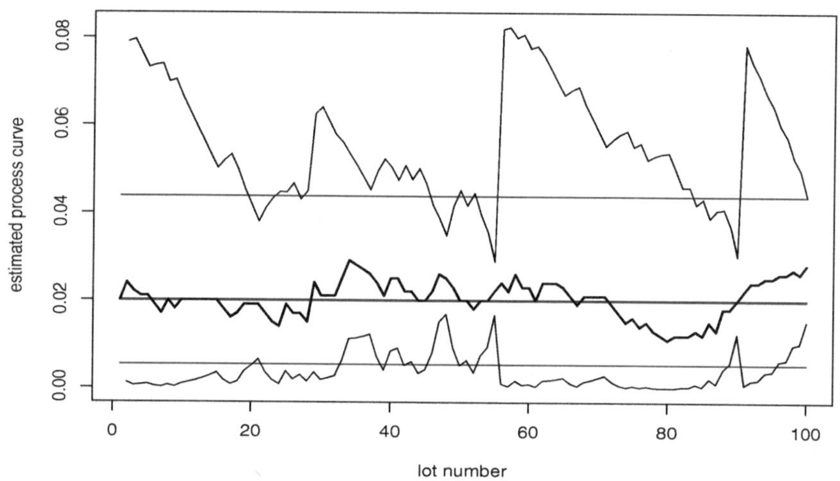

Figure 2. Process average (bold), 2.5% quantile and 97.5% quantile of the true (blue) and the estimated process curve (black) of scenario 1

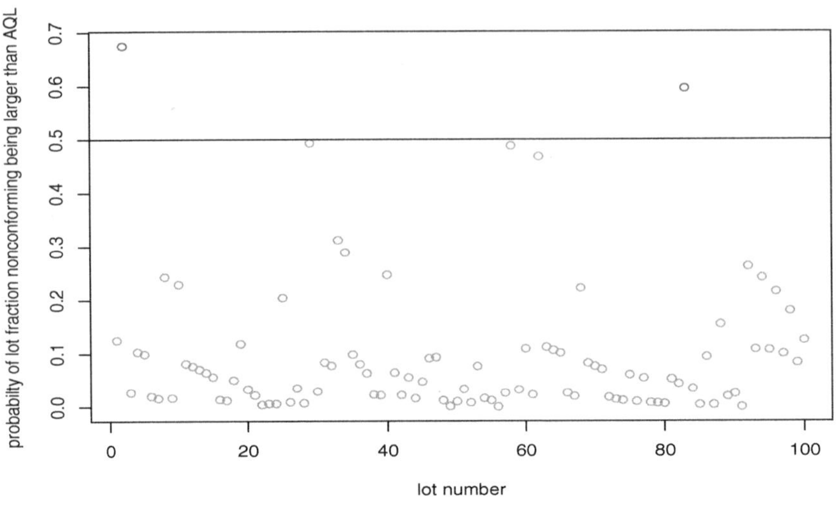

Figure 3. *A posteriori* probability of the fraction of nonconforming items in the lot being larger than p_{AQL}, for scenario 1

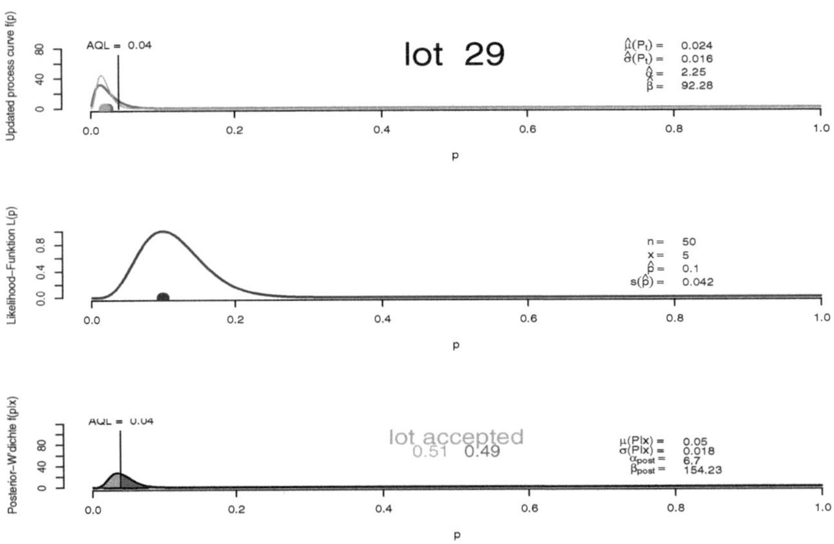

Figure 4. Bayesian analysis for lot 29. The graph on top shows the true (green) and the updated (red) process curve, the graph in the middle the likelihood, the graph on the bottom the *a posteriori* distribution of the fraction of nonconforming items in lot 29 of scenario 1

forming items that gives the classical estimate $\hat{p} = x/n = 0.10$ of the fraction of nonconforming items in lot 29; the true fraction of nonconforming items in this lot (which is known in the simulation study but would be unknown in practice) is $p_{29} = 0.025$. The graph on the bottom is the *a posteriori* distribution of the fraction of nonconforming items in lot 29. The *a posteriori* estimate of the fraction of nonconforming items in lot 29 is $\hat{\mu}(P) = 0.05$. The *a posteriori* probability of this fraction being larger than p_{AQL}, $PA_{p_{k+1,posteriori}} = 0.49$, is visualized by the red area under the *a posteriori* beta distribution. Since it is smaller than $L = 0.5$, the lot is accepted. The graph on top shows the true process curve (green) and the updated estimate of the process curve (red). The updated estimates of process average and process standard deviation are $\hat{\mu}(P_{29}) = 0.024$ and $\hat{\sigma}(P_{29}) = 0.016$, respectively. (true values: $\mu(P_{29}) = 0.02$ and $\sigma(P_{29}) = 0.01$).

Scenario 2: Sudden shifts of the process average

Like scenario 1, scenario 2 starts with a beta process curve with process average $\mu(P) = 0.02$ and process standard deviation $\sigma(P) = 0.01$. At lot 51 the process average is shifted to $\mu(P) = 0.04$ and at lot 51 it is again shifted to $\mu(P) = 0.06$. The simulation ends with lot 150.

Figure 5 corresponds to Figure 1 of scenario 1. The blue curves indicate the process curve, and the points indicate the fractions of nonconforming items in the lots together with the acceptance/rejection decisions (green/red) by Bayes (large) and ISO (small). ISO remains on normal inspection and accepts 145 of the 150 lots whereas Bayes accepts only 90 lots. The 5 lots rejected by ISO are also rejected by Bayes. Concerning the requirement to detect a process average above p_{AQL}, Bayes performs much better than ISO. As in Figure 2 for scenario 1, Figure 6 shows the updated (black) and the true (blue) process curve.

Scenario 3: Drift of the process curve

In this scenario, the beta process curve has a linearly increasing process average (from 0.02 at lot 1 to 0.06 at lot 100) and a linearly increasing standard deviation (from 0.01 at lot 1 to 0.03 at lot 100). Figure 7 corresponds to Figures 1 and 5, and Figure 8 corresponds to Figures 2 and 6. ISO starts to switch to tightened inspection at lot 83. There are 9 rejections by ISO and 24 by Bayes. Bayes reacts much better to the increasing process average than ISO. The estimated process curve follows roughly the actual process curve.

Scenario 4: A process with outlier lots

As in scenario 1, the process curve is a beta distribution with constant process average $\mu(P) = 0.02$ and constant process standard deviation $\sigma(P) =$

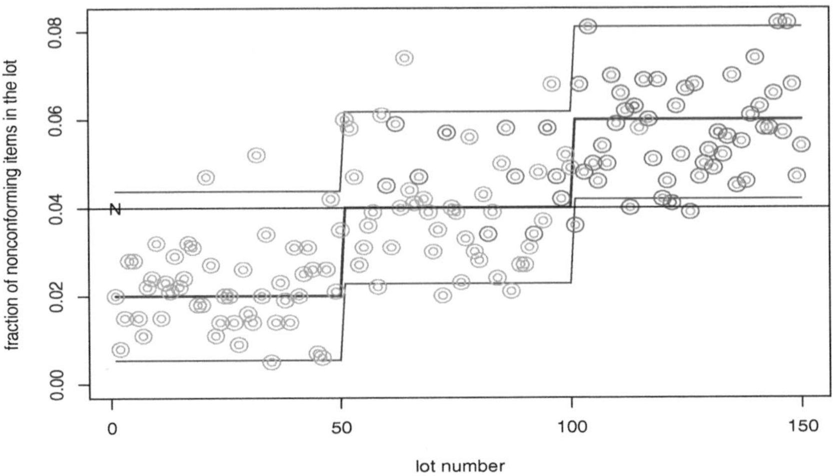

Figure 5. Fractions of nonconforming items in the lots and acceptance decisions of scenario 2; green indicates acceptance, red rejection. Larger circles indicate the Bayes decision, smaller circles the ISO 2859 decision

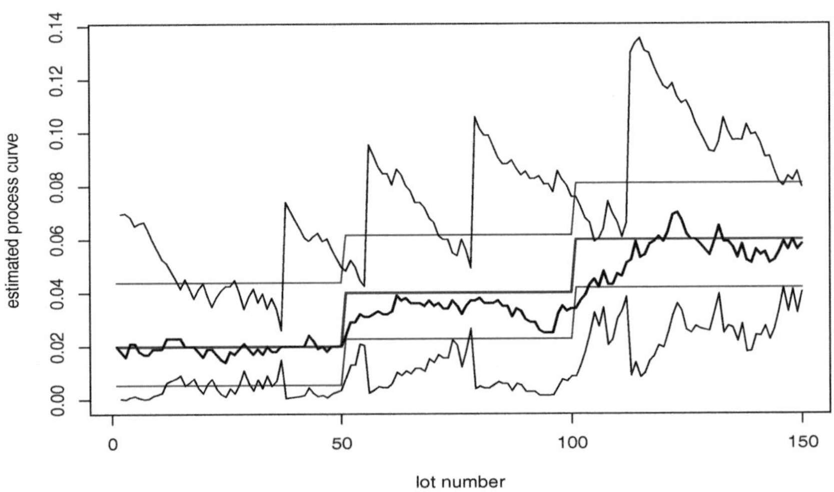

Figure 6. Process average (bold), 2.5% quantile and 97.5% quantile of the true (blue) and the estimated process curve (black) of scenario 2

0.01. However, there is a probability of 0.2 that the lot will be an outlier lot with a fraction of nonconforming items chosen from a uniform distribution between 0.1 and 0.2. Figure 9 corresponds to Figures 1, 5 and 7, and Figure 10 corresponds to Figures 2, 6 and 8. ISO switches to tightened inspection after the fifth lot and discontinues inspection after the 23rd lot. Neglecting the discontinuation of inspection, ISO would switch permanently between normal and tightened inspection. Bayes rejects all outlier lots, ISO all but two. On the other hand, Bayes rejects 6 lots with a fraction nonconforming below p_{AQL}, but ISO does not. Altogether, ISO rejects 17, Bayes rejects 25 lots of the lot series with about 20% outlier lots. Figure 10 shows that, due to some of the outlier lots included in the update of the estimate of the process curve, the estimated process standard deviation and hence, the 95% range of the process curve are overestimated. However, the estimates of the process average are (almost) unbiased.

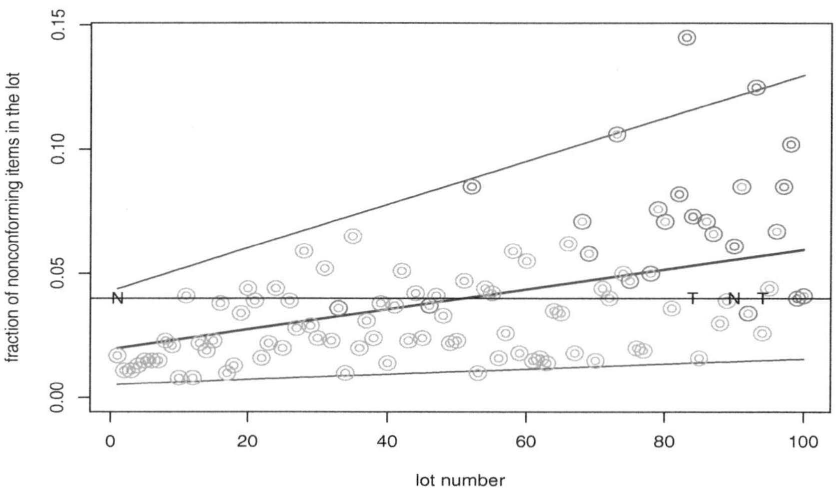

Figure 7. Fractions of nonconforming items in the lots and acceptance decisions of scenario 3; green indicates acceptance, red rejection. Larger circles indicate the Bayes decision, smaller circles the ISO 2859 decision

5 Conclusions

Compared with ISO 2859 the Bayesian sampling inspection has some advantages:

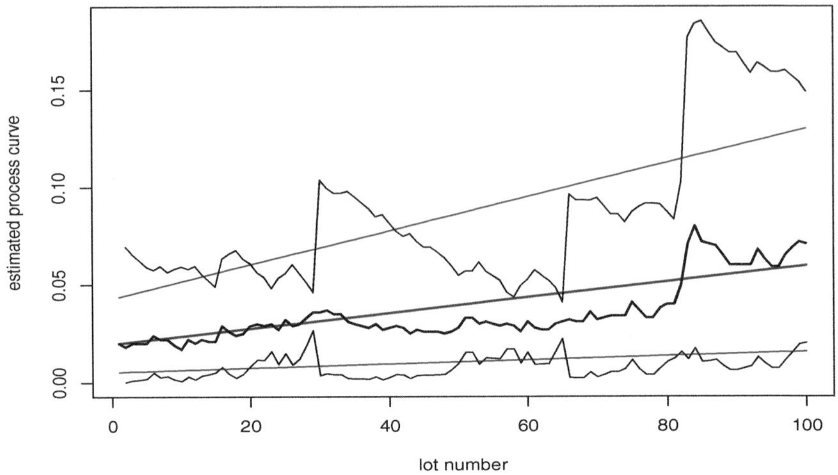

Figure 8. Process average (bold), 2.5% quantile and 97.5% quantile of the true (blue) and the estimated process curve (black) of scenario 3

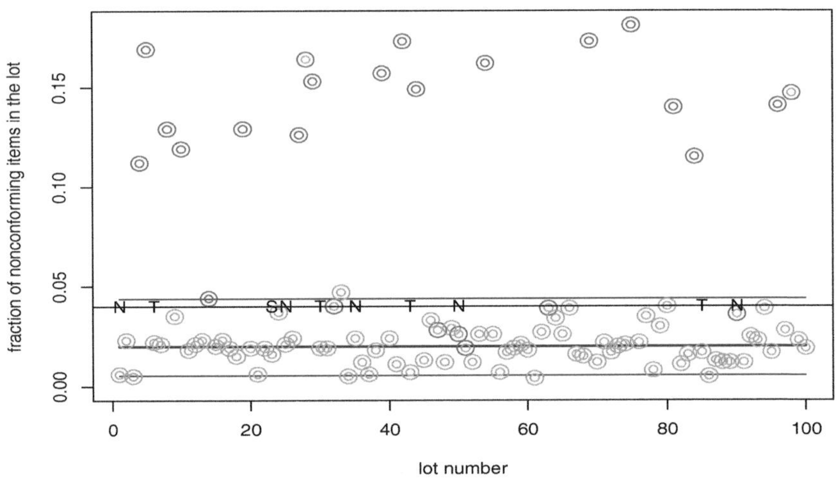

Figure 9. Fractions of nonconforming items in the lots and acceptance decisions of scenario 4; green indicates acceptance, red rejection. Larger circles indicate the Bayes decision, smaller circles the ISO 2859 decision

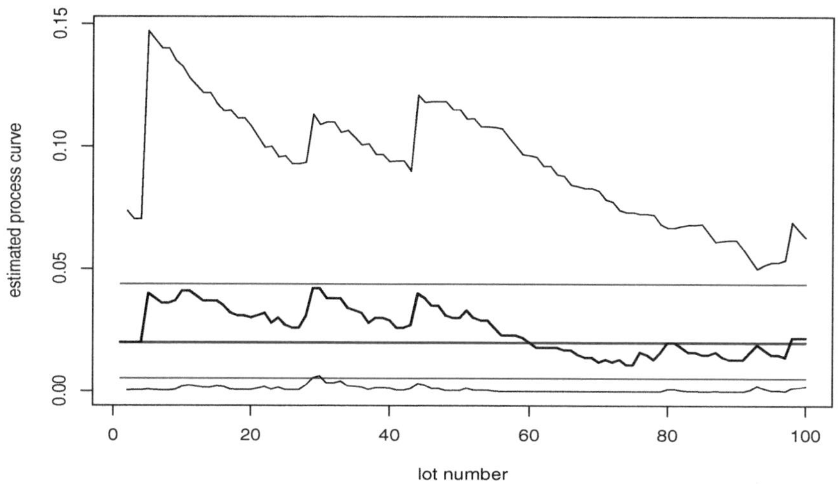

Figure 10. Process average (bold), 2.5% quantile and 97.5% quantile of the true (blue) and the estimated process curve (black) of scenario 4

- The acceptance decision does not end only with the decision result "lot accepted" or "lot rejected" but in addition with the estimated probability of the fraction of nonconforming items in the lot being larger than p_{AQL} (or another established value). Hence, the user gets an idea of the reliability of the decision. In addition, he can establish other border lines for this estimated probability, in order to sort lots into two or more groups with higher or lower probability of fractions nonconforming larger than p_{AQL} being used as different grades of product.
- After having inspected only a few lots, the user will have a reliable knowledge of the process curve, especially its process average. This quality history facilitates steps towards quality improvements. In addition, it can be used for switches to reduced inspection with smaller sample size.

The Bayesian method works irrespective of the sampling plan chosen from ISO 2859 or having been designed as cost or risk optimal in the Bayesian sense.

References

1. Hald, A.: Statistical Theory of Sampling Inspection by Attributes. Institute of Mathematical Statistics of the University of Copenhagen, Part 1 (1976), Part 2 (1978)

2. Rendtel, U., Lenz, H.-J.: Adaptive Bayes'sche Stichprobensysteme für die Gut-Schlecht-Prüfung. Heidelberg: Physica-Verlag 1990
3. ISO 2859–1: Sampling procedures for inspection by attributes – Part 1: Sampling schemes indexed by acceptance quality limit (AQL) for lot-by-lot inspection. Geneva: International Standardisation Organisation 1999
4. Stange, K.: Bayes-Verfahren. Schätz- und Testverfahren bei Berücksichtigung von Vorinformationen. Berlin – Heidelberg – New York: Springer 1977

Part III

Off-line Control

Stochastic Modelling as a Tool for Quality Assessment and Quality Improvement Illustrated by Means of Nuclear Fuel Assemblies

Elart von Collani[1] and Karl Baur[2]

[1] University Würzburg, Sanderring 2, D-97070 Würzburg, Germany
collani@mathematik.uni-wuerzburg.de
[2] Stochastikon GmbH, Schießhausstr. 15, D-97072 Würzburg, Germany
baur@stochastikon.com

Summary. The design of fuel rods is essential for a safe and economic operation of nuclear power plants. Recent considerations have led to the idea of an increase of the average fuel rod burn-up, which would lead to a better efficiency and smaller environmental exposure by reducing the number of fuel assemblies to be disposed of. However, any further increase of the burn-up means an additional stress for the fuel rods, in particular for the cladding tube material, and necessitates a new verification of fuel rod integrity. This is not so easy with the traditional approach and, thus, requires a more realistic description of the processes in the reactor core, enabling more reliable and more accurate predictions of the performance of the fuel rods. The predictions are the foundation of a good fuel rod design and constitute the proof of its integrity.

Instead of the commonly used deterministic models, in this paper a stochastic model is proposed, which would provide new possibilities for showing the fuel rod integrity. The proposed stochastic model removes the well-known weaknesses of the traditional approach and applies directly all existing knowledge and uncovers any remaining ignorance in order to realistically model the existing uncertainty.

1 Introduction

The performance of fuel rods depends on many different factors, which can be classified in certain groups:

- Fuel rod design.

H.-J. Lenz et al. (eds.), *Frontiers in Statistical Quality Control 9*,
DOI 10.1007/978-3-7908-2380-6_14,

- Fuel rod properties and condition.
- Nuclear power plant.
- Operational conditions.

The rod design is well known. The properties and conditions of the produced rods can be described, if a verified stochastic model is available for the production process. Their is plenty of information about the operational conditions for each of the used rods and as well as about the power plant in question.

The problem is to use all the available knowledge for developing a stochastic model reflecting the existing uncertainty, which materializes in performance variability. Subsequently, the stochastic model can be used for making reliable predictions about the performance of fuel rods, under the desired conditions with respect to the considered burn-up.

2 Developing a Stochastic Model

A stochastic model for the operation of fuel rods refers to one or more performance characteristics denoted by X, which are of interest for a safe and efficient operation. The stochastic model describes the future, indeterminate behavior of the fuel rods, which is characterized by a certain variability. Such a description is necessary for quality assessment, quality comparison and quality improvement. Variability is characterized by the range of possible values the random variable X may adopt and the corresponding probabilities.

2.1 Qualitative Properties

Any performance characteristic of a specified fuel rod behaves differently under different operating conditions, for instance with respect to time and amount of irradiation, irradiation history, state of burn-up, etc. Performance characteristics of interest are, for example, the fission gas release, the creep behavior, the fuel temperature, the internal rod pressure, and others. The main difficulty in describing the performance is the fact that under the same conditions the performance can be quite different. Thus, for making reliable and at the same time accurate predictions, one has to describe the variability of the future performance.

For the purpose of illustration, the internal rod pressure is taken here as the characteristic of interest. It is considered for a certain fuel rod design in a specified situation of increased high burn-up. The random variable X stands for the maximum internal pressure in the specified situation. From experience, we can immediately list the following qualitative properties with respect to the future variability of the maximum internal pressure X:

- The maximum internal pressure X cannot fall below a finite lower bound.
- The maximum internal pressure X cannot exceed a finite upper bound.
- For specified initial conditions, the form of the probability mass function of any random variable is either constant, monotonic, or uni-modal. In the case of the maximum internal pressure X, the extreme values will seldom occur under regular operation, while some values in between will occur more frequently. This implies that the first two alternatives can be abandoned with certainty, leaving a uni-modal probability distribution P_X.

From the above consideration we conclude that for developing a quantitative stochastic model for the given situation, a lower and an upper bound and a uni-modal probability mass function for X have to be determined.

2.2 Quantitatively Modelling Uncertainty

A quantitative stochastic model must be developed based only on the available knowledge, as otherwise no valid statement about the uncertainty on hand can be made. If there is no complete knowledge, an approximation must be used which covers the existing uncertainty. Uncertainty about X is described by the probability distribution P_X and its magnitude can be measured by the corresponding entropy denoted by $H(P_X)$. An approximation covers the true probability distribution, if both agree in all important properties and, additionally, the entropy of the approximation is not smaller than that of the true probability distribution.

Consequently, the following items have to be determined based on available observations:

- A lower bound ℓ for the range of variability must be determined, which meets the requirement that it does not exceed the true, but unknown lower bound of X.
- An upper bound u for the range of variability must be determined, which meets the requirement that it is not smaller than the true, but unknown upper bound of X.

- A probability distribution $\hat{P}_X$ for X must be determined, which has the same relevant properties concerning uncertainty as the true probability distribution P_X, and has an entropy which is not smaller than that of the true true probability distribution. Let F_i, $i = 1, \ldots, r$, denote the relevant properties of P_X. Then, the two requirements are represented by the following two conditions:

$$F_i[\hat{P}_X] = F_i[P_X] \quad \text{for } i = 1, \ldots, r \tag{1}$$

$$H\left[\hat{P}_X\right] \geq H\left[P_X\right] \tag{2}$$

The requirement (1) means that the approximation and the true distribution have the same relevant properties, while not meeting (2) would mean underestimation of the magnitude of the existing uncertainty and, thus, would be tantamount to assuming more knowledge about the probability distribution P_X than what is actually available.

Determination of the above specified quantities requires a certain amount of quantitative information. This is represented by a sample $(X_1, \ldots, X_n)$ of X, where the sample elements are independent copies of X.

2.3 Relevant Features of P_X

As a first step the uni-modality of P_X must be utilized. The relevant properties of a uni-modal probability distribution for a given range of variability of X are represented by its first two moments (for details see [1]). Less information about P_X would lead to an inappropriate probability distribution; more information would generally reduce the uncertainty represented by the probability distribution only marginally. Therefore, we conclude that for meeting condition (1), the values of the first moment $E[X]$ and of the variance $V[X]$ of X or P_X are needed.

In order to be certain that condition (2) is satisfied, the probability distribution must be selected that has maximum entropy for the given range of variability of the X values and for the given first two moments of X. In [1] it is shown that the corresponding maximum entropy distribution is uniquely determined and uni-modal.

The resulting probability distribution covers the existing uncertainty in the sense of (1) and (2). Note that in a given situation the relevant properties do not refer to the effects of single factors, like temperature or pressure, but to the properties of the probability distribution, which are the outcome of all known and unknown factors.

Thus, the approach proposed here does not depend on any knowledge about the effects of single factors, but describes the joint effect of all involved factors.

Stochastic Measurement Procedure for the First Moment $E[X]$

For a uni-modal probability distribution, the probability distribution of the sample mean may be approximated for sufficiently large sample size n by the normal distribution, which has the important advantage of being independent of the unknown range of variability of X. Assuming that n is large enough, we can use the well-known procedure based on the t-distribution with n-1 degrees of freedom for measuring[3] the unknown value of $E[X]$.

The measurement procedure for determining the value of the first moment $E[X]$ of X denoted $C^{(\beta)}_{E[X]}$ is based on the sample mean and the sample variance:

$$\overline{X} = \frac{1}{125}\sum_{i=1}^{125} X_i \tag{3}$$

$$S^2 = \frac{1}{124}\sum_{i=1}^{125} (X_i - \overline{X})^2 \tag{4}$$

The measurement procedure $C^{(\beta)}_{E[X]}$ assigns to each possible observed event $\{(\bar{x}, s^2)\}$ a measurement interval for $E[X]$, which contains all those values μ, which are consistent with $\bar{x}$ and s^2. The upper index β specifies the reliability level of the measurement procedure, which constitutes a reliability requirement. Applying a measurement procedure with confidence level β means that a correct result will be obtained with a probability of at least β. In other words the risk of getting a wrong result when applying $C^{(\beta)}_{E[X]}$ is less than $1 - \beta$.

The measurement intervals of $C^{(\beta)}_{E[X]}$ are given as follows:

$$C^{(\beta)}_{E[X]}\left(\{(\bar{x}, s^2)\}\right) = \left\{\mu \,\middle|\, \bar{x} - t^{(n-1)}_{\frac{1+\beta}{2}}\frac{s}{\sqrt{n}} \le \mu \le \bar{x} + t^{(n-1)}_{\frac{1+\beta}{2}}\frac{s}{\sqrt{n}}\right\} \tag{5}$$

[3]Note that in stochastics two classes of procedures are distinguished. The first one is known as the prediction class, as it refers to the indeterminate future, while the second one is called measurement class, as it refers to the determinate past.

Stochastic Measurement Procedure for the Variance of $V[X]$

Similar as in the case of $E[X]$, there is a powerful measurement procedure in traditional statistics based on the sample variance and the χ^2-distribution. This well-known procedure is denoted by $C^{(\beta)}_{V[X]}$ here. Because only an upper bound for the unknown value of $V[X]$ is needed, the measurement intervals of $C^{(\beta)}_{V[X]}$ assigns to each possible observed event $\{s^2\}$ of S^2 the following measurement interval:

$$C^{(\beta)}_{V[X]}\left(\{s^2\}\right) = \left\{\sigma^2 \,\middle|\, 0 < \sigma^2 \leq \frac{n-1}{\chi^{2\,(n-1)}_{\;1-\beta}} s^2\right\} \tag{6}$$

Applying the measurement procedure $C^{(\beta)}_{V[X]}$ yields a correct result with a probability of at least β.

2.4 Range of Variability of X

The range of variability of X contains all those values that X may adopt. In general, the range of variability depends on the actual value μ of the first moment and the value σ^2 of the variance of X. Therefore, a stochastic measurement procedure is needed for determining the lower and upper bound of the range of variability of X for given values μ and σ^2.

In [5] to [8] a measurement procedure denoted by $C^{(\beta)}_{L,U}$ for the bounds L and U of X is developed. The procedure is based on the sample functions

$$\min(X_1, \ldots, X_n)$$
$$\max(X_1, \ldots, X_n)$$

and assigns for given (μ, σ^2) to each observed event $\{(x_{min}, x_{max})\}$ a measurement set for the unknown lower and upper bounds of X. Again the value β specifies the procedure's reliability level, which guarantees that the procedure will yield a correct result with a probability of at least β. A result is correct if the measurement set contains the true but unknown bounds of X.

Note that, in contrast to stochastics, there are no methods in statistics for determining the bounds of variability of a random variable. Instead, in statistics the range of variability is generally assumed to be unbounded.

2.5 The Probability Distribution $\hat{P}_{X|\{(\mu,\sigma^2)\}}$

Applying the above outlined measurement procedures yields the following knowledge about the true probability distribution P_X of X, where β_1, β_2 and β_3 specify the corresponding reliability levels:

- A set $C_{E[X],V[X]} = C^{(\beta_1}_{E[X]}(\{(\bar{x}, s^2)\}) \times C^{(\beta_2)}_{V[X]}(\{s^2\})$ containing all those values of the first two moments of X that are consistent with the sample observations $(\bar{x}, s^2)$:

$$C_{E[X],V[X]}(\{(\bar{x}, s^2)\}) = \{(\mu, \sigma^2) \mid \mu_\ell \leq \mu \leq \mu_u, \sigma^2 \leq \sigma_u\} \quad (7)$$

- For each possible element (μ, σ^2) a set

$$C_{L,U|\{(\mu,\sigma^2)\}}(\{(x_{min}, x_{max})\}) = C^{(\beta_3)}_{L,U|\{(\mu,\sigma^2)\}}(\{(x_{min}, x_{max})\})$$

containing all those values for the bounds that are consistent with the sample observations (x_{min}, x_{max}) for given $(\mu, \sigma^2) \in C_{E[X],V[X]}$. Let $\ell(\mu, \sigma^2)$ be the smallest lower bound and $u(\mu, \sigma^2)$ the largest upper bound being consistent with (x_{min}, x_{max}), then the measurement result for given (μ, σ^2) takes the following form:

$$C_{L,U|\{(\mu,\sigma^2)\}}(\{(x_{min},x_{max})\})=\{(\ell, u) \mid \ell(\mu,\sigma^2)\leq\ell,u\leq u(\mu, \sigma^2)\} \quad (8)$$

Each element of (7) describes one possible situation and, therefore, has to be taken in account when making statements about the future development. As shown in [1], each of these values together with corresponding values of the bounds $(\ell(\mu, \sigma^2), u(\mu, \sigma^2))$ of X given as elements of $C_{L,U|\{(\mu,\sigma^2)\}}$ determines uniquely a probability distribution $\hat{P}_{X|\{(\mu,\sigma^2)\}}$ which has maximum entropy of all probability distribution of random variables having a range of variability given by $\{x \mid \ell(\mu, \sigma^2) \leq x \leq u(\mu, \sigma^2)\}$ and having a first moment $E[X]$ with value μ and a variance $V[X]$ with value σ^2.

The corresponding maximum entropy probability distributions are determined by the values of three distribution parameters

$$\lambda_0(\mu, \sigma^2), \lambda_1(\mu, \sigma^2), \lambda_2(\mu, \sigma^2) \quad (9)$$

with $\lambda_1(\mu, \sigma^2), \lambda_2(\mu, \sigma^2)$ being the solution of the following two equations:

$$\int_{\ell}^{u} x e^{\lambda_0(\mu,\sigma^2)+\lambda_1(\mu,\sigma^2)x+\lambda_2(\mu,\sigma^2)x^2} dx = \mu$$
$$\int_{\ell}^{u} (x-\mu)^2 e^{\lambda_0(\mu,\sigma^2)+\lambda_1(\mu,\sigma^2)x+\lambda_2(\mu,\sigma^2)x^2} dx = \sigma^2 \tag{10}$$

and $\lambda_0(\mu, \sigma^2)$ is given by:

$$\lambda_0(\mu,\sigma^2) = -\ln\left(\int_{\ell}^{u} e^{\lambda_1(\mu,\sigma^2)x+\lambda_2(\mu,\sigma^2)x^2} dx\right) \tag{11}$$

For each $(\mu, \sigma^2) \in C_{E[X],E[V]}$ and each $(\ell, u) \in C_{L,U|\{(\mu,\sigma^2)\}}$ the above system of equations has to be solved numerically. The solutions transform the elements of $C^{(0.95)}_{E[X],V[X]}$ and $C_{L,U|\{(\mu,\sigma^2)\}}$ onto a set of probability distributions which cover the true, but unknown, probability distribution of the maximum internal pressure X.

3 The Stochastic Model

In [1] the *Bernoulli-Space* is introduced as the general stochastic model. Any *Bernoulli-Space* refers to a pair of variables (X, D), where the first one represents the future aspect of interest, while the second one represents the initial conditions. In our case, the aspect of interest X is the maximum internal pressure for a specified situation. The initial conditions $D = (E[X], V[X])$ are given by the values of the first two moments. The true values are unknown and, therefore, all those values which cannot be excluded constitute the set of possible situations with respect to the internal pressure.

The *Bernoulli-Space* is denoted $\mathbb{B}_{X,(E[X],V[X])}$ and contains three components:

- The first one is called the *ignorance space*. It reflects the existing ignorance about the relevant initial conditions represented here by $(E[X], V[X])$. The true value of $(E[X], V[X])$ is unknown except for the measurement result:

$$C_{E[X],V[X]} \tag{12}$$

  which, therefore, constitutes the *ignorance space* with respect to the initial conditions.

- The second component, denoted $\mathcal{X}$, is called the *variability function*. It maps subsets of the *ignorance space* onto the corresponding range of variability of X. In the example, the variability function is given for $(\mu, \sigma^2) \in C_{E[X],V[X]}$ as follows:

$$\mathcal{X}(\{(\mu, \sigma^2)\}) = \{x \mid \ell(\mu, \sigma^2) \leq x \leq u(\mu, \sigma^2)\} \tag{13}$$

- The third component of a *Bernoulli-Space* is called the *random structure function*, denoted $\mathcal{P}$. It maps subsets of the *ignorance space* on to the corresponding probability distribution. In our case, we obtain for the singletons:

$$\mathcal{P}(\{(\mu, \sigma^2)\}) = P_{X|\{(\mu,\sigma^2)\}} \quad \text{for } \{(\mu, \sigma^2)\} \in C_{[X],V[X]} \tag{14}$$

where $P_{X|(\mu,\sigma^2)}$ represents an approximation meeting the conditions (1) and (2).

With (12), (13) and (14), we obtain the following *Bernoulli-Space* for the maximum internal pressure:

$$\mathbb{B}_{X,(E[X],V[X])} = \left(C_{E[X],V[X]}, \mathcal{X}, \mathcal{P}\right) \tag{15}$$

A *Bernoulli-Space* (15) comprises a stochastic model that utilizes the entire available knowledge, but which is not based on any unfounded assumption. A *Bernoulli-Space* contains a set of probability distributions that cover the true but unknown probability distribution in the sense of (1) and (2).

Any *Bernoulli-Space* describes explicitly the existing ignorance and the prevailing randomness and is a necessary basis for making reliable and accurate predictions about the future performance of X, which is here the the maximum internal pressure.

4 Illustrative Example

For convenience, the reliability specification for the three measurement procedures is set to be equal, i.e., $\beta = \beta_1 = \beta_2 = \beta_3 = 0.95$. A sample[4] of size $n = 125$ is used for the measurement procedures. The sample experiment is performed and yields the result given in Table 1:

[4]Note that the data are fictitious and only used for illustrating the process of developing a stochastic model. Real data for such an investigation could, however, be taken from the Halden Reactor Project.

n	$\bar{x}$	s^2	$\min x_i$	$\max x_i$
125	27.05	4.36	24.90	30.15

Table 1: Numerical sample results for the maximum internal pressure.

Measuring the First Moment

For a reliability specification of $\beta = 0.95$, the measurement procedure $C_{E[X]}^{(0.95)}$ yields the following measurement interval for the unknown value of $E[X]$:

$$C_{E[X]}^{(0.95)}(\{(27.05, 4.36)\}) = \{\mu \,|\, 26.99 \leq \mu \leq 27.11\} \tag{16}$$

After the measurement has been performed, the measurement interval $C_{E[X]}^{(0.95)}(\{(27.05, 4.36)\})$ represents the available knowledge about the true value of $E[X]$.

Measuring the Variance

Again, the reliability specification is set to be $\beta = 0.95$. The measurement procedure $C_{V[X]}^{(0.95)}$ yields the following measurement interval, which, analogously to the case of the first moment, represents the knowledge about the true value of the variance:

$$C_{V[X]}^{(0.95)}(\{4.36\}) = \{\sigma^2 \,|\, 3.51 \leq \sigma^2 \leq 5.55\} \tag{17}$$

Measuring the Lower and an Upper Bounds

The measurement procedure for the bounds is based on the lower and upper quantiles of the sample functions $\min(X_1, \ldots, X_n)|\{(\mu, \sigma^2)\}$ and $\max(X_1, \ldots, X_n)|\{(\mu, \sigma^2)\}$. For the reliability level $\beta = 0.95$ and a given pair (μ, σ^2), the measurement procedure $C_{(L,U)|\{(\mu,\sigma^2)\}}^{(0.95)}$ yields a set of possible values for the bounds. For example, this set for given (μ, σ^2) and observed sample event $\{(24.9, 30.15)\}$ could be as follows:

$$\begin{aligned} C_{L,U|\{(\mu,\sigma^2)\}}^{(0.95)}(\{(24.90, 30.15)\}) &= \\ \{(\ell, u) \,|\, 22.83 \leq \ell \leq 24.90, 30.15 \leq u \leq 33.78\} \end{aligned} \tag{18}$$

Each value of $C_{L,U|\{(\mu,\sigma^2)\}}^{(0.95)}(\{(24.9, 30.15)\})$ represents potentially possible bounds of the range of variability of X for the case (μ, σ^2).

Any statement about the maximum internal pressure must be valid for each of the possible situations given by the values of the first two moments and the corresponding values of the bounds of X.

Determining the Probability Distributions

By solving the system of equation (10) for each of the possible pairs (μ, σ^2) and the corresponding values of the bounds, we obtain a set of uniquely defined maximum entropy distributions, where each element has the specified values of the first two moments and the specified range of variability:

$$\left\{ P_{X|\{\mu,\sigma^2)\}} \middle| (\mu, \sigma^2) \in C_{E[X],V[X]} \right\} \tag{19}$$

In Figure 1 the *Bernoulli-Space* for the maximum internal pressure X is illustrated by the density functions of some of the probability distribution obtained as images of the random structure function. The flat density functions are those with a large variance, while the more peaked ones have a smaller variance.

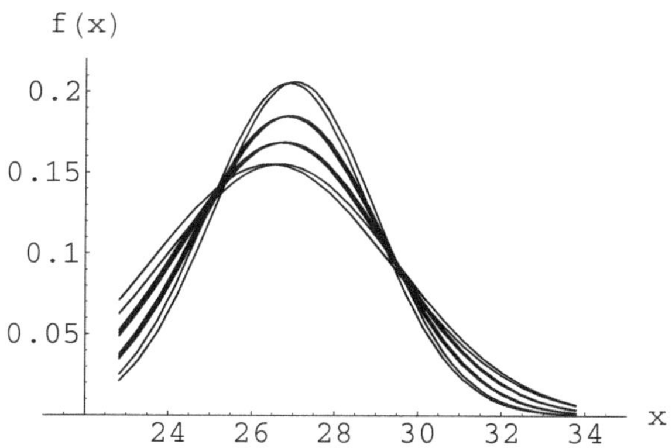

Figure 1: Density function of some of the potential probability distribution of the maximum internal pressure X.

5 Stochastic Prediction Procedures

A stochastic prediction procedure is based on a *Bernoulli-Space*. A prediction procedure is a set function that maps subsets of the *ignorance space*, which represent different levels of ignorance, on to subsets of the corresponding range of variability of the considered random variable X, where the latter subsets represent the set of possible predictions.

A prediction procedure is denoted by $A_X^{(\beta)}$, where the reliability level β constitutes a specification for the procedure. If the specification is met, the procedure produces predictions that will occur with a probability of at least β.

In the example considered here, let the system of subsets of the *ignorance space* be given by the singletons:

$$\mathcal{T}_{(E[X],V[X])}\left(C_{E[X],V[X]}\right) = \left\{\{(\mu,\sigma^2)\} \,\middle|\, (\mu,\sigma^2) \in C_{E[X],V[X]}\right\} \tag{20}$$

The possible ranges of variability of X for specified (μ,σ^2) are given by the variability function $\mathcal{X}(\{(\mu,\sigma^2)\})$, while the corresponding probability distributions are obtained by means of the random structure function $\mathcal{P}$.

$$A_X^{(\beta)} : \mathcal{T}_{(E[X],V[X])}\left(C_{E[X],V[X]}\right) \to \mathcal{T}_X\left(\mathcal{X}\left(C_{E[X],V[X]}\right)\right) \tag{21}$$

where $\mathcal{T}_X$ is an appropriate system of subsets of the overall range of variability of X, where each element of $\mathcal{T}_X$ represents a possible prediction.

Let $C_0 \in \mathcal{T}_{(E[X],V[X])}\left(C_{E[X],V[X]}\right)$ be a level of ignorance, and $A_X^{(\beta)}(C_0)$ the corresponding predicted event. Then

$$P_{X|\{(\mu,\sigma^2)\}}\left(A_X^{(\beta)}(C_0)\right) \geq \beta \quad \text{for } (\mu,\sigma^2) \in C_0 \tag{22}$$

i.e., the predicted event under the condition of C_0 occurs for any $(\mu,\sigma^2) \in C_0$ with a probability that is at least β.

Of course, a prediction procedure $A_X^{(\beta)}$ should be derived in such a way that the predictions are most accurate, where the accuracy is measured by the geometrical size of the predicted event. Therefore, an optimal prediction procedure ${}^*A_X^{(\beta)}$ produces predictions, which have minimum size:

$$\left|{}^*A_X^{(\beta)}(C_0)\right| = \min \tag{23}$$

Illustrative Example

In the case considered here, an upper bound for the maximum internal pressure in the given situation is sought. Therefore, as a first step, the worst case given by the first moment $E[X]$ and the variance $V[X]$ with respect to the maximum internal pressure X is identified.

The worst case refers to the maximum possible value of $E[X]$ and the maximum possible value of $V[X]$. If the true values are given by $\mu_{max} = 27.11$ and $\sigma^2_{max} = 5.55$, then the corresponding prediction will contain the largest values compared with all other possible values of $E[X]$ and $V[X]$.

As only an upper bound for the maximum internal pressure is of interest, the following system $\mathcal{T}_X$ of one-sided predictions is selected:

$$\mathcal{T}_X\left(\mathcal{X}\left(C_{E[X],V[X]}\right)\right) = \left\{\{x \,|\, x \le b\} \,\middle|\, \ell < b \le u\right\} \tag{24}$$

The corresponding values of the distribution parameters for the worst case $\mu_{max} = 27.11$ and $\sigma^2_{max} = 5.55$ and the corresponding bounds are given by

$$\begin{aligned} \lambda_0(27.11, 5.55) &= -45.8746113827101 \\ \lambda_1(27.11, 5.55) &= 3.301795856164792 \\ \lambda_2(27.11, 5.55) &= -0.06192663171432966 \end{aligned} \tag{25}$$

The density function of the resulting probability distribution with respect to the internal pressure $X|\{(27.11, 5.55)\}$ is displayed in Figure 2 below:

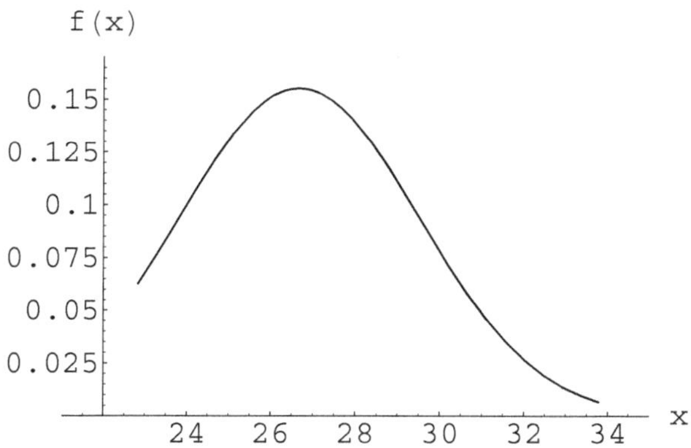

Figure 2: Density function of the worst-case probability distribution of the maximum internal pressure X.

Denoting the reliability specification for the prediction procedure by β, then the optimal prediction procedure yields the following prediction:

$$A_X^{(\beta)}(\{(27.11, 5.55)\}) = \{x \mid x \leq Q_{X|\{(27.11,5.55)\}}(\beta)\} \qquad (26)$$

where $Q_{X|\{(27.11,5.55)\}}$ is the upper quantile function of the random variable $X|\{(27.11, 5.55)\}$.

The prediction (26) has the maximum bound among all predictions based on any of the probability distributions contained in the *Bernoulli-Space*. Thus, it actually represents the worst possible case, i.e., it contains or covers each of the possible cases with respect to the initial conditions.

For $\beta = 0.995$ we obtain the following prediction:

$$A_X^{(0.995)}(\{(27.11, 5.55)\}) = \{x \mid x \leq 33.204\} \qquad (27)$$

The predicted event will occur with a probability that is at least 0.995. Thus, it is guaranteed (with a known risk not larger then 0.005) that the bound 33.204 will not be exceeded.

6 Conclusions

Particularly in areas where safety is involved, quality assessment and quality improvement are important issues. Therefore, adequate means for measuring quality should be applied in order to arrive at reliable and at the same time accurate results.

Clearly, an adequate description of the uncertainty involved cannot be given by the traditional way in physics, which assumes deterministic phenomena and cannot model the inherent variability of real processes. More details of how to consider and model uncertainty with respect to the design of fuel rods are given in [3] and [4], in which the problem of fuel rod design and the technical details of an appropriate mathematical modelling of the operational behavior of the fuel rods are elaborated.

In this paper, a stochastic model – the *Bernoulli-Space* – is used for describing uncertainty, without the necessity of postulating a deterministic world as in physics or of making assumptions which violate reality as, for example, in the statistical approaches. The *Bernoulli-Space* allows one to guarantee the predicted performance of fuel rods for a given design and, thus, it can be used to investigate a safe increase of the burn-up of nuclear fuel.

References

1. v. Collani E (2004) Theoretical Stochastics. In: v. Collani E (ed) Defining the Science of Stochastics. Heldermann Verlag, Lemgo, 147-174.
2. v. Collani E (2004) Empirical Stochastics. In: v. Collani E (ed) Defining the Science of Stochastics. Heldermann Verlag, Lemgo, 175-213.
3. v. Collani E, Baur K (2004) Fuel rod design and modeling of fuel rods - Part 1 (In German). Kerntechnik 69: 253-260.
4. v. Collani E, Baur K (2005) Fuel rod design and modeling of fuel rods - Part 2 (In German). Kerntechnik 70: 158-166.
5. Schelz F, Sans W, v. Collani E (2007) Improving the Variability Function in Case of a Uni-Modal Probability Distribution. Economic Quality Control 22: 19-39.
6. Sans W, Stübner D, Sen S, v. Collani E (2005) Improving the Variability Function in Case of a Uniform Distribution. iapqr transactions 30: 1-12.
7. Sans W, Zhai X, Stübner D, Sen S, v. Collani E (2005) Improving the Variability Function in Case of a Monotonic Probability Distribution. Economic Quality Control 20: 121-142.
8. Stübner, D., Sans, W., Zhai, Xiaomin and v. Collani, E. (2004): Measurement Procedures for Improving the Variability Function. *Economic Quality Control* 19, 215-228.

Hierarchical Modeling for Monitoring Defects

Christina M. Mastrangelo[1], Naveen Kumar[2], and David Forrest [3]

[1]University of Washington, Box 352650, Seattle, WA 98195-2650,
mastr@u.washington.edu
[2]Intel Hillsboro, OR
nkchawla@yahoo.com
[3]Virginia Institute of Marine Science, Gloucester Point, VA
drf5n@maplepark.com

Summary. In semiconductor manufacturing, discovering the processes that are attributable to defect rates is a lengthy and expensive procedure. This paper proposes a approach for understanding the impact of process variables on defect rates. By using a process-based hierarchical model, we can relate sub-process manufacturing data to layer-specific defect rates. This paper demonstrates a hierarchical modeling method using process data drawn from the Gate Contact layer, Metal 1 layer, and Electrical Test data to produce estimates of defect rates. A benefit of the hierarchical approach is that the parameters of the high-level model may be interpreted as the relative contributions of the sub-models to the overall yield. Additionally, the output from the sub-models may be monitored with a control chart that is 'oriented' toward yield.

1 Introduction

Manufacturing systems can produce such large quantities of data that an understanding of the interrelations between subsystems is difficult. Yet, the large quantities of data from these processes may be a valuable resource if information can be extracted from it. High-dimensional complex systems, such as the semiconductor manufacturing process, is an example of this and present a number of unique challenges in data analysis for understanding and prediction. In order to understand and improve chip production in semiconductor manufacturing, the manufacturing and process data already recorded during processing may be used to more fully understand the system and provide avenues for improvement.

Figure 1 summarizes this scenario. Manufacturing systems have embraced statistical process control to understand and monitor problem areas in manufacturing. Traditionally, these are univariate, and maybe even multivariate, and bound to a specific process to monitor key processing and electrical parameters (Montgomery 2005). These methods, however, do not readily enable a process engineer to analytically answer questions such as "what characteristics are present that result in a 'good lot'?" 'Good lot' is typically defined in terms of high yield. The dashed lines in Figure 1 represent the relationships required to address

H.-J. Lenz et al. (eds.), *Frontiers in Statistical Quality Control 9*,
DOI 10.1007/978-3-7908-2380-6_15, © Springer-Verlag Berlin Heidelberg 2010

this question. These questions go beyond the process level; they need to be addressed at a higher or operational level.

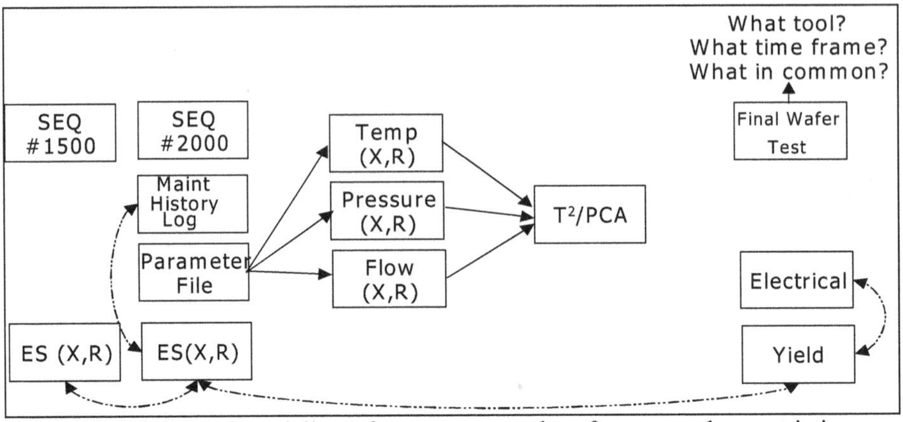

Figure 1. Conceptual modeling of parameters and performance characteristics.

The goal of this work is to understand the influences and contributions of lower-level processes (i.e. sub-processes) on *performance characteristics*, such as yield, conformance to specifications, defectivity, customer returns, or other quality measures. Characterizing the relationships at the operational level will also improve productivity and enable the evaluation of competing process models. In order to examine the relationships between sub-processes and performance characteristics, a hierarchical model of sub-process outputs can provide support for decision making in allocating resources between sub-processes.

This paper demonstrates a hierarchical modeling method using process data to produce estimates of defect rates. In addition, it demonstrates how control charts may be used to monitor sub-process outputs where these outputs have been oriented towards a performance characteristic such as yield and compares these to multivariate control (or univariate) charts on the sub-processes. The next section discusses the challenges of modeling at the operational level and summarizes the structure of hierarchical models. An application of this approach with data drawn from the Gate Contact layer, Metal 1 layer, and Electrical Test and Probe is given in the third section.

2 Background

Yield modeling is a key goal of semiconductor manufacturers, and improvements in yield and yield prediction provide direct financial gain to manufacturers. Yield models depend primarily on the combination of defect rates from the various sub-processes to predict yield rates. Unfortunately, the defect rate information attributed to specific sub-processes is expensive and slow to collect (Van Zant 2004; Horton 1998; Nurani et al. 1998; and Hess and Weiland 1999).

In addition to yield modeling, there has been work in modeling and classifying defect patterns after final test to more fully understand low yield (Cunningham and MacKinnon 1998; Hansen et al., 1997; Friedman et al., 1997). While the previous papers analyze yield and defect patterns, Skinner et al. (2002) use multivariate statistical methods to analyze relationships, at the wafer level, between the parameters of a process and yield to potentially determine the cause of low-yield wafers. However, the process is the electrical test, or probe operations, that occur *after* fabrication. This work links yield estimates to process variables much earlier in production. See Kumar et al. (2005) for a review of empirical modeling methods in semiconductor manufacturing.

Data stored in process, engineering and metrology databases detail the manufacturing processes involved in the production process. Analysis of the data is challenging because of the large number of variables, interactions between subprocesses and relatively small number of observations. For example, a memory device involving 22 layers of semiconductor can involve 300 processing steps over 3 months with 21710 process variables. Figure 2 shows a sample of 90 days of lot-level production data for one product. Note the misalignments between separate data tables and that the number of tractable lots, or observations (n=221), may be much less than the number of variables (p=21710).

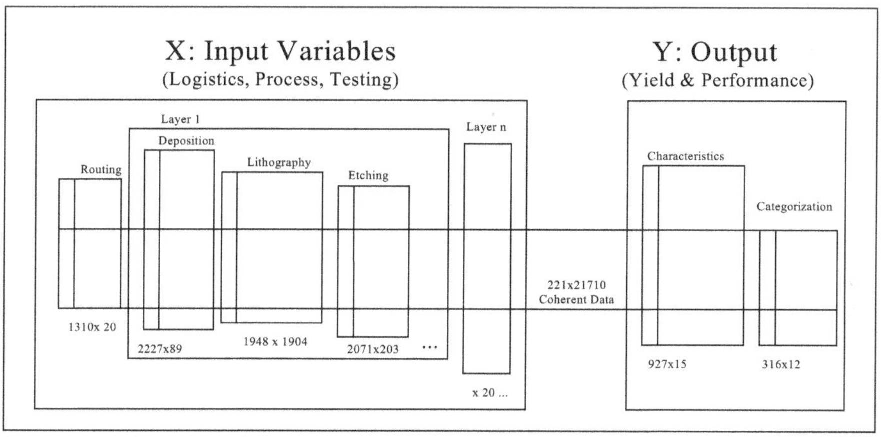

Figure 2. Schematic of processing and quality data, X, and testing and yield data, Y. Note that the number of consistent and completed lots is less than the number of variables, or $n<<p$.

The domain has a very high dimensionality compared to the number of observations. Many tools exist for dealing with high dimensional data; however, these methods assume a well-formed data matrix (Johnson and Wichern 2004; Hastie et al. 2001). Combining heterogeneous data sets produces increased problems with missing data and dimensionality. Forming a larger data matrix requires joins of datasets, and the large number of variables exacerbates missing data problems. Consider, for example, Metal 1 layer production data. Suppose that only 7% of the data in each set was missing at random in each of 10 sub-processes. The resulting table join would have 52% of the rows with missing data. Established analysis techniques, such as regression, clustering, and principal component analysis, perform a case-wise delete of data with missing elements. Thus, combining datasets with missing data results in larger missing data challenges and missed modeling opportunities.

In addition, the data may be inconsistent for several reasons: frequent process changes which produce a dynamic process structure; missing data due to structure changes (e.g. using a different machine may produce a different set of process variables); recording and transmission errors resulting from manual entry that erroneously report values against improper lots. Individually, these data characteristics are well-studied. However, these issues combine making the problem difficult to solve. Since complex systems do not often generate data as a simple data matrix, but instead as a set of distinct subsystem-specific datasets, common methods such as standard regression, feature selection, and feature extraction are not easily applied without significant preprocessing. Changes in the structure of the system, due to equipment or policy changes, makes construction of a full rank data matrix impossible.

3 Hierarchical Modeling

Analyzing high dimensional, low sample size data by using a hierarchical model to combine sub-process variables utilizes the many variables to produce meaningful intermediate variables. Figure 3 shows a two-level model comprised of several sub-models each estimating and intermediate variable (e.g. an electrical test parameter), and the high-level model which combines the estimates from the subsystems to produce an overall estimate of the system.

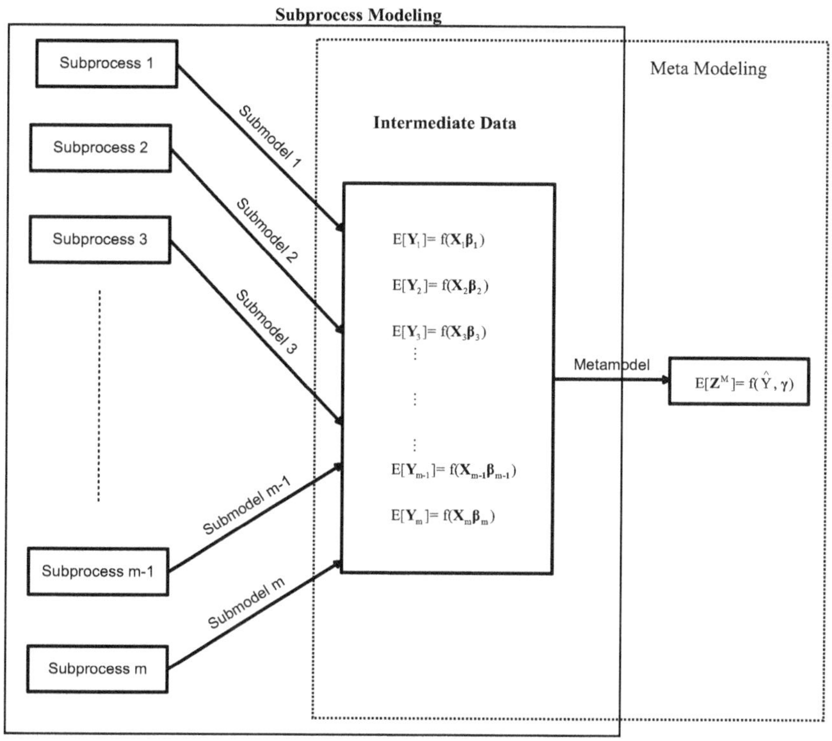

Figure 3: A schematic representation of a two-level model.

Using intermediate data (i.e. the output from the sub-models) provides two important benefits: it can summarize a large number of input variables into a smaller, more manageable set, and it can also provide a mechanism for interpreting the interaction between subsystems. Correlations between the intermediate models represent correlations between the sub-processes. The weights of the standardized intermediate data represent the significance of the sub-processes to the variable of interest in the high-level system. By using the "natural" divisions in the process, data collection and storage system, the entire system can be broken down into sub-processes, each examined and modeled separately, and recombined in a hierarchy of interacting models.

Equally contributing sub-models will individually will have low predictive power, but taken together, will have stronger predictive power. Suppose that the estimate of the defect rate due to some sub-model, m: $\hat{Y}_m = f(X_m b)$ where X_m is a vector of variables, b is a vector of estimated parameters, and $f(\,)$ is an estimation model. The model, $\hat{Y}_m$, will be able to predict the system output with some level of accuracy. Supposing further that $m = 10$ independent subsystems and that each is responsible for $1/20^{th}$ of the variance observed in the output, with the remaining ½ of the variance due to other sources. Each of the 10 models can explain at most 5% of variation. This may seem insignificant, but a combination of the subsystems has toe potential to explain ½ of the variance of the entire system.

The three elements to hierarchical modeling are the identification of sub-processes, the sub-process and meta- models, and the outputs (intermediate and system data). In developing the sub-process models shown in Figure 2, consider m sub-processes separately each modeling a response variable, Y. Taken together, $Y_m = X_m \beta_m + \varepsilon_m$, where Y_m represents vector of response variables associated with m sub-processes, ε_m is a vector representing the error terms, and X_m is a matrix of regressors representing a set of sub-process input variables. The m sub-models are developed in the sub-process modeling stage to provide an estimate of intermediate data.

The predicted values from the sub-process models are referred to as intermediate data. The intermediate data becomes the inputs in the meta-modeling stage to estimate the system output. In other words, the predicted response from each of the sub-models is used to estimate the system output:

$$Z = \hat{Y}_1\gamma_1 + \hat{Y}_2\gamma_2 + \ldots + \hat{Y}_m\gamma_m + \varepsilon_{mm}$$

$$E[Z] = \hat{Y}_1\gamma_1 + \hat{Y}_2\gamma_2 + \ldots + \hat{Y}_m\gamma_m$$

$$E[Z] = \hat{Y}\gamma$$

where vector Z is the meta-model response variable, ε_{mm} is an error term associated with the meta-model, and γ is a matrix of parameters representing each sub-model contribution towards the meta-model. Kumar et al. (2007) provide further discussion on the properties of the hierarchical linear models.

Finding $Y_m = f(X_m)$ is a well studied problem for many sorts of $f(X_m)$, however, issues in the construction and usage of the data matrices and interpretation of the resultant model make the practical application of standard tools on large and complex data sets problematic. A simple model of one output variable would be a linear regression - a linear model with Gaussian error terms.

This work uses Generalized Linear Models (GLMs). GLMs relax the assumptions of the least-squares linear model and extend it to a larger class of problems than linear, Gaussian models. The generalized linear model consists of three components: a linear model or linear predictor that involves the independent variables in the system, the link function $g[(E(y)]$, which relates the mean of the response variable to the linear predictor, and the response distribution (sometimes called the error structure). Myers, Montgomery, and Vining (2002) write the GLM as

$$g[E(y)] = f(\vec{X}, \beta)$$

where g is a monotonic differentiable link function and the function f is the linear predictor. The vector of model parameters, β, is estimated using maximum likelihood. The computational procedure involves an iteratively reweighted least squares algorithm. Very thorough descriptions of the GLM are in Myers, Montgomery, and Vining (2002) and McCullagh and Nelder (1989). Model adequacy is tested through a variety of methods. McCullagh and Nelder (1989) suggest some informal checks using plots to check the model: scaling the standardized deviance residuals to a constant information scale, and plotting against the predicted and explanatory variables produces a method for identifying systematic patterns. Hosmer and Lemeshow (2000) develop several model diagnostic plots for examining influence, outliers, and leverage to identify interesting data points. GLMs have already found application in semiconductor manufacturing as the basis of process monitoring techniques for multistage processes. See Skinner, Montgomery, and Runger (2003) and Jearkpaporn, Montgomery, Runger, and Borror (2003).

The intermediate data from the sub-process models, in essence, provides a quality control variable for monitoring the process.

4 Application

A semiconductor device is the result of many distinct processes applied to many different layers on a silicon wafer layers over a long period of time to produce a complex product with measurable performance characteristics. The manufacturing data this methodology was applied to comes from a 200mm fabrication line producing 250 chips per wafer, in lots of 25 wafers. Each lot undergoes 270 individual processing steps, in which different parts of the circuits are etched in thin layers of material grown or deposited on the working surface of the wafers. Each process step must be tightly controlled to ensure dimensional tolerances measured in nanometers. Fabrication of a single lot requires approximately three months. While the base unit of production may be a small lot of wafers (25 wafers) each process can operate on different units of production. For instance, an oven can anneal multiple lots per baking operation, while some processes act on smaller groups of wafers, individual wafers, or at the sub-wafer level. Examples of sub-wafer operations are electrical testing of an individual chip, and reticle level masking of groups of chips.

Six process datasets drawn from key manufacturing steps such as the gate contact layer and the metal 1 layer are used to model the in-line electrical test parameters. The final performance measure, i.e. the number of functional die in different bins is the final dataset. The die of every lot are assigned to one of six bins based on their performance in the sort process. In this study, we focus on three bins, i.e. bin 11, 4, and 10, which represent DC Fail Count, Function Fail Count, and Marginal Fail Count, respectively. The objective of the hierarchical modeling is to predict the number of die in these three bins.

In this section, we report the model building and diagnostic steps used for predicting one of the in-line electrical test results (AVG6055) using lithography process data (LXMASK). The response variable is continuous, and so a GLM with an identity link function is used. We tested the normality assumption and the validity of identity link function in the model diagnostic stage. Note that due to the identity link function, the GLM is equivalent to multiple linear regression and produces the same output as multiple linear regression. As a result, the model diagnostics such as raw residuals and deviance residuals are equal when identity link is used. The diagnostic plots for the model are shown in Figure 4 and appear acceptable.

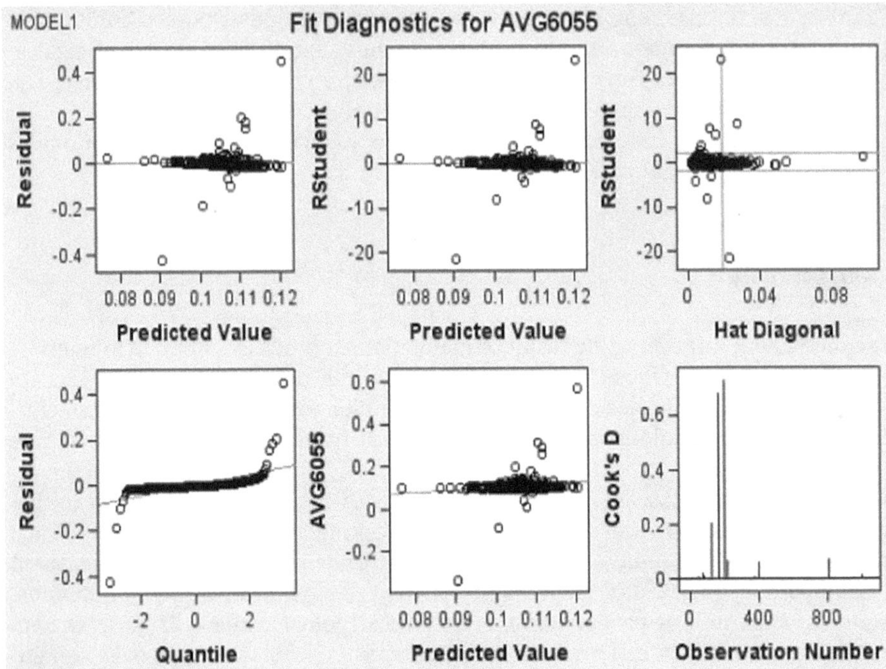

Figure 4. Diagnostic plots of the LXMASK model.

Figure 5 shows the fitting results of the sub-process model LXMASK. As a result of iterative model building, six of the original variables are retained in the final model. Models were fit for five additional sub-processes from the gate contact and metal 1 layers. Next, the meta-model was fit using the estimated values of the sub-process models as the inputs to predict the number of die that will be sorted into a specific bin. Figure 6 shows the results of using a negative binomial GLM with a log link to predict the number of die in Bin 4, DC Fail Count.

Criteria for Assessing Goodness of Fit

Criterion	DF	Value	Value/DF
Deviance	1036	0.6025	0.0006
Scaled Deviance	1036	1043.0000	1.0068
Pearson Chi-Square	1036	0.6025	0.0006
Scaled Pearson X2	1036	1043.0000	1.0068

Analysis of Parameter Estimates

Parameter	DF	Estimate	Standard Error	Wald 85% Confidence Limits		Chi-Square	Pr > ChiSq
Intercept	1	-1.5873	1.9381	-4.3772	1.2026	0.67	0.4128

NC00139	1	-0.0488	0.0241	-0.0834	-0.0142	4.11	0.0426
NC0057	1	-0.0068	0.0038	-0.0123	-0.0013	3.16	0.0754
NC0069	1	0.0113	0.0060	0.0027	0.0198	3.57	0.0589
NC00184	1	0.1007	0.0718	-0.0026	0.2041	1.97	0.1607
NC00116	1	-0.0152	0.0041	-0.0212	-0.0093	13.54	0.0002
NC00117	1	-0.0174	0.0069	-0.0273	-0.0076	6.47	0.0110
Scale	1	0.0240	0.0005	0.0233	0.0248		

Figure 5. Partial GLM Output for the LXMASK process and AVG 6055 electrical test parameter.

The GENMOD Procedure
Model Information

Data Set	WORK.MERGE7
Distribution	Negative Binomial
Link Function	Log
Dependent Variable	YLD0004

Criteria for Assessing Goodness of Fit

Criterion	DF	Value	Value/DF
Deviance	701	722.4246	1.0306
Scaled Deviance	701	722.4246	1.0306
Pearson Chi-Square	701	615.3860	0.8779
Scaled Pearson X2	701	615.3860	0.8779
Log Likelihood		526916.1127	

Analysis of Parameter Estimates

Parameter	DF	Estimate	Standard Error	Wald 85% Confidence Limits		Chi-Square	Pr > ChiSq
Intercept	1	3.0039	0.5651	2.1904	3.8173	28.26	<.0001
OVERPred	1	14.4474	2.6602	10.6180	18.2769	29.50	<.0001
MASKPred	1	-6.9074	3.2984	-11.6555	-2.1593	4.39	0.0362
LQPred	1	-1.2256	0.3779	-1.7695	-0.6816	10.52	0.0012
LXPred	1	16.0875	4.4065	9.7443	22.4307	13.33	0.0003
VDPred	1	-1.9153	4.7355	-8.7322	4.9016	0.16	0.6859
Dispersion	1	0.0463	0.0028	0.0422	0.0503		

Figure 6. GLM Output for the Bin 4 Meta-Model.

The results are validated by partitioning the data into a training and test set. The training data was used to fit the above models and the test set validates the findings. The performance of each model is evaluated based on the change in Mean Squared Prediction Error (MSPE) of training dataset with respect to test dataset. With the proposed model, overall predictive capability has increased by ~15% over the naive models (Table 1). Apply this model to the 322 lots of data results in average predicted number of die per lot is 371 and the average actual number of die is 242. Since information from only five sub-processes (out of 270 process operations) and one in-line electrical test variable (out of 143) are used, the increase in prediction power by 15% can be treated as a potentially significant improvement.

Table 1. Bin 4 meta-model results.

Bin 4 Yield	MSPE Training Set	MSPE Test Set	Variance (Training	R^2
AVG6055	1324.063365	21015.15957	1495.077468	14.87172

5 Defect Monitoring

The results from the previous modeling activity can provide for attribution of defect rates to physical subsystems. From Figure 6, 'MASKPred' has the largest coefficient and corresponds to the LXMASK sub-process model in Figure 5. Given the magnitude of the coefficient, 16.09, relative to the other variables, the LXMASK sub-process appears to be one of the drivers of the Bin 4 yield. Using SPC methods to monitor the variables of the process may provide an opportunity earlier in the fabrication process to detect conditions that result in an undesirable higher Bin 4 yield. However, after pre-processing, the LXMASK operation has 94 process variables. As a result of the modeling in the previous section, the variables in Table 1 may be monitored univariately (Figure 7) or multivariately (Figure 8).

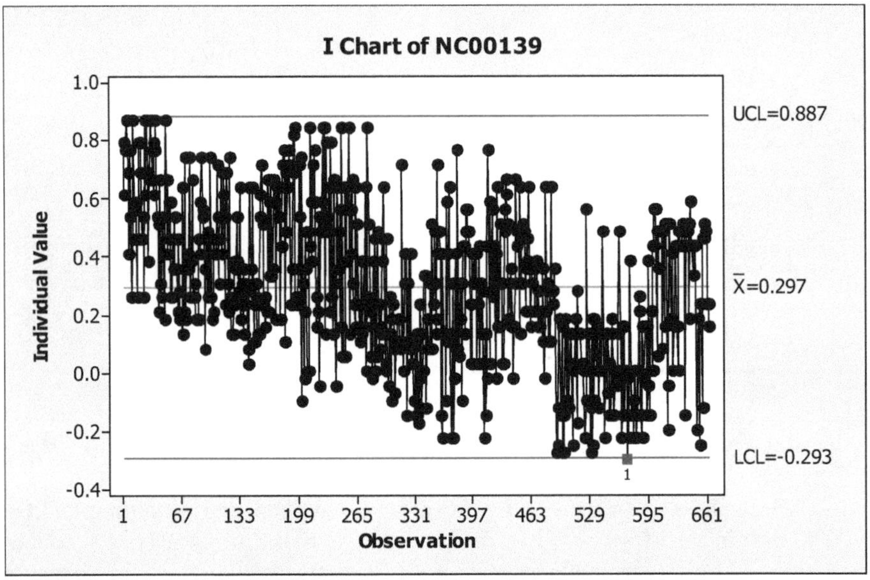

Figure 7. Representative individuals control chart of a variable from the LXMASK process.

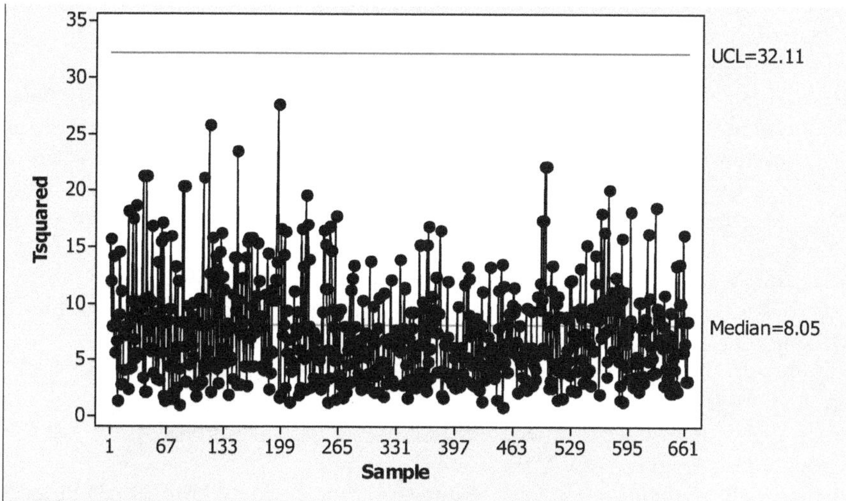

Figure 8. Multivariate control chart of the six LXMASK variables.

6 Summary

This modeling approach leads to several results: empirical estimates of the defective chip count using process data, estimates of the relative importance of the subcomponent, and a mechanism to monitor relative process variables. The meta-model on the scaled intermediate data (i.e. the scaled output from the sub-models) provides a means of ranking and assessing the relative importance of the subsystem to the top-level system. This provides an inexpensive alternative to estimate defect rates and monitor variables through the existing use of automated data collection and enables a process engineer to more fully understand the system and provide avenues for improvement.

Characterizing the relationships at the systems or operational level will enable verification of quality control parameters, improve the utilization of technical resources, and provide the means to validate the effectiveness of current operating methods and metrics.

References

Cunningham, S.P. and MacKinnon, S. (1998) Statistical methods for visual defect methodology. IEEE Transactions on Semiconductor Manufacturing 11(1): 48-53.

Friedman, D.J., Hansen, M.H., Nair, V.N. and James, D.A. (1997) Model-free estimation of defect clustering in integrated circuit fabrication. IEEE Transactions in Integrated Circuit Manufacturing 10(3): 344-359.

Hansen, M.H., Nair, V.N. and Friedman, D.J. (1997) Monitoring wafer map data from integrated circuit fabrication processes for spatially clustered defects. Technometrics 39(3): 241-254.

Hastie, T., Tibshirani, R. and Friedman, J. (2001) The elements of statistical learning: data mining, inference, and prediction. Springer-Verlag, Berlin New York.

Hess, C. and Weiland, L.H. (1999) Extraction of wafer-level defect density distributions to improve yield prediction. IEEE Transactions on Semiconductor Manufacturing 12(2): 175-183.

Horton, D. (1998) Modeling the yield of mixed-technology die. Solid State Technology 41(9): 109-119.

Hosmer, D. W. and Lemeshow, S. (2000). Applied logistic regression, 2nd ed., John Wiley & Sons, New York.

Jearkpaporn, D., Montgomery, D.C., Runger, G. and Borror, C.M. (2003) Process monitoring for correlated gamma distributed data using generalized linear model based control charts. Quality and Reliability Engineering International 19(6): 477-491.

Johnson, R.A. and Wichern, D.W. (1992). Applied multivariate statistical analysis, 3rd ed., Prentice Hall, Englewood Cliffs, New Jersey.

Kumar N, Kennedy K, Gildersleeve K, Abelson R, Mastrangelo CM, and Montgomery DC (2006) A review of yield modeling techniques for semiconductor manufacturing. International Journal of Production Research 44(23): 5019-5036.

McCullagh P and Nelder JA (1989) Generalized linear models. Chapman & Hall, London.

Montgomery DC (2005) Introduction to statistical quality control 5th ed. John Wiley & Sons, NY.

Myers RH, Montgomery DC, and Vining GG (2002). Generalized linear models with applications in engineering and the sciences, John Wiley & Sons, NY.

Nurani RK, Strojwas AJ, Maly WP, Ouyang C, Shindo W, Akella R, McIntyre MG, and Derrett J (1998) In-line yield prediction methodologies using patterned wafer inspection information. IEEE Transactions on Semiconductor Manufacturing 11(1): 40- 47.

Skinner KR, Montgomery DC, Runger GC, Fowler JW, McCarville DR, Rhoads TR and Stanley JD (2002) Multivariate statistical methods for modeling and analysis of wafer probe test data. IEEE Transactions on Semiconductor Manufacturing 15(4): 523-530.

Skinner KR, Montgomery DC, and Runger GC (2003) Process monitoring for multiple count data using generalized linear model-based control charts. International Journal of Production Research 41(6): 1167-1180.

Van Zant P (2004) Microchip fabrication: a practical guide to semiconductor processing. McGraw-Hill, New York.

Conformance Analysis of Population Means under Restricted Stratified Sampling

Rainer Göb and Arne Müller

Institut für Angewandte Mathematik und Statistik, Universität Würzburg,
Sanderring 2, D-97070 Würzburg, Germany, goeb@mathematik.uni-wuerzburg.de

Summary. Risk based auditing has recently been promoted by national and international auditing standards, e. g., SAS 10 through SAS 111, ISA 315 and ISA 330, IDW PS 261. These schemes have created a broadening demand for stochastic methods of risk quantification and risk control in the auditing community. In particular, risk based auditing requires the specification and control of the detection risk, i. e., the probability that the auditor erroneously approves the conformance of the financial statements of the audited entity with the de facto financial status. Confronted with large volumes of accounting and inventory data, auditors are interested in sampling based methods of conformance testing. Customary sampling schemes are not completely in line with the requirements of conformance testing in auditing. The paper introduces *restricted stratified sampling* as a viable scheme for auditing purposes.

1 Introduction: Audit Risk and Conformance Testing

Audit risk is the risk of not detecting material misstatements on the financial situation of a company in course of the auditing process, see the definitions in the professional standards [2], AU 312, §02 and §11, or [1], AU 350, Appendix. Risk based auditing is a scheme which requires the definition, analysis, quantification, and control of all audit risks. Risk based auditing has recently been promoted by national and international auditing standards, e. g., SAS 104 through SAS 111 (*Statements on Auditing Standards* promulgated by the Auditing Standards Board of the American Institute of Certified Public Accountants, AICPA) and the corresponding *Professional Standards* (AUs) of AICPA, ISA 315 and ISA 330 (*International Standard on Auditing* maintained by the International Auditing and Assurance Standards Board, IAASB), IDW PS 261 (developed by the German *Institut der Wirtschaftsprüfer in Deutschland*, IDW). An important motive for these developments were serious auditing attestation errors which ended up in public scandals. Known worldwide is the collapse of Enron in 2001. Revelations on irregular

H.-J. Lenz et al. (eds.), *Frontiers in Statistical Quality Control 9*,
DOI 10.1007/978-3-7908-2380-6_16, © Springer-Verlag Berlin Heidelberg 2010

accounting procedures led to bankruptcy of the company and to the dissolution of Arthur Andersen which had served as Enron's accounting auditor. The loss of Enron stock values amounted to around 60 billion US dollars. A major German scandal was the collapse of FlowTex in 2000. The company alleged book values from selling 3000 drilling machines at a price of 1.5 million DM per item. Only 10% of the machines really existed. FlowTex's business and accounting activities had been approved by KPMG. The total indictable loss of FlowTex creditors amounted to around 2.6 billion Euro. In an extra-judicial agreement, KPMG paid around 50 million Euro to the creditors.

Risk based auditing schemes distinguish two types of audit risks. i) Risk resulting from the type and behaviour of the audited entity, e. g., a company, public organization, or a part thereof. ii) Risk resulting from procedures of investigative and assessment performed by the auditor. All these risks are modeled as probabilities. The definition of these risk probabilities in auditing standards are by far not totally clear. We adopt the consistent view that all probabilities are conditional probabilities under the condition that a financial misstatement exists in a company's accounts. For further discussion of the interpretation of audit risk probabilities, the reader is referred to auditing literature, e. g., [7] or [8].

The risks of type i) are differentiated into two components: the *inherent risk* (IR) as a function of the character of the audited entity, e. g., field, branch, type, magnitude, diversification, internationalisation, financial status, liquidity position, and the *control risk* (CR) as a function of the entity's internal control system. The risk of type ii) residing in the auditing activities is the *detection risk* (DR), i. e., the probability that the auditing process does not reveal existing material misstatements. The total *auditing risk* (AR) is the probability of erroneously approving the totality of financial statements of an entity. Inherent risk factors, control risk factors, and detection risk factors are assumed to be independent which leads to the equation $\mathrm{AR} = \mathrm{IR} \times \mathrm{CR} \times \mathrm{DR}$, see [1].

Risk based auditing schemes have created a broadening demand for stochastic methodology in the auditing community, essentially in two respects: 1) Risk modeling, in particular risk quantification schemes. 2) Risk control procedures to restrict the detection risk DR. The latter respect will be considered in the sequel.

The key interest of auditing in the detection area is *data conformance testing*, i. e., investigating the conformance of financial *book values* kept in lists or data bases with financial *de facto values* existing in reality, e. g., comparing the values of articles reported in an inventory list with the de facto values of the warehoused articles, or comparing the sales transaction figures stored in an enterprise resource planning (ERP) base with the de facto transaction volumes occurred in reality. The book values can easily determined with negligible expense in time or cost, whereas determining de facto values is often considerably laborious and expensive. Hence data conformance testing is often not based on exhaustive 100 % comparisons of book values with associated de facto values, but on *sample* comparisons.

Many sampling techniques are designed for *exploring* unknown populations, e. g., survey sampling in social or market research. The conformance problem in auditing requires a different design. The following aspects are important. 1) The book values are not in complete accordance with the de facto values, but nevertheless they provide some partially reliable prior information on the de facto values. 2) Simple random sampling can be misleading if aggregates of book values and true values are to be compared. 3) Auditing considers finite populations, e. g., an inventory and the corresponding inventory list. Methods from standard statistics are rather designed for random phenomena with an infinite horizon. Such methods cannot readily be used for the analysis of finite populations.

The present paper is arranged in the following parts. Paragraph 2 introduces the data conformance problem in processes (superpopulations) and lots (populations). Paragraph 3 discusses the problems of statistical tests for data conformance in finite populations. Paragraphs 4 and 5 introduce a formal model for superpopulations (processes), and finite populations (lots), and sampling therefrom. The restricted stratified sampling scheme is formally defined by paragraph 6. Paragraphs 7 and 8 describe the implementation of restricted sampling for testing conformance of book and de facto values. Paragraph 9 discusses properties of the restricted stratified sampling scheme.

2 Data Conformance in Processes and Lots

Consider discrete items i, e. g., retail articles , invoices, transaction records in an ERP data base. Each item i is assigned two characteristic quantities W_i and V_i, mostly in monetary terms. W_i is the *de facto value*, e. g., the true monetary value of article i in an inventory, the true amount of money invoiced from a customer, the true value of a sales transaction. The *book value* V_i allegedly reports the true value, e. g., the monetary value of article i reported in an inventory list, the invoice amount reported in accounting records, the sales transaction amount kept in an ERP data base.

The items and item characteristics can be considered in two different respects. i) In the framework of a *process* or *superpopulation* where the items and their characteristics are repetitive realisations of a continuing random business phenomenon, e. g., updating an inventory and the corresponding inventory list, delivering an invoice to a customer and booking the invoice amount in an accounting system, closing a sales transaction and recording the transaction amount in the ERP base. ii) As a finite *population* or *lot* where the items $1, ..., N$ and their characteristics are a completed segment of the underlying process or superpopulation, e. g., an inventory and the corresponding list on a certain reference day, the invoices delivered over a certain period and the corresponding accounting entries, the transactions closed over a certain period and the corresponding ERP records.

Testing for *data conformance*, i. e., the conformance of book values to de facto values, is the essential purpose of auditing. Conformance can be defined in various ways. The subsequent interpretations are most popular. i) *Itemwise conformance*: Data for item i are conforming, if $V_i = W_i$, and otherwise nonconforming. For process auditing of itemwise conformance, the target parameter is the probability $\mathrm{P}(V_i = W_i)$ of the itemwise identity of book and de facto values. For lot auditing of itemwise conformance, the target parameter is the relative frequency of items $i \in \{1, ..., N\}$ with $V_i = W_i$. ii) *Conformance of means*: Data are conforming, if the mean of the book values is sufficiently close to the mean of the de facto values. For process auditing, the target parameters are the process book value expectation $E[V_i] = \mu_V$ and the process de facto value expectation $E[W_i] = \mu_W$. For lot auditing, the target parameters are the lot book value mean $\overline{V} = \frac{1}{N}\sum_{i=1}^{N} V_i$ and the lot de facto value mean $\overline{W} = \frac{1}{N}\sum_{i=1}^{N} W_i$.

Subsequently, we consider auditing for conformance of means. The meaning of "sufficiently close book and de facto means" is specified by imposing a bound τ for a *tolerable misstatement*, i. e., a tolerable deviation of the book mean from the de facto mean. Hence auditing for conformance of process means amounts to testing the process conformance hypothesis $H\colon |\mu_V - \mu_W| \leq \tau$, and auditing for conformance of lot means amounts to testing the lot conformance hypothesis $H\colon |\overline{V} - \overline{W}| \leq \tau$. The tolerance parameter τ is often calculated as a percentage of the known book value mean $\overline{V}$, ranging from 0.5% to 10.0% of $\overline{V}$. $\tau = 0.02\overline{V}$, i. e., 2% of $\overline{V}$, is often used.

3 Statistical Testing of the Data Conformance Hypothesis

We consider statistical testing of the lot conformance hypothesis $H\colon |\overline{V} - \overline{W}| \leq \tau$ based on paired samples $(X_1, Y_1), ..., (X_n, Y_n)$ of book values $X_l = V_{\nu_l}$ and de facto values $Y_l = W_{\nu_l}$ of randomly sampled items $\nu_1, ..., \nu_n$.

Risk based auditing schemes contain a guideline on how to specify the error bounds for a statistical test $H\colon |\overline{V} - \overline{W}| \leq \tau$. The detection risk DR is identified with the probability of erroneously accepting H. DR can be calculated from the basic equation AR $=$ IR $\times$ CR $\times$ DR by specifying AR, IR, CR, and then solving for DR $=$ AR/(IR $\times$ CR). However, it is a serious problem that the existing statistical methodology is rather inappropriate for the empirical analysis of the lot conformance hypothesis $H\colon |\overline{V} - \overline{W}| \leq \tau$. Standard statistics provides two approaches.

i) The *explorative appoach* considers $H\colon |\overline{V} - \overline{W}| \leq \tau$ as a hypothesis on two *unknown* means $\overline{V}$, $\overline{W}$. This approach amounts to testing H by comparing the empirical mean of the pairwise sample differences $Y_i - X_i$ of de facto and book values with 0.

ii) The *comparative appoach* considers $H\colon |\overline{V} - \overline{W}| \leq \tau$ as a *reference* hypothesis with the prescribed and known book value mean $\overline{V}$ as the reference value. This approach amounts to testing H by comparing the empirical mean $\overline{Y}$ of the de facto sample observations Y_i with the known book value mean $\overline{V}$.

Both approaches have advantages and disadvantages, but in final analysis both are not fully adequate for the analysis of the lot conformance hypothesis $H\colon |\overline{V} - \overline{W}| \leq \tau$. The advantage of the explorative approach i) is that by considering $Y_i - X_i$ the positive correlation between the paired sample variables X_i and Y_i is used implicitly to reduce the sample variance. The conceptual problem is that $\overline{V}$ is actually not unknown, but completely known as the parameter of a finite population. The latter aspect is accurately reflected by the comparative approach ii), whereas the method of comparing the sample de facto mean $\overline{Y}$ with the population book value mean $\overline{V}$ makes no adequate use of the correlation between book and de facto values. In auditing practice, the deficiencies of either of the two approaches may lead to irritations, as illustrated by the subsequent example 1.

Example 1 (testing conformance of means). Consider a population of $N = 17$ items with book values V_i and de facto values W_i given by the following table:

i	1	2	3	4	5	6	7	8	9	10	11	12	13	14	15	16	17
V_i	0	0	0	0	0	0	50	50	50	50	50	100	100	100	100	100	100
W_i	0	0	0	0	0	0	$\star$	$\star$	$\star$	$\star$	$\star$	100	100	100	100	100	100
$V_i - W_i$	0	0	0	0	0	0	$\star$	$\star$	$\star$	$\star$	$\star$	0	0	0	0	0	0

We consider the population as given, without recurring to an underlying process model. The book value mean is $\overline{V} = 50.00$. Let $n = 2$ items be selected at random without replacement from the $N = 15$ items. Then for all samples (i, j) constituted from the set $1, ..., 6$ we have $\overline{Y} = 0 = \overline{X}$, $\overline{Y} - \overline{X} = 0$, and for all samples (i, j) constituted from the set $12, ..., 17$ we have $\overline{Y} = 100 = \overline{X}$, $\overline{Y} - \overline{X} = 0$. Hence $\mathrm{P}(\overline{Y} \in \{0, 100\})$ and $\mathrm{P}(\overline{Y} - \overline{X} = 0)$ both have the lower bound $2\binom{6}{2}/\binom{17}{2} = 0.221$.

Consider the tolerance parameter $\tau = 1.00$ which amounts to 2% of the book value mean.

The explorative approach i) interprets $H\colon |\overline{V} - \overline{W}| \leq 1.00$ as a hypothesis on unknown means, and tests by comparing $\overline{Y} - \overline{X}$ with 0. Any reasonable method would have to accept $H\colon |\overline{V} - \overline{W}| \leq 1.00$ under the result $\overline{Y} - \overline{X} = 0$. However, in view of the actually known book value mean $\overline{V} = 50$, this decision seems implausible. The client would feel fine, whereas the auditor would feel uneasy to attest a de facto mean of $49.00 \leq \overline{W} \leq 51.00$ based on sample averages strongly deviating from 50.00. The implausible result occurs with a probability of at least 0.221.

The comparative approach ii) considers the book value mean to be known and considers the conformance hypothesis in the form $H: 49.00 \leq \overline{W} \leq 51.00$. H is tested by comparing the sample de facto average $\overline{Y}$ with the book value mean $\overline{V} = 50$. Any reasonable method would have to reject $H: 49.00 \leq \overline{W} \leq 51.00$ under $\overline{Y} = 0$ or $\overline{Y} = 100$. This decision is implausible rather from the client's perspective. The auditor will feel on the safe side, whereas the client will complain not to be approved although his books are in complete accordance with the de facto values. The implausible result occurs with a probability of at least 0.221. •

A method is required which preserves the advantages both of the explorative and comparative approach without incurring their inadequacies and irritating implications as illustrated by example 1. In particular, the following requirements should be observed. 1) The lot book value mean $\overline{V}$ is known. Hence it is basically better to follow the comparative approach, i. e., to estimate the de facto mean $\overline{W}$ by the sample de facto mean $\overline{Y}$, and then to compare $\overline{V}$ and $\overline{Y}$. 2) The use of the correlation between the book and de facto values, which is implicit in the explorative approach, should be introduced in the comparative approach. In particular, the correlation should be used according to two objectives: i) to avoid the implausible rejection of the conformance hypothesis on grounds of samples which exhibit good itemwise conformance between book and de facto values, but bad conformance of the sample de facto mean $\overline{Y}$ with the lot book value mean $\overline{V}$, see example 1; ii) to reduce the variation through sampling, and in consequence to reduce the amount of sampling.

The guidelines 1) and 2) lead to the methods of *restricted random sampling* and *restricted stratified sampling*. The subsequent paragraphs 4 and 5 provide the basis for a formal definition and technical analysis of the restricted random sampling policy on grounds of an appropriate process, lot, and sampling model. Restricted random sampling is formally defined by paragraph 6 and further investigated by paragraphs 7 through 9.

4 Process Model, Lot Model, Sampling Model

Paragraph 2 distinguishes an underlying *process* or *superpopulation* and the resulting finite *lot* or *population*. The present paragraph gives a formal and more refined account of this distinction.

The process and the resulting finite population are classified into *strata*, i. e., particular processes and particular resulting populations of different origin or properties, e. g., articles of different price range or from different storage locations, invoices of different order of magnitude, transactions of different size or from different branches or departments. In some cases, the strata are predefined in a natural way, in particular by location characteristics like storage, branch, department. The strata can also be defined by the sampling design to minimize the sampling variance, see for instance the allocation policies by [12] and [9], and the exhaustive survey by [4]. Table 1 shows an arrangement into k strata with k underlying processes and corresponding resulting stratum populations of sizes $N_1, ..., N_k$. The stratum population l consists of items labeled $(l, 1), ..., (l, N_l)$. Each item (l, j) is associated with a book value V_{lj} and a de facto value W_{lj}. These values are arranged into the *book value vector* $\boldsymbol{V}_l = (V_{l1}, ..., V_{l,N_l})^\top$ and the *de facto value vector* $\boldsymbol{W}_l = (W_{l1}, ..., W_{lN_l})^\top$. The total process and, respectively, the total population is constituted by the weighted mixture of the strata with stratum weights $q_1, ..., q_k$, $q_1 + ... + q_k = 1$. The finite total population consists of $N = N_1 + ... + N_k$ items, where the stratum weights can be expressed as $q_l = N_l/N$.

Table 1. Process strata and population strata.

	infinite stratum process	population stratum
stratum 1	$..., V_{11}, V_{12}, ...$ $..., W_{11}, W_{12}, ...$	$\boldsymbol{V}_1 = (V_{11}, ..., V_{l,N_1})^\top$ $\boldsymbol{W}_1 = (W_{11}, ..., W_{1N_1})^\top$
$\vdots$	$\vdots$	$\vdots$
stratum k	$..., V_{k1}, V_{k2}, ...$ $..., W_{k1}, W_{k2}, ...$	$\boldsymbol{V}_k = (V_{k1}, ..., V_{k,N_k})^\top$ $\boldsymbol{W}_k = (W_{k1}, ..., W_{kN_k})^\top$

Table 2. Process strata characteristics and population strata characteristics under strata weights $q_1 + ... + q_k = 1$.

	process stratum characteristics	population stratum characteristic
stratum 1	$\mathrm{E}[V_{1i}] = \mu_{1,V}, \ \mathrm{V}[V_{1i}] = \sigma^2_{1,V}$ $\mathrm{E}[W_{1i}] = \mu_{1,W}, \ \mathrm{V}[W_{1i}] = \sigma^2_{1,W}$	$\overline{V}_1 = \frac{1}{N_1}\sum_{j=1}^{N_1} V_{1j}$ $\overline{W}_1 = \frac{1}{N_1}\sum_{j=1}^{N_1} W_{1j}$
$\vdots$	$\vdots$	$\vdots$
stratum k	$\mathrm{E}[V_{ki}] = \mu_{k,V}, \ \mathrm{V}[V_{ki}] = \sigma^2_{k,V}$ $\mathrm{E}[W_{ki}] = \mu_{k,W}, \ \mathrm{V}[W_{ki}] = \sigma^2_{k,W}$	$\overline{V}_k = \frac{1}{N_k}\sum_{j=1}^{N_k} V_{kj}$ $\overline{W}_k = \frac{1}{N_k}\sum_{j=1}^{N_k} W_{kj}$
total	$\mathrm{E}[V_i] = \sum_{l=1}^{k} q_l \mu_{V,l}$ $\mathrm{E}[W_i] = \sum_{l=1}^{k} q_l \mu_{W,l},$	$\overline{V} = \sum_{l=1}^{k} q_l \overline{V}_l$ $\overline{W} = \sum_{l=1}^{k} q_l \overline{W}_l$

The relevant process and lot parameters are displayed by table 2. The underlying process has an infinite horizon. Hence process characteristics have to be defined as parameters of the probability distribution of the random item characteristics V and W. The process is assumed to be stable, such that the distribution remains invariant over time. The book process mean in stratum l is defined as the invariant expectation $\mu_{V,l} = \mathrm{E}[V_{lt}]$ of the book values V_{lt} generated by the lth stratum process. The de facto process mean in stratum l is defined as the invariant expectation $\mu_{W,l} = \mathrm{E}[W_{lt}]$ of the de facto values W_{lt} generated by the lth stratum process. The total process means are the mixtures $\sum_{l=1}^{k} q_l \mu_{V,l}$, $\sum_{l=1}^{k} q_l \mu_{W,l}$ of the stratum means. The finite population means are empirical means, e. g., the average value of articles in an inventory on a specific reference date, the average amount of all invoices billed in a specified reference period, the average volume of all transactions stored in an ERP base during a specified reference period. The book mean in stratum l is $\overline{V}_l = \frac{1}{N_l} \sum_{j=1}^{N_l} V_{lj}$, the de facto mean in stratum l is $\overline{W}_l = \frac{1}{N_l} \sum_{j=1}^{N_l} W_{lj}$. The total population means are the mixtures $\overline{V} = \sum_{l=1}^{k} q_l \overline{V}_l$, $\overline{W} = \sum_{l=1}^{k} q_l \overline{W}_l$ of the stratum means.

The value *variance* is a secondary characteristic of processes and populations. In accounting and auditing, the value variance is not a target characteristic, and appears only as a nuisance parameter of statistical methods. The process variance is defined as the stable variance of the item characteristics as explained by table 2. In analogy to the definition of the population mean, a population variance may be introduced as an empirical variance of the item characteristics in the population. However, in view of the status of the variance as a mere nuisance parameter, introducing a separate population variance is redundant. The technical involvements can be handled with the process variance in a satisfactory way.

The auditing targets are the population means $\overline{V}_l$, $\overline{W}_l$, and not the individual population vectors $\boldsymbol{V}_l$, $\boldsymbol{W}_l$. All populations with the same values $\overline{V}_1 = \overline{v}_1, ..., \overline{V}_k = \overline{v}_k$, $\overline{W}_1 = \overline{w}_1, ..., \overline{W}_k = \overline{w}_k$ are equivalent. Hence the stochastic analysis of finite population auditing is concerned with the analysis of the *conditional distribution* of the population vectors $\boldsymbol{V}_1, ..., \boldsymbol{V}_k$, $\boldsymbol{W}_1, ..., \boldsymbol{W}_k$ under conditions $\overline{V}_1 = \overline{v}_1, ..., \overline{V}_k = \overline{v}_k$, $\overline{W}_1 = \overline{w}_1, ..., \overline{W}_k = \overline{w}_k$.

The notations and quantities associated with sampling are explained by table 10 in the appendix D. A sample of n_l items from the population stratum l is represented by a sequence of numbers $\nu_{l1}, ..., \nu_{ln_l}$ picked from the item labels $1, ..., N_l$ in the stratum. The corresponding book values are denoted by $X_{l1} = V_{\nu_{l1}}, ..., X_{ln_l} = V_{\nu_{ln_l}}$, and the corresponding de facto values are denoted by $Y_{l1} = W_{\nu_{l1}}, ..., Y_{ln_l} = W_{\nu_{ln_l}}$.

For auditing practice, the only reasonable sampling technique is sampling without replacement, i. e., the sampled numbers $\nu_{l1}, ..., \nu_{ln_l}$ are pairwise different. The distribution of the sample number vector $\boldsymbol{\nu}_l = (\nu_{l1}, ..., \nu_{ln_l})$ determines the *sampling design*. The design of *random sampling without replacement* is characterized by the following properties (S1) and (S2). (S1) All sequences $j_1, ..., j_{n_l}$ of pairwise different numbers from $1, ..., N_l$ have the same probability $\mathrm{P}(\nu_{l1} = j_1, ..., \nu_{ln_l} = j_l) = \frac{(N_l - n_l)!}{N_l!}$ of constituting the sample. (S2) The sample number vector $\boldsymbol{\nu}_l$ is independent of the vector $\boldsymbol{V}_l$ of book values and of the vector $\boldsymbol{W}_l$ of de facto values.

5 Conditional Sampling Distributions

We analyse the conditional distribution of the sample vectors $\boldsymbol{X}_1, ..., \boldsymbol{X}_k$, $\boldsymbol{Y}_1, ..., \boldsymbol{Y}_k$ under conditions $\overline{V}_1 = \overline{v}_1, ..., \overline{V}_k = \overline{v}_k$, $\overline{W}_1 = \overline{w}_1, ..., \overline{W}_k = \overline{w}_k$ under random sampling without replacement. For this analysis, we need two types of assumptions on the generating strata processes: i) Assumptions on the relations between different strata processes. ii) Assumptions on the inherent properties of each individual strata process. The relevant assumptions in the respects i) and ii) are summarised by table 3. The assumptions of independent strata and of an i.i.d. stratum process are not problematic. Assuming a normal distribution within a stratum is more controversial. Many business data are rather skewed than symmetric and have no normal shape. However, the basic objective is to analyse the sample means $\overline{X}_l, \overline{Y}_l$. For these quantities, a normal distribution is at least a good approximation.

Table 3. Process properties.

stratum relation model	strata independent, i. e., pairs $(\boldsymbol{V}_1, \boldsymbol{W}_1), ..., (\boldsymbol{V}_k, \boldsymbol{W}_k)$ independent
basic stratum model	stratum variables $\begin{pmatrix} V_{l1} \\ W_{l1} \end{pmatrix}, ..., \begin{pmatrix} V_{lN_l} \\ W_{lN_l} \end{pmatrix}$ i.i.d., each with density $f_{V,W}^{(l)}$
normal stratum model	stratum variables $\begin{pmatrix} V_{l1} \\ W_{l1} \end{pmatrix}, ..., \begin{pmatrix} V_{lN_l} \\ W_{lN_l} \end{pmatrix}$ i.i.d., each with normal distribution $N(\boldsymbol{\mu}_l, \boldsymbol{\Sigma}_l)$ where $\boldsymbol{\mu}_l = \begin{pmatrix} \mu_{V,l} \\ \mu_{W,l} \end{pmatrix}$, $\boldsymbol{\Sigma}_l = \begin{pmatrix} \sigma_{V,l}^2 & \kappa_l \\ \kappa_l & \sigma_{W,l}^2 \end{pmatrix}$

Under the basic stratum model of table 3, the conditional sampling distributions can be analysed by formulae for the conditional densities. The results are displayed by table 4. The proof of these results is based on the fact that for distribution calculations we can assume $\overline{X} = \frac{1}{n}\sum_{i=1}^{n} V_{li}$, $\overline{Y} = \frac{1}{n}\sum_{i=1}^{n} W_{li}$, see [6] for a formal proof.

The conditional sampling distributions under the normal stratum model of table 3 are normal distributions as displayed by table 5. The proof of these results is given by appendix A.

Consider the sequence $(X_{l1}, Y_{l1})^\top, ..., (X_{ln_l}, Y_{ln_l})^\top$ of pairs of book values X_{li} and de facto values Y_{li} in the sample asymptotically for large stratum size N_l. The first row of table 5 shows that asymptotically the pairs $(X_{l1}, Y_{l1})^\top, ..., (X_{ln_l}, Y_{ln_l})^\top$ are i.i.d., each with bivariate normal distribution $N(\boldsymbol{\mu}_l, \boldsymbol{\Sigma}_l)$. From normal distribution theory, see [3], p. 53, it follows that asymptotically the pair $(\overline{X}_l, \overline{Y}_l)^\top$ of sample means and the sample variance-covariance matrix $\widehat{\boldsymbol{\Sigma}}_l$ introduced by table 10 are independent.

Table 4. Conditional stratum sampling densities under the basic stratum model. h_k is the joint density of arithmetical averages $\frac{1}{m}\sum_{i=1}^m V_{li}$, $\frac{1}{m}\sum_{i=1}^m W_{li}$.

variables	conditional density under $\overline{V}_l = \overline{v}_l$, $\overline{W}_l = \overline{w}_l$
sample vectors $\boldsymbol{X}_l$, $\boldsymbol{Y}_l$	$f_{\boldsymbol{X}_l, \boldsymbol{Y}_l \mid \overline{V}_l = \overline{v}_l, \overline{W}_l = \overline{w}_l}(\boldsymbol{x}, \boldsymbol{y}) = \left(\frac{N_l}{N_l - n_l}\right)^2 \frac{\left(\prod_{i=1}^{n_l} f_{V,W}^{(l)}(x_i, y_i)\right) h_{N_l - n_l}\left(\frac{N_l \overline{v}_l - \sum x_i}{N_l - n_l}, \frac{N_l \overline{w}_l - \sum y_i}{N_l - n_l}\right)}{h_{N_l}(\overline{v}_l, \overline{w}_l)}$
sample means $\overline{X}_l$, $\overline{Y}_l$	$f_{\overline{X}_l, \overline{Y}_l \mid \overline{V}_l = \overline{v}_l, \overline{W}_l = \overline{w}_l}(x, y) = \left(\frac{N_l}{N_l - n_l}\right)^2 \frac{h_{n_l}(x, y) h_{N_l - n_l}\left(\frac{N_l \overline{v}_l - n x}{N_l - n}, \frac{N_l \overline{w}_l - n_l y}{N_l - n_l}\right)}{h_{N_l}(\overline{v}_l, \overline{w}_l)}$
sample mean $\overline{Y}_l$	$f_{\overline{Y}_l \mid \overline{V}_l = \overline{v}_l, \overline{W}_l = \overline{w}_l}(y) = \frac{N_l}{N_l - n_l} \frac{h_{n_l}(y) h_{N_l - n_l}\left(\overline{v}_l, \frac{N_l \overline{w}_l - n_l y}{N_l - n_l}\right)}{h_{N_l}(\overline{v}_l, \overline{w}_l)}$

6 Restricted Random Sampling and Restricted Stratified Sampling

Restricted random sampling from a stratum l is defined by a *restriction parameter* δ and the following steps (RS1), (RS2).

(RS1) Select a random sample without replacement from the stratum items $1, ..., N_l$.

(RS2) If $|\overline{X}_l - \overline{v}_l| \le \delta \sigma_{V,l} \sqrt{e_l}$ where $e_l = \frac{1}{n_l} - \frac{1}{N_l}$: Evaluate $\overline{Y}_l$, terminate. If $|\overline{X}_l - \overline{v}_l| > \delta \sigma_{V,l} \sqrt{e_l}$: Discard the sample, return to step (RS1).

Table 5. Conditional stratum sampling distributions under the normal stratum model.

variables	conditional distribution under $\overline{V}_l = \overline{v}_l$, $\overline{W}_l = \overline{w}_l$
sample vectors $\boldsymbol{X}_l$, $\boldsymbol{Y}_l$	normal distribution of dimension $2n_l$ with $\mathrm{E}[(\boldsymbol{X}_l, \boldsymbol{Y}_l)^\top \mid \overline{V}_l = \overline{v}_l, \overline{W}_l = \overline{w}_l] = \begin{pmatrix} \overline{v}_l \mathbf{1}_{n_l} \\ w_l \mathbf{1}_{n_l} \end{pmatrix}$ $\mathrm{Cov}[(\boldsymbol{X}_l, \boldsymbol{Y}_l)^\top \mid \overline{V}_l = \overline{v}_l, \overline{W}_l = \overline{w}_l] = \begin{pmatrix} \sigma^2_{V,l}\mathbf{I}_{n_l} & \kappa_l \mathbf{I}_{n_l} \\ \kappa_l \mathbf{I}_{n_l} & \sigma^2_{W,l}\mathbf{I}_{n_l} \end{pmatrix} - \frac{1}{N} \begin{pmatrix} \sigma^2_{V,l}\mathbf{1}_{n_l,n_l} & \kappa_l \mathbf{1}_{n_l,n_l} \\ \kappa_l \mathbf{1}_{n_l,n_l} & \sigma^2_{W,l}\mathbf{1}_{n_l,n_l} \end{pmatrix}$
sample means $\overline{X}_l$, $\overline{Y}_l$	bivariate normal distribution with $\mathrm{E}[\begin{pmatrix} \overline{X}_l \\ \overline{Y}_l \end{pmatrix} \mid \overline{V}_l = \overline{v}_l, \overline{W}_l = \overline{w}_l] = \begin{pmatrix} \overline{v}_l \\ \overline{w}_l \end{pmatrix}$ $\mathrm{Cov}[\begin{pmatrix} \overline{X}_l \\ \overline{Y}_l \end{pmatrix} \mid \overline{V}_l = \overline{v}_l, \overline{W}_l = \overline{w}_l] = e_l \boldsymbol{\Sigma}_l, \quad e_l = \frac{1}{n_l} - \frac{1}{N_l}$
sample mean $\overline{Y}_l$	univariate normal distribution with $\mathrm{E}[\overline{Y}_l \mid \overline{V}_l = \overline{v}_l, \overline{W}_l = \overline{w}_l] = \overline{w}_l, \quad \mathrm{V}[\overline{Y}_l \mid \overline{V}_l = \overline{v}_l, \overline{W}_l = \overline{w}_l] = e_l \sigma^2_{W,l}$

Applying the algorithm defined by the steps (RS1) and (RS2) to each stratum $l = 1, ..., k$ we obtain *restricted stratified sampling*. We give a brief survey of the properties and potential advantages of restricted sampling. A more detailed discussion follows in subsequent paragraphs 7 through 9.

1) In comparison with simple random sampling, restricted random sampling involves no additional difficulties or costs. The only additional requirement is the knowledge of the book values in the lot so as to calculate $\overline{X}_l$, $\overline{v}_l$, and $\sigma_{V,l}$. This requirement is fulfilled without additional expenses.

2) The implementation of restricted random sampling requires the specification of the restriction parameter δ. The numerical analysis of real auditing data and of simulated data shows that $0.5 \leq \delta \leq 1.0$ are reasonable choices.

3) Restricted sampling conforms to the objective i) established by paragraph 3: Implausible inferences as illustrated by example 1 are avoided.

4) Restricted sampling conforms to the objective ii) established by paragraph 3: By using prior information, restricted sampling achieves lower sample sizes than unrestricted sampling.

7 Inference on the Population De Facto Mean under Restricted Random Sampling

The theoretical analysis of restricted random sampling in the lot model amounts to considering the conditional sampling distribution under the additional condition $|\overline{X}_l - \overline{v}_l| \leq \delta\sigma_{V,l}\sqrt{e_l}$ or, equivalently, $\overline{v}_l - \delta\sigma_{V,l}\sqrt{e_l} \leq \overline{X}_l \leq \overline{v}_l + \delta\sigma_{V,l}\sqrt{e_l}$. The main properties of this distribution are collected in table 6. The proofs of the results are provided by appendix B. For the analysis of the distributions it has to be noted that the basic stratum model of table 3 postulates independent strata.

In the conditional population situation, the book value stratum means $\overline{v}_1, ..., \overline{v}_k$ and the book value population mean $\overline{v} = q_1\overline{v}_1 + ... + q_k\overline{v}_k$ are known in a strict sense. The model of paragraph 4 considers only the process stratum variances $\sigma^2_{1,V}, ..., \sigma^2_{k,V}$ without population counterparts. Since the numbers N_l of stratum book values is large, the process variances $\sigma^2_{1,V}, ..., \sigma^2_{k,V}$ can be estimated with high precision, and can also be assumed to be known. The de facto parameters $\overline{w}_1, ..., \overline{w}_k$, $\overline{w} = q_1\overline{w}_1 + ... + q_k\overline{w}_k$, $\sigma^2_{1,W}, ..., \sigma^2_{k,W}$ have to be estimated from the stratum samples of sizes $n_1, ..., n_k$.

The inference on the de facto population mean $\overline{w}$ under the normal stratum model is described by table 7. The variances and correlations are involved as nuisance parameters. The properties of the stratified sample mean $\overline{Y}_{\text{st}}$ are immediately evident from the facts given by table 6. If the variances $\sigma^2_{W,l}$ and correlations ρ_l are all known, the approximate normality of $\overline{Y}_{\text{st}}$ allows to calculate an approximate confidence interval for $\overline{w}$ based on quantiles of the normal distribution $N(0,1)$ as displayed by table 7.

Table 6. Conditional sampling distributions under the conditioning event $C = \{\overline{V}_l = \overline{v}_l, \overline{W}_l = \overline{w}_l, \overline{v}_l - \delta\sigma_{V,l}\sqrt{e_l} \leq \overline{X}_l \leq \overline{v}_l + \delta\sigma_{V,l}\sqrt{e_l} \text{ for } l = 1, ..., k\}$ in the normal stratum model of table 3.

moments of $\overline{Y}_l$	$\mathrm{E}[\overline{Y}_l \mid C] = \overline{w}_l, \quad \mathrm{V}[\overline{Y}_l \mid C] = \sigma^2_{W,l} e_l h_l$ $e_l = \frac{1}{n_l} - \frac{1}{N_l}, \quad h_l = 1 - \rho_l^2 \frac{2\delta\varphi(\delta)}{2\Phi(\delta) - 1}$
asymptotic distribution of $S^2_{Y,l}$	the asymptotic (N_l large) distribution of $S^2_{Y,l}$ under C is the same as under $\overline{V}_l = \overline{v}_l$, $\overline{W}_l = \overline{w}_l$, i. e., $\frac{n_l - 1}{\sigma^2_{Y,l}} S^2_{Y,l}$ is asymptotically distributed by $\chi^2(n_l - 1)$ with $E[S^2_{Y,l} \mid C] \approx \sigma^2_{Y,l}$
asymptotic relation of $\overline{Y}_l$, $S^2_{Y,l}$	$\overline{Y}_l$, $S^2_{Y,l}$ are asymptotically (for N_l large) independent under C

Table 7. Inference on the population de facto mean $\overline{w} = q_1\overline{w}_1 + ... + q_k\overline{w}_k$ under the conditioning event $C = \{\overline{V}_l = \overline{v}_l, \overline{W}_l = \overline{w}_l, \overline{v}_l - \delta\sigma_{V,l}\sqrt{e_l} \leq \overline{X}_l \leq \overline{v}_l + \delta\sigma_{V,l}\sqrt{e_l}$ for $l = 1, ..., k\}$ in the normal stratum model.

coefficients	$q_l = \frac{N_l}{N}$, $h_l = 1 - \rho_l^2 \frac{2\delta\varphi(\delta)}{2\Phi(\delta) - 1}$, $e_l = \frac{1}{n_l} - \frac{1}{N_l}$
estimator of w	$\widehat{w} = \overline{Y}_{\mathrm{st}} = \sum_{l=1}^k q_l \overline{Y}_l$ (stratified sample mean) $E[\overline{Y}_{\mathrm{st}} \mid C] = w$, $V[\overline{Y}_{\mathrm{st}} \mid C] = \sum_{l=1}^k q_l^2 e_l h_l \sigma_{W,l}^2$ under C, $\overline{Y}_{\mathrm{st}}$ is approximately normally distributed for large k
estimator of $V[\overline{Y}_{\mathrm{st}} \mid C]$ when $\sigma_{W,1}^2, ..., \sigma_{W,k}^2$ unknown, $\rho_1, ..., \rho_k$ known	$V[\overline{Y}_{\mathrm{st}} \mid C]$ is estimated by $S_{\mathrm{st},Y}^2 = \sum_{l=1}^k q_l^2 e_l h_l S_{Y,l}^2$ under C, $S_{\mathrm{st},Y}^2$ is approximately unbiased when $N_1, ..., N_k$ large
estimator of ρ_l^2 when $\sigma_{W,l}^2$ unknown, ρ_l unknown	$\widehat{\rho_l^2} = \frac{n_l(n_l - 1)}{n_l^2 + 1} \frac{S_{X,Y,l}^2}{\sigma_{V,l}^2 S_{Y,l}^2} - \frac{n_l - 1}{n_l^2 + 1}$
estimator of $V[\overline{Y}_{\mathrm{st}} \mid C]$ when $\sigma_{W,1}^2, ..., \sigma_{W,k}^2$ unknown, $\rho_1, ..., \rho_k$ unknown	h_l is estimated by $\widehat{h}_l = 1 - \widehat{\rho}_{l,2}^2 \frac{2\delta\varphi(\delta)}{2\Phi(\delta) - 1}$ $V[\overline{Y}_{\mathrm{st}} \mid C]$ is estimated by $S_{\mathrm{st},Y}^2 = \sum_{l=1}^k q_l^2 e_l \widehat{h}_l S_{Y,l}^2$ under C, $S_{\mathrm{st},Y}^2$ is approximately unbiased for $V[\overline{Y}_{\mathrm{st}} \mid C]$ when $N_1, ..., N_k$ large
approximate level γ confidence interval for w	$\left(\overline{Y}_{\mathrm{st}} - z\widehat{\sigma}_{\mathrm{st},Y};\ \overline{Y}_{\mathrm{st}} + z\widehat{\sigma}_{\mathrm{st},Y}\right)$ where $z = z\left(\frac{1+\gamma}{2}\right)$ is the $100\frac{1+\gamma}{2}$%-quantile of the inference distribution
$\widehat{\sigma}_{\mathrm{st},Y}$ when $\sigma_{W,1}^2, ..., \sigma_{W,k}^2$, $\rho_1, ..., \rho_k$ known	$\widehat{\sigma}_{\mathrm{st},Y} = \sum_{l=1}^k q_l^2 e_l h_l \sigma_{W,l}^2$, and the inference distribution is $N(0,1)$
$\widehat{\sigma}_{\mathrm{st},Y}$ when $\sigma_{W,1}^2, ..., \sigma_{W,k}^2$ unknown	$\widehat{\sigma}_{\mathrm{st},Y} = S_{\mathrm{st},Y}^2$ and the inference distribution is the central t distribution $t(\nu)$, $\nu = \left(\sum q_l^2 e_l h_l S_{Y,l}^2\right)^{-2} \sum (q_l^2 e_l h_l)^2 \frac{S_{Y,l}^4}{n_l - 1}$ $\widehat{h}_l$ is inserted for h_l if $\rho_1, ..., \rho_k$ are unknown

Consider the case of unknown stratum variances $\sigma_{W,1}^2, ..., \sigma_{W,k}^2$, but known stratum correlations $\rho_1, ..., \rho_k$. In this case, the approximate unbiasedness of the stratified sample variance $S_{\mathrm{st},Y}^2 = \sum_{l=1}^k q_l^2 e_l h_l S_{Y,l}^2$ follows from $E[S_{Y,l}^2 \mid C] \approx$

$\sigma^2_{Y,l}$, see table 6, and the definition of $S^2_{\mathrm{st},Y}$. An approximation of the distribution of $\sum q_l^2 e_l h_l \sigma^2_{W,l} S^2_{\mathrm{st},Y} / (\sum (q_l^2 e_l h_l)^2 \frac{\sigma^4_{W,l}}{n_l-1})$ can be constructed following the approximation scheme suggested by [10], [11] and [13]. The distribution of

$$Z_\nu \;=\; \nu \frac{\sum_l q_l^2 e_l h_l S^2_{Y,l}}{E[\sum_l q_l^2 e_l h_l S^2_{Y,l} | C]} \;\approx\; \nu \frac{\sum_l q_l^2 e_l h_l S^2_{Y,l}}{(\sum_l q_l^2 e_l h_l \sigma^2_{W,l})}$$

is approximated by the $\chi^2(\nu)$ distribution with the same variance as Z_ν. Hence ν is determined from

$$2\nu \;\stackrel{!}{=}\; V[Z_\nu|C] \;\approx\; \frac{\nu^2}{(\sum_l q_l^2 e_l h_l \sigma^2_{W,l})^2} \, 2 \sum_l (q_l^2 e_l h_l)^2 \frac{\sigma^4_{W,l}}{n_l - 1} \,.$$

Solving for ν provides

$$\nu \;=\; \frac{(\sum_l q_l^2 e_l h_l \sigma^2_{W,l})^2}{\sum_l (q_l^2 e_l h_l)^2 \frac{\sigma^4_{W,l}}{n_l-1}} \,. \tag{1}$$

By the results of table 6, $\overline{Y}_{\mathrm{st}}$ and $S^2_{\mathrm{st},Y}$ are asymptotically independent for large stratum sizes $N_1, ..., N_k$. For a large number k of strata, the weighted sum $\overline{Y}_{\mathrm{st}}$ is under C approximately normally distributed. Hence the ratio

$$\frac{\frac{Y_{\mathrm{st}} - \overline{w}}{\sqrt{\sum_l b_l^2 \sigma^2_{W,l}}}}{\sqrt{\frac{\sum b_l \sigma^2_{W,l}}{\sum b_l^2 \frac{\sigma^4_{W,l}}{n_l-1}} S^2_{\mathrm{st},Y}}} \sqrt{\frac{\left(\sum b_l \sigma^2_{W,l}\right)^2}{\sum b_l^2 \frac{\sigma^4_{W,l}}{n_l-1}}} \;=\; \frac{Y_{\mathrm{st}} - \overline{w}}{S_{\mathrm{st},Y}}$$

is approximately distributed by the central t distribution $t(\nu)$. To calculate the degree of freedom ν, in formula (1) the unknown $\sigma^2_{W,l}$ are replaced by the estimates $S^2_{Y,l}$. This justifies the approximate confidence interval for w.

In the case of both unknown stratum variances $\sigma^2_{W,1}, ..., \sigma^2_{W,k}$ and unknown stratum correlations $\rho_1, ..., \rho_k$, the squared correlation estimators $\widehat{\rho}_l^2$ suggested by table 7 are plugged into the formula for $S^2_{\mathrm{st},Y}$, which provides the estimate $S^2_{\mathrm{st},Y} = \sum_{l=1}^k q_l^2 e_l \widehat{h}_l S^2_{Y,l}$, and into formula (1) for the degree of freedom ν. The approximate unbiasedness of the estimate $S^2_{\mathrm{st},Y}$ for $V[\overline{Y}_{\mathrm{st}}|C]$ under large $N_1, ..., N_k$ follows from the results of the appendix C: $\widehat{h}_l S^2_{Y,l}$ is approximately unbiased for $h_l \sigma^2_{W,l}$; $S^2_{\mathrm{st},Y} = \sum_{l=1}^k q_l^2 e_l h_l S^2_{Y,l}$ is apprimately unbiased for $V[\overline{Y}_{\mathrm{st}}|C]$ under known correlations $\rho_1, ..., \rho_l$, see table 6 and the related explanations; hence $S^2_{\mathrm{st},Y} = \sum_{l=1}^k q_l^2 e_l \widehat{h}_l S^2_{Y,l}$ is apprimately unbiased for $V[\overline{Y}_{\mathrm{st}}|C]$ under unknown correlations $\rho_1, ..., \rho_l$.

Table 7 considers confidence intervals for the unknown population de facto mean $\overline{w}$ whereas paragraph 2 places special emphasis on a test of the lot conformance hypothesis $H\colon |\overline{v} - \overline{w}| \le \tau$, or equivalently $H\colon \overline{v} - \tau \le \overline{w} \le \overline{v} + \tau$. The confidence interval approach is more general, more flexible, and more informative. A test for $H: \overline{v} - \tau \leq \overline{w} \leq \overline{v} + \tau$ can easily be defined on grounds of a two-sided confidence interval B for $\overline{w}$ by the following rule: H is accepted, if $B \subset [\overline{v} - \tau; \overline{v} + \tau]$; H is rejected, if $B \cap [\overline{v} - \tau; \overline{v} + \tau] = \emptyset$; otherwise the decision amounts to indifference. Under the confidence level γ, the probability of erroneous acceptance and the probability of erroneous rejection are both bounded by $1 - \gamma$. Hence, following the discussion in paragraph 2, the confidence level should be chosen as $\gamma = 1 - \mathrm{DR} = 1 - \mathrm{AR}/(\mathrm{IR} \times \mathrm{CR})$. The confidence interval conveys much more information than the test. The auditor obtains an information about the location of the unknown population de facto mean $\overline{w}$ in comparison with the population book mean $\overline{v}$.

8 Selection of Sampling Weights and Total Sample Size

Previous paragraphs considered the the total sample size n and the sampling weights $d_l = n_l/n$ as prescribed. For applications, we need an algorithm for planning the sample size and the sampling weights. We frame such rules by adapting techniques from classical survey sampling. As in classical survey sampling, the planning phase assumes the de facto variances $\sigma^2_{W,l}$ and the correlations ρ_l to be known. In practice, $\sigma^2_{W,l}$ is estimated by the book value variance $\sigma^2_{V,l}$. Assigning values to the correlations ρ_l is more difficult. An auditor can use three sources of knowledge to determine the correlations: 1) Estimators from previous audits. 2) Small trial samples to get rough initial estimates. 3) Rules of thumb to express the auditor's confidence in the auditee be assigning appropriate values of the correlations.

For unrestricted stratified sampling, [12] and [9] calculate sampling weights $d_l = n_l/n$ which minimize the variance of the stratified sample mean $\overline{Y}_{\mathrm{st}}$ under a sampling cost function $C = c_0 + \sum_{l=1}^{k} c_l n_l$. The same approach can be used under restricted random sampling, with results displayed by table 8. The derivation is completely analogous to the derivation under unrestricted random sampling, see [4], pp. 97-98.

The purpose of sampling is to obtain precise information on the de facto population mean $w = q_1\overline{w}_1 + \ldots + q_k\overline{w}_k$. A reasonable measure for the precision of sampling inference is the length of the confidence interval for w calculated according to table 7. To determine a lower limit for the sample size n, an upper limit Δ is prescribed for the length $2z\sqrt{\sum_{l=1}^{k} q_l^2 e_l h_l \sigma^2_{W,l}}$ of the confidence interval, see table 7. The resulting lower limit for n is displayed by table 9. The derivation of the result is based on simple algebraic manipulations.

The number k of strata can be chosen by algorithms familiar from classical survey sampling, see [4].

Table 8. Neyman-Tschuprow sampling weights $d_l = n_l/n$ under restricted sampling. Coefficients q_l and h_l defined by table 7.

cost function $C = c_0 + \sum_{l=1}^{k} c_l n_l$	$d_l \;=\; \dfrac{\frac{q_l \sigma_{W,l}\sqrt{h_l}}{\sqrt{c_l}}}{\sum_{m=1}^{k} \frac{q_m \sigma_{W,m}\sqrt{h_m}}{\sqrt{c_m}}}$
cost function $C = c_0 + cn$	$d_l \;=\; \dfrac{q_l \sigma_{W,l}\sqrt{h_l}}{\sum_{m=1}^{k} q_m \sigma_{W,m}\sqrt{h_m}}$
cost function $C = c_0 + cn$, identical correlations $\rho_1 = ... = \rho_k$	$d_l \;=\; \dfrac{q_l \sigma_{W,l}}{\sum_{m=1}^{k} q_m \sigma_{W,m}}$

Table 9. Choice of total sample size n under restricted sampling. Coefficients q_l and h_l defined by table 7.

prescription	length of level γ confidence interval for the de facto population mean w smaller than Δ, i. e., $2z\sqrt{\sum_{l=1}^{k} q_l^2 e_l h_l \sigma_{W,l}^2} \overset{!}{\leq} \Delta$ where $z = z\left(\frac{1+\gamma}{2}\right)$ is the $100\frac{1+\gamma}{2}$%-quantile of $N(0,1)$
sample size	$n \;\geq\; \dfrac{\sum_{l=1}^{k} q_l^2 \frac{h_l}{d_l} \sigma_{W,l}^2}{\frac{\Delta^2}{4z^2} + \frac{1}{N}\sum_{l=1}^{k} q_l h_l \sigma_{W,l}^2}$

9 Discussion of Restricted Sampling

We discuss the aspects 2), 3), and 4) mentioned after the definition of restricted random sampling in paragraph 6: 2) the choice of the restriction parameter δ, 3) the potential of restricted sampling to avoid implausible inferences as illustrated by example 1, 4) the potential of restricted sampling to achieve lower sample sizes than unrestricted sampling by using prior information.

Aspects 2) and 4) interfere. The prior information is conveyed by the correlations ρ_l: Increasing ρ_l amounts to increasing confidence of the auditor in the quality of book values. The correlations affect the sample size formula of table 9 through the coefficient h_l, where the latter is defined by table 7. h_l is decreasing in ρ_l and increasing in δ. From table 9 it is easy to see that consequently the sample size lower limit is also decreasing in ρ_l and increasing in δ. Too large values of δ counteract the reducing effect of the confidence parameter ρ_l. As an illustration, figure 9 displays the lower limit of the sample size according to table 9 as a function $n(\rho)$ of a homogeneous correlation $\rho = \rho_1 = ... = \rho_k$ for $\delta = 0.75$ and $\delta = 1.50$. The figure is based on a set of real inventory book data. The number of strata was prescribed with $k = 10$. The strata boundaries and the resulting strata means and variances were determined from the real book values by the algorithm of [5]. As usual in planning stratified sampling, see [4], the unknown de facto variances were identified with the book value variances. Figure 9 shows a strong effect of the confidence parameter ρ for $\delta = 0.75$ and a much smoother effect for $\delta = 1.50$. For auditing practice, a strong effect is desirable. Hence $0.5 \leq \delta \leq 1.0$ are reasonable choices.

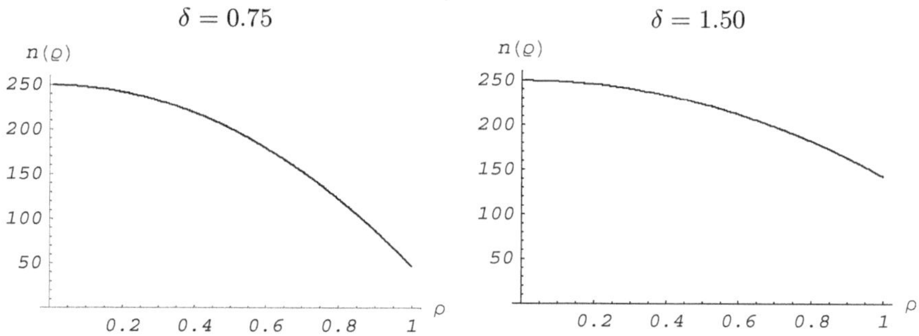

Fig. 1. Lower limit of the sample size according to table 9 as a function $n(\rho)$ of a homogeneous correlation $\rho = \rho_1 = ... = \rho_k$ in the strata $l = 1, ..., k$.

To discuss the potential of restricted sampling to avoid implausible inferences as illustrated by example 1, consider the data of example 1 as one stratum. We have $\overline{v} = 50.00$, $\sigma_V = 43.30$. For sample size $n = 2$ we have $e = \frac{1}{n} - \frac{1}{N} = 0.44$. For moderate restriction parameter $\delta = 1.5$ we obtain $\delta\sigma_V\sqrt{e} = 43.14$. According to the restricted sampling rule, samples with $|X - \overline{v}| > \delta\sigma_V\sqrt{e} = 43.14$ are discarded. For samples constituted either from numbers $1, ..., 6$ or from numbers $12, ..., 17$ we have $|X - \overline{v}| = 50.00 > 43.14$. Hence the samples entailing irritating inferences are discarded.

10 Conclusion

Restricted stratified sampling is a method specifically designed for the purpose of sampling based analysis of the conformance of book value averages and de facto value averages in finite populations. Used for this purpose, standard survey sampling techniques may run into intolerable conclusions. Although the theory of statistical inference from restricted sampling is more involved than for standard stratified sampling we have demonstrated that the method can be arranged into a practically manageable and simple inference scheme.

The planning of restricted sampling with respect to stratification and sample size requires the prior specification of the correlation between book and de facto values, see paragraph 8, in particular tables 8 and 9. This is an important practical issue which, however, goes beyond the scope of the present introductory paper. Further research should concentrate on providing practically useful and usable solutions to this problem.

A Proof of the Results of Table 5

The conditional sampling distributions displayed by table 5 can be obtained by inserting into the density formulae of table 4, or by using the following well-known theorem on conditional distributions under the multivariate normal, see [3], p. 29.

Theorem 1 (Conditional Distributions under the Multivariate Normal). *Let* $\boldsymbol{U}_1$ *be a* k_1*-dimensional,* $\boldsymbol{U}_2$ *a* k_2*-dimensional random vector and let* $\boldsymbol{U}^\top = (\boldsymbol{U}_1^\top, \boldsymbol{U}_2^\top)$*. Let* $\boldsymbol{U}$ *have the* (k_1+k_2)*-dimensional normal distribution* $N(\boldsymbol{\mu}, \boldsymbol{\Sigma})$ *where*

$$\boldsymbol{\mu} = \begin{pmatrix} \boldsymbol{\mu}_1 \\ \boldsymbol{\mu}_2 \end{pmatrix}, \quad \boldsymbol{\Sigma} = \begin{pmatrix} \boldsymbol{\Sigma}_1 & \boldsymbol{\Gamma} \\ \boldsymbol{\Gamma}^\top & \boldsymbol{\Sigma}_2 \end{pmatrix},$$

$\boldsymbol{\mu}_i = E[\boldsymbol{U}_i]$, $\boldsymbol{\Sigma}_i = Cov[\boldsymbol{U}_i]$. *Then we have:*

(a) The random vectors $\boldsymbol{U}_1 - \boldsymbol{\Gamma}\boldsymbol{\Sigma}_2^{-1}\boldsymbol{U}_2$ *und* $\boldsymbol{U}_2$ *are independent.*

(b) For any $\boldsymbol{u}_2 \in \mathbb{R}^{k_2}$ *the conditional distribution of* $\boldsymbol{U}_1$ *under the condition* $\boldsymbol{U}_2 = \boldsymbol{u}_2$ *is the* k_1*-dimensional normal distribution with*

$$\begin{aligned} E[\boldsymbol{U}_1|\boldsymbol{U}_2 = \boldsymbol{u}_2] &= \boldsymbol{\mu}_1 + \boldsymbol{\Gamma}\boldsymbol{\Sigma}_2^{-1}(\boldsymbol{u}_2 - \boldsymbol{\mu}_2), \\ Cov[\boldsymbol{U}_1|\boldsymbol{U}_2 = \boldsymbol{u}_2] &= \boldsymbol{\Sigma}_1 - \boldsymbol{\Gamma}\boldsymbol{\Sigma}_2^{-1}\boldsymbol{\Gamma}^\top. \end{aligned} \qquad \bullet$$

The following multivariate distributions are easily found.

1) The distribution of $\boldsymbol{G} = (\boldsymbol{X}, \boldsymbol{Y}, \overline{V}, \overline{W})^\top$ is the multivariate normal of dimension $2n+2$ with

$$\mathrm{E}[\boldsymbol{G}] = \begin{pmatrix} \mu_V \mathbf{1}_n \\ \mu_W \mathbf{1}_n \\ \mu_V \\ \mu_W \end{pmatrix}, \quad \mathrm{Cov}[\boldsymbol{G}] = \begin{pmatrix} \sigma_V^2 \mathbf{I}_n & \kappa \mathbf{I}_n & \frac{\sigma_V^2}{N}\mathbf{1}_n & \frac{\kappa}{N}\mathbf{1}_n \\ \kappa \mathbf{I}_n & \sigma_W^2 \mathbf{I}_n & \frac{\kappa}{N}\mathbf{1}_n & \frac{\sigma_W^2}{N}\mathbf{1}_n \\ \frac{\sigma_V^2}{N}\mathbf{1}_n^\top & \frac{\kappa}{N}\mathbf{1}_n^\top & \frac{\sigma_V^2}{N} & \frac{\kappa}{N} \\ \frac{\kappa}{N}\mathbf{1}_n^\top & \frac{\sigma_W^2}{N}\mathbf{1}_n^\top & \frac{\kappa}{N} & \frac{\sigma_W^2}{N} \end{pmatrix}. \tag{2}$$

2) The distribution of $\boldsymbol{H} = (\overline{X}, \overline{Y}, \overline{V}, \overline{W})^\top$ is the multivariate normal of dimension 4 with

$$\mathrm{E}[\boldsymbol{H}] = (\mu_V, \mu_W, \mu_V, \mu_W)^\top, \quad \mathrm{Cov}[\boldsymbol{H}] = \begin{pmatrix} \frac{1}{n}\boldsymbol{\Sigma} & \frac{1}{N}\boldsymbol{\Sigma} \\ \frac{1}{N}\boldsymbol{\Sigma} & \frac{1}{N}\boldsymbol{\Sigma} \end{pmatrix}. \tag{3}$$

3) The distribution of $\boldsymbol{K} = (\overline{X}, \overline{V}, \overline{W})^\top$ is the multivariate normal of dimension 3 with

$$\mathrm{E}[\boldsymbol{K}] = (\mu_V, \mu_V, \mu_W)^\top,\ \mathrm{Cov}[\boldsymbol{K}] = \begin{pmatrix} \frac{\sigma_W^2}{n} & \gamma^\top \\ \gamma & \frac{1}{N}\boldsymbol{\Sigma} \end{pmatrix},\ \gamma = \frac{1}{N}\begin{pmatrix} \kappa \\ \sigma_W^2 \end{pmatrix}. \quad (4)$$

From these results and from theorem 1 we obtain the assertions of table 5.

B Derivation of the Results of Table 6

The moments of $\overline{X}_l$ under $\overline{V}_l = \overline{v}_l$, $\overline{W}_l = \overline{w}_l$, $\overline{v}_l - \delta\sigma_{V,l}\sqrt{e_l} \le \overline{X}_l \le \overline{v}_l + \delta\sigma_{V,l}\sqrt{e_l}$ are obtained by applying the subsequent proposition 1 to the results of table 5 on conditional distributions under $\overline{V}_l = \overline{v}_l$, $\overline{W}_l = \overline{w}_l$.

Proposition 1. *Let the random pair $\boldsymbol{U} = (U_1, U_2)^\top$ have a bivariate normal distribution $N(\boldsymbol{\mu}, \boldsymbol{\Sigma})$ with*

$$\boldsymbol{\Sigma} = \begin{pmatrix} \sigma_1^2 & \kappa \\ \kappa & \sigma_2^2 \end{pmatrix}, \quad \text{i. e., with correlation } \rho = \frac{\kappa}{\sigma_1\sigma_2}.$$

Let $a, b \in \mathbb{R}$, $a < b$. Then we have:

(a) The conditional density $f_{U_1|a\le U_2\le b}$ of U_1 under the condition $a \le U_2 \le b$ is

$$f_{U_1|a\le U_2\le b}(u_1) = \frac{1}{\sigma_1}\varphi\left(\frac{u_1-\mu_1}{\sigma_1}\right)\frac{\Phi\left(\frac{x-\mu_2-\frac{\kappa}{\sigma_1^2}(y-\mu_1)}{\sqrt{\sigma_2^2-\frac{\kappa^2}{\sigma_1^2}}}\right)\Bigg|_{x=a}^{x=b}}{\Phi\left(\frac{b-\mu_2}{\sigma_2}\right)-\Phi\left(\frac{a-\mu_2}{\sigma_2}\right)} \quad (5)$$

(b) The conditional first and second moments of U_1 under the condition $a \le U_2 \le b$ are

$$E[U_1|a\le U_2\le b] = \mu_1 + \frac{\kappa}{\sigma_2}\frac{\varphi\left(\frac{a-\mu_2}{\sigma_2}\right)-\varphi\left(\frac{b-\mu_2}{\sigma_2}\right)}{\Phi\left(\frac{b-\mu_2}{\sigma_2}\right)-\Phi\left(\frac{a-\mu_2}{\sigma_2}\right)}, \quad (6)$$

$$E[U_1^2|a\le U_2\le b] =$$

$$\sigma_1^2+\mu_1^2 - \frac{2\kappa\mu_1}{\sigma_2}\frac{\varphi(x)\Big|_{\frac{a-\mu_2}{\sigma_2}}^{\frac{b-\mu_2}{\sigma_2}}}{\Phi\left(\frac{b-\mu_2}{\sigma_2}\right)-\Phi\left(\frac{a-\mu_2}{\sigma_2}\right)} - \frac{\kappa^2}{\sigma_2^2}\frac{x\varphi(x)\Big|_{\frac{a-\mu_2}{\sigma_2}}^{\frac{b-\mu_2}{\sigma_2}}}{\Phi\left(\frac{b-\mu_2}{\sigma_2}\right)-\Phi\left(\frac{a-\mu_2}{\sigma_2}\right)} \quad (7)$$

(c) The conditional variance of U_1 under the condition $a \le U_2 \le b$ is

$$V[U_1|a \leq U_2 \leq b] \;=\; \sigma_1^2 \left\{ 1 - \rho^2 \left[\frac{x\varphi(x)\Big|_{\frac{a-\mu_2}{\sigma_2}}^{\frac{b-\mu_2}{\sigma_2}}}{\Phi\left(\frac{b-\mu_2}{\sigma_2}\right) - \Phi\left(\frac{a-\mu_2}{\sigma_2}\right)} + \left(\frac{\varphi(x)\Big|_{\frac{a-\mu_2}{\sigma_2}}^{\frac{b-\mu_2}{\sigma_2}}}{\Phi\left(\frac{b-\mu_2}{\sigma_2}\right) - \Phi\left(\frac{a-\mu_2}{\sigma_2}\right)} \right)^2 \right] \right\}. \tag{8}$$

(d) In case of $a = \mu_2 - \xi$, $b = \mu_2 + \xi$ we have $E[U_1|a \leq U_2 \leq b] = \mu_1$, and

$$V[U_1|a \leq U_2 \leq b] \;=\; \sigma_1^2 \left\{ 1 - \rho^2 \frac{\frac{2\xi}{\sigma_2}\varphi\left(\frac{\xi}{\sigma_2}\right)}{2\Phi\left(\frac{\xi}{\sigma_2}\right) - 1} \right\}. \tag{9}$$

Proof of assertion (a) of theorem 1. By theorem 1, the conditional distribution of U_2 under $U_1 = u_1$ is the normal distribution with $\mathrm{E}[U_2|U_1 = u_1] = \mu_2 + \frac{\kappa}{\sigma_1^2}(u_1 - \mu_1)$, $\mathrm{V}[U_2|U_1 = u_1] = \sigma_2^2 - \frac{\kappa^2}{\sigma_1^2}$. Hence we have

$$f_{U_1|a \leq U_2 \leq b}(u_1) \;=\; \frac{\int_a^b f_{U_1,U_2}(u_1,u_2)\,\mathrm{d}u_2}{\mathrm{P}(a \leq U_2 \leq b)} \;=\; \frac{\int_a^b f_{U_2|U_1=u_1}(u_2)\,\mathrm{d}u_2 f_{U_1}(u_1)}{\mathrm{P}(a \leq U_2 \leq b)}.$$

Assertion (a) follows immediately from the latter formula.

Proof of assertions (b) and (c) of theorem 1. By theorem 1, the conditional distribution of U_1 under $U_2 = u_2$ is the normal distribution with $\mathrm{E}[U_1|U_2 = u_2] = \mu_1 + \frac{\kappa}{\sigma_2^2}(u_2 - \mu_2)$, $\mathrm{V}[U_1|U_2 = u_2] = \sigma_1^2 - \frac{\kappa^2}{\sigma_2^2}$, and with $\mathrm{E}[U_1^2|U_2 = u_2] = \sigma_1^2 - \frac{\kappa^2}{\sigma_2^2} + \mu_1^2 + \frac{2\kappa\mu_1}{\sigma_2^2}(u_2 - \mu_2) + \frac{\kappa^2}{\sigma_2^4}(u_2 - \mu_2)^2$. Hence we have

$$\mathrm{E}[U_1|a \leq U_2 \leq b] \;=\; \frac{\int\limits_{u_1 \in \mathbb{R}} u_1 \int\limits_{a \leq u_2 \leq b} f_{U_1,U_2}(u_1,u_2)\,\mathrm{d}u_2\,\mathrm{d}u_1}{\mathrm{P}(a \leq U_2 \leq b)} \;=$$

$$\frac{\int\limits_{a \leq u_2 \leq b} \int\limits_{u_1 \in \mathbb{R}} u_1 f_{U_1|U_2=u_2}(u_1)\,\mathrm{d}u_1\, f_{U_2}(u_2)\,\mathrm{d}u_2}{\mathrm{P}(a \leq U_2 \leq b)} \;=$$

$$\frac{\int_a^b \left(\mu_1 + \frac{\kappa}{\sigma_2^2}(u_2 - \mu_2)\right) f_{U_2}(u_2)\,\mathrm{d}u_2}{\mathrm{P}(a \leq U_2 \leq b)} \;=$$

$$\mu_1 + \frac{\kappa}{\sigma_2} \frac{\int_a^b \frac{1}{\sigma_2}\frac{u_2-\mu_2}{\sigma_2}\varphi\left(\frac{u_2-\mu_2}{\sigma_2}\right)\mathrm{d}u_2}{\mathrm{P}(a \leq U_2 \leq b)} \;=\; \mu_1 + \frac{\kappa}{\sigma_2} \frac{-\varphi(x)\Big|_{\frac{a-\mu_2}{\sigma_2}}^{\frac{b-\mu_2}{\sigma_2}}}{\mathrm{P}(a \leq U_2 \leq b)},$$

and

$$\mathrm{E}[U_1^2|a \le U_2 \le b] = \frac{\int\limits_{u_1 \in \mathbb{R}} u_1^2 \int\limits_{a \le u_2 \le b} f_{U_1,U_2}(u_1,u_2)\,\mathrm{d}u_2\,\mathrm{d}u_1}{\mathrm{P}(a \le U_2 \le b)} =$$

$$\frac{\int\limits_{a \le u_2 \le b} \int\limits_{u_1 \in \mathbb{R}} u_1^2 f_{U_1|U_2=u_2}(u_1)\,\mathrm{d}u_1\, f_{U_2}(u_2)\,\mathrm{d}u_2}{\mathrm{P}(a \le U_2 \le b)} =$$

$$\frac{\int_a^b \left(\sigma_1^2 - \frac{\kappa^2}{\sigma_2^2} + \mu_1^2 + \frac{2\kappa\mu_1}{\sigma_2^2}(u_2-\mu_2) + \frac{\kappa^2}{\sigma_2^4}(u_2-\mu_2)^2\right) f_{U_2}(u_2)\,\mathrm{d}u_2}{\mathrm{P}(a \le U_2 \le b)} =$$

$$\sigma_1^2 - \frac{\kappa^2}{\sigma_2^2} + \mu_1^2 + \frac{2\kappa\mu_1}{\sigma_2}\frac{-\varphi(x)\Big|_{\frac{a-\mu_2}{\sigma_2}}^{\frac{b-\mu_2}{\sigma_2}}}{\mathrm{P}(a \le U_2 \le b)} + \frac{\kappa^2}{\sigma_2^2}\frac{\int_{\frac{a-\mu_2}{\sigma_2}}^{\frac{b-\mu_2}{\sigma_2}} x^2\varphi(x)\,\mathrm{d}x}{\mathrm{P}(a \le U_2 \le b)} =$$

$$\sigma_1^2 - \frac{\kappa^2}{\sigma_2^2} + \mu_1^2 - \frac{2\kappa\mu_1}{\sigma_2}\frac{\varphi(x)\Big|_{\frac{a-\mu_2}{\sigma_2}}^{\frac{b-\mu_2}{\sigma_2}}}{\mathrm{P}(a \le U_2 \le b)} + \frac{\kappa^2}{\sigma_2^2}\frac{\Big(\Phi(x) - x\varphi(x)\Big)\Big|_{\frac{a-\mu_2}{\sigma_2}}^{\frac{b-\mu_2}{\sigma_2}}}{\mathrm{P}(a \le U_2 \le b)} =$$

$$\sigma_1^2 + \mu_1^2 - \frac{2\kappa\mu_1}{\sigma_2}\frac{\varphi(x)\Big|_{\frac{a-\mu_2}{\sigma_2}}^{\frac{b-\mu_2}{\sigma_2}}}{\mathrm{P}(a \le U_2 \le b)} - \frac{\kappa^2}{\sigma_2^2}\frac{x\varphi(x)\Big|_{\frac{a-\mu_2}{\sigma_2}}^{\frac{b-\mu_2}{\sigma_2}}}{\mathrm{P}(a \le U_2 \le b)}\,. \qquad \bullet$$

The results of table 6 on the asymptotic (N_l large) conditional distribution of $S_{Y,l}^2$ and the conditional independence of $\overline{Y}_l$, $S_{Y,l}^2$ follow from the asymptotic independence of $(\overline{X}_l, \overline{Y}_l)^\top$ and $S_{Y,l}^2$ under $\overline{V}_l = \overline{v}_l$, $\overline{W}_l = \overline{w}_l$, and from the resulting asymptotic equation

$$\mathrm{P}_{\overline{V}_l=\overline{v}_l,\overline{W}_l=\overline{w}_l}(\overline{Y}_l \in A, S_{Y,l}^2 \in B|\overline{X}_l \in C) \simeq$$

$$\frac{\mathrm{P}_{\overline{V}_l=\overline{v}_l,\overline{W}_l=\overline{w}_l}(\overline{Y}_l \in A, \overline{X}_l \in C)}{\mathrm{P}_{\overline{V}_l=\overline{v}_l,\overline{W}_l=\overline{w}_l}(X_l \in C)}\,\mathrm{P}_{\overline{V}_l=\overline{v}_l,\overline{W}_l=\overline{w}_l}(S_{Y,l}^2 \in B) \simeq$$

$$\mathrm{P}_{\overline{V}_l=\overline{v}_l,\overline{W}_l=\overline{w}_l}(\overline{Y}_l \in A|\overline{X}_l \in C)\,\mathrm{P}_{\overline{V}_l=\overline{v}_l,\overline{W}_l=\overline{w}_l}(S_{Y,l}^2 \in B|\overline{X}_l \in C)\,.$$

C Derivation of the Asymptotically Unbiased Estimator of $h_l\sigma_{W,l}^2$

Consider the estimator $\widehat{h}_l = 1 - \widehat{\rho}_{l,2}^2 \frac{2\delta\varphi(\delta)}{2\Phi(\delta)-1}$ introduced in table 7. We demonstrate the asymptotic (N_l large) unbiasedness of $\widehat{h}_l S_{Y,l}^2$ for $h_l\sigma_{W,l}^2$. The essential step to this result is contained in the following proposition 2.

Proposition 2. *Let $\boldsymbol{Z}_1, ..., \boldsymbol{Z}_n$ be a sequence of i.i.d. random vectors of dimension K with $E[\boldsymbol{Z}_m] = \boldsymbol{\xi}_m$, $Cov[\boldsymbol{Z}_l] = \boldsymbol{\Gamma}$, $\nu_{abrs} = E[Z_{ma}Z_{mb}Z_{mr}Z_{ms}]$ for each $m = 1, ..., n$, $1 \leq a, b, r, s \leq K$. Let*

$$\widehat{\boldsymbol{\Gamma}} = \frac{1}{n-1}\sum_{m=1}^{n}(\boldsymbol{Z}_m - \overline{\boldsymbol{Z}})(\boldsymbol{Z}_m - \overline{\boldsymbol{Z}})^\top = \frac{1}{n-1}\Big\{\sum_{l=m}^{n}\boldsymbol{Z}_m\boldsymbol{Z}'_m - n\overline{\boldsymbol{Z}}\overline{\boldsymbol{Z}}^\top\Big\}.$$

Then

$$E[\widehat{\gamma}_{ab}\widehat{\gamma}_{rs}] = \frac{n^2-2n+3}{n(n-1)}\gamma_{ab}\gamma_{rs} + \frac{1}{n}\nu_{abrs} \quad \textit{for } 1 \leq a, b, r, s \leq K \tag{10}$$

and if each $\boldsymbol{Z}_m$ has the K-dimensional normal distribution $N(\boldsymbol{0}, \boldsymbol{\Gamma})$ we have

$$E[\widehat{\gamma}_{ab}^2] = \frac{n^2+1}{n(n-1)}\gamma_{ab}^2 + \frac{1}{n}\gamma_{aa}\gamma_{bb} \quad \textit{for } 1 \leq a, b \leq K\,. \tag{11}$$

Proof of proposition 2. Obviously one can assume $\boldsymbol{\xi}_m = \boldsymbol{0}$. Because of the independence of $\boldsymbol{Z}_1, ..., \boldsymbol{Z}_n$ we have for $1 \leq a, b, r, s \leq K$

$$E[Z_{ia}Z_{jb}Z_{lr}X_{ms}] = 0, \quad \text{if one of the indices } i, j, l, m \text{ differs from all others,}$$

$$E[Z_{ia}Z_{jb}Z_{lr}Z_{ms}] = \begin{cases} \gamma_{ab}\gamma_{rs}, \text{ if } i = j,\ l = m,\ i \neq l, \\ \nu_{abrs}, \quad \text{if } i = j = l = m. \end{cases}$$

Hence for $1 \leq a, b, r, s \leq K$

$$(n-1)^2 E[\widehat{\gamma}_{ab}\widehat{\gamma}_{rs}] =$$

$$E\Big[\Big(\sum_{i=1}^{n} Z_{ia}Z_{ib} - \frac{1}{n}\sum_{j=1}^{n}\sum_{l=1}^{n} Z_{ja}Z_{lb}\Big)\Big(\sum_{m=1}^{n} Z_{mr}Z_{ms} - \frac{1}{n}\sum_{p=1}^{n}\sum_{t=1}^{n} Z_{pr}Z_{ts}\Big)\Big] =$$

$$\sum_{i=1}^{n}\sum_{m=1}^{n} E[Z_{ia}Z_{ib}Z_{mr}Z_{ms}] + \frac{1}{n^2}\sum_{j=1}^{n}\sum_{l=1}^{n}\sum_{p=1}^{n}\sum_{t=1}^{n} E[Z_{ja}Z_{lb}Z_{pr}Z_{ts}]$$

$$-\frac{1}{n}\sum_{i=1}^{n}\sum_{p=1}^{n}\sum_{t=1}^{n} E[Z_{ia}Z_{ib}Z_{pr}Z_{ts}] - \frac{1}{n}\sum_{m=1}^{n}\sum_{j=1}^{n}\sum_{l=1}^{n} E[Z_{mr}Z_{ms}Z_{ja}Z_{lb}] =$$

$$\Big[n(n-1)\gamma_{pb}\gamma_{rs} + n\nu_{abrs}\Big] + \frac{1}{n^2}\Big[3n(n-1)\gamma_{ab}\gamma_{rs} + n\nu_{abrs}\Big] =$$

$$-\frac{2}{n}\Big[n(n-1)\gamma_{ab}\gamma_{rs} + n\nu_{abrs}\Big] =$$

$$(n-1)\frac{n^2-2n+3}{n}\gamma_{ab}\gamma_{rs} + \frac{(n-1)^2}{n}\nu_{abrs}.$$

Dividing the latter expression by $n-1$ we obtain the equation (10).

Let each $\boldsymbol{Z}_l$ have the K-dimensional normal distribution $N(\boldsymbol{0}, \boldsymbol{\Gamma})$. Then from [3], pp. 38-39,

$$\nu_{abab} = 2\gamma_{ab}^2 + \gamma_{aa}\gamma_{bb},$$

and the second result expressed by equation (11) follows from inserting into equation (10). •

Table 5 shows that for large N_l in the conditional distribution the vectors $\boldsymbol{Z}_{lm} = (X_{lm}, Y_{lm})^\top$, $m = 1, ..., n_l$, satisyfy the assumptions of proposition 2 with $K = 2$, $n = n_l$, $\boldsymbol{\xi}_l = (\overline{v}_l, \overline{w}_l)^\top$, $\boldsymbol{\Gamma} = \boldsymbol{\Sigma}_l$. Hence for N_l large

$$E[S_{X,Y,l}^2] \simeq \frac{n_l^2+1}{n_l(n_l-1)}\kappa_l^2 + \frac{1}{n_l}\sigma_{V,l}^2\sigma_{W,l}^2$$

and consequently

$$E\left[\frac{n_l(n_l-1)}{n_l^2+1}S_{X,Y,l}^2 - \frac{(n_l-1)\sigma_{V,l}^2}{n_l^2+1}S_{Y,l}^2\right] \simeq \kappa_l^2. \tag{12}$$

Hence

$$E[\widehat{h}_l S_{Y,l}^2] = E\left[S_{Y,l}^2 - \widehat{\rho}_{l,2}^2\frac{2\delta\varphi(\delta)}{2\Phi(\delta)-1}S_{Y,l}^2\right] \simeq$$

$$\sigma_{W,l}^2 - E\left[\frac{n_l(n_l-1)}{n_l^2+1}S_{X,Y,l}^2 - \frac{(n_l-1)\sigma_{V,l}^2}{n_l^2+1}S_{Y,l}^2\right]\frac{2\delta\varphi(\delta)}{2\Phi(\delta)-1} \simeq_{(12)}$$

$$\sigma_{W,l}^2 - \frac{\kappa_l^2}{\sigma_{V,l}^2}\frac{2\delta\varphi(\delta)}{2\Phi(\delta)-1} = \sigma_{W,l}^2 h_l.$$

D Sampling Notation

The subsequent table 10 explains the notations associated with stratified sampling, see paragraph 4.

Table 10. Description of samples.

formal notation	interpretation
$n_1, ..., n_k$	sample sizes in strata $1, ..., k$
$n = n_1 + ... + n_k$	total sample size
$\frac{n_1}{n}, \ldots, \frac{n_k}{n}$	weights of strata $1, ..., k$ in the total sample of size n
$\nu_{l1}, ..., \nu_{ln_l}$	numbers of items sampled from stratum l
$X_{l1} = V_{\nu_{l1}}, ..., X_{ln_l} = V_{\nu_{ln_l}}$	book values of the n_l items sampled from stratum l
$\boldsymbol{X}_l = (X_{l1}, ..., X_{ln_l})^\top$	vector of book values sampled from stratum l
$\overline{X}_l = \frac{1}{n_l} \sum_{i=1}^{n_l} X_{li}$	sample mean of book values in the sample from stratum l
$Y_{l1} = W_{\nu_{l1}}, ..., W_{ln_l} = W_{\nu_{ln_l}}$	de facto values of the n_l items sampled from stratum l
$\boldsymbol{Y}_l = (Y_{l1}, ..., Y_{ln_l})^\top$	vector of de facto values sampled from population stratum l
$\overline{Y}_l = \frac{1}{n_l} \sum_{i=1}^{n_l} Y_{li}$	sample mean of de facto values in the sample from stratum l
$S_{Y,l}^2 = \frac{1}{n_l - 1} \sum_{i=1}^{n_l} (Y_{li} - \overline{Y}_l)^2$	sample variance of de facto values in the sample from stratum l
$S_{X,Y,l} = \frac{1}{n_l - 1} \sum_{i=1}^{n_l} (X_{li} - \overline{X}_l)(Y_{li} - \overline{Y}_l)$	sample covariance of book and de facto values in the sample from stratum l
$\widehat{\boldsymbol{\Sigma}}_l = \begin{pmatrix} S_{X,l}^2 & S_{X,Y,l} \\ S_{X,Y,l} & S_{Y,l}^2 \end{pmatrix}$	sample variance-covariance matrix of the sample from stratum l

References

1. AICPA (2006) AU Section 350 – Audit Sampling. American Institute of Certified Public Accountants, New York.
2. AICPA (2007) AU Section 312 – Audit Risk and Materiality in Conducting an Audit. American Institute of Certified Public Accountants, New York.
3. Anderson, T. W. (1958) An Introduction to Multivariate Statistical Analysis. John Wiley & Sons, Inc., New York, London, Sydney.
4. Cochran, W. G. (1977) Sampling Techniques. Third edition. John Wiley & Sons, Inc., New York, London, Sydney.
5. Dalenius, T., and Hodges, J. L. (1959) Minimum variance stratification. Journal of the American Statistical Association, Vol. 54, pp. 88-101.
6. Göb, R. (1996) An Elementary Model for Statistical Lot Inspection and Its Application to Sampling by Variables. Metrika, Vol. 44, pp. 135-163.
7. Guy, D. M., Carmichael, D. R., and Whittington, O. R. (2002) Audit Sampling: An Introduction. Fifth Edition. John Wiley & Sons, Inc., New York, London, Sydney.
8. Hayes, R., Dassen, R., Schilder, A., an Wallage, P. (2005) Principles of Auditing: An Introduction to International Standards on Auditing. Second Edition. Prentice Hall, Harlow, England.
9. Neyman, J. (1934) On two different aspects of the representative method: the method of stratified sampling and the method of purposive selection. Journal of the Royal Statistical Society, Vol. 97, pp. 558-606.
10. Satterthwaite, F. E. (1941) Synthesis of variance. Psychometrika, Vol. 6, pp. 309-316.
11. Satterthwaite, F. E. (1946) An approximate distribution of estimates of variance components. Biometrics Bulletin, Vol. 2., No. 6, pp. 110-114.
12. Tschuprow, A. A. (1923) On the Mathematical Expectation of the Moments of Frequency Distributions in the Case of Correlated Observations. Metron 2, pp. 461-493, 646-683.
13. Welch, B. L. (1947) The generalization of "student's" problem when several different population variances are involved. Biometrika, Vol. 34, pp. 28-35.

Data Quality Control Based on Metric Data Models

Veit Köppen[1] and Hans - J. Lenz[2]

[1]Institute of Production, Information Systems and Operations Research, Freie Universität Berlin, Garystr. 21, D-14195 Berlin, Germany, koeppen@wiwiss.fu-berlin.de

[2]Institute of Statistics and Econometrics, Freie Universität Berlin, Garystr. 21, D-14195 Berlin, Germany, hjlenz@wiwiss.fu-berlin.de

Summary. We consider statistical edits defined on a metric data space spanned by the non-key attributes (variables) of a given database. Integrity constraints are defined on this data space based on definitions, behavioral equations or a balance equation system. As an example think of a set of business or economic indicators. The variables are linked by the four basic arithmetic operations only. Assuming a multivariate Gaussian distribution and an error in the variables model estimation of the unknown (latent) variables can be carried out by a generalized least-squares (GLS) procedure. The drawback of this approach is that the equations form a non-linear equation system due to multiplication and division of variables, and that generally one assumes independence between all variables due to a lack of information in real applications. As there exists no finite parameter density family which is closed under all four arithmetic operations we use MCMC-simulation techniques, cf. Smith and Gelfand (1992) and Chib (2004) to derive the "exact" distributions in the non-normal case and under cross-correlation. The research can be viewed as an extension of Köppen and Lenz (2005) in the sense of studying the robustness of the GLS approach with respect to non-normality and correlation.

1 Introduction

Fellegi and Holt (1976) published a break-through paper on automatic editing and imputation. Wetherill and Gerson (1987) put together the methodology about edits as validation rules on symbolic, logical, probabilistic and relational data sets. Lenz and Rödel (1991) extended validation rules to statistical edits, and Lenz and Müller (2000) to fuzzy edits in the case of metric data spaces. Liepins and Uppuluri (1990) published their view and pragmatics on data quality control. Aitchison (1986) considered non-negative measurements that sum up to unity.

H.-J. Lenz et al. (eds.), *Frontiers in Statistical Quality Control 9*,
DOI 10.1007/978-3-7908-2380-6_17, © Springer-Verlag Berlin Heidelberg 2010

Recently, Batini and Scannapieco (2006) put together the methodology about data quality known in both areas 'Statistics' and 'Database Theory'.

In the following we are concerned with statistical edits, i.e. validation rules based on a fully specified model and defined on a metric data space. The model is assumed to be correctly specified. As an example think of sales = profit + costs as a linear equation which is true due to definition. Note, that the fundamental economic equation "sales = sold_quantity * unit_price" is a non-linear relation. Such relations can be represented by an error-in-the-variables model. Let $\boldsymbol{\xi}$ be a p-dimensional vector of error-free variables and $\boldsymbol{x}$ the corresponding observation vector with superimposed measurement errors $\boldsymbol{u}$, i.e. we have $\boldsymbol{\xi} = \boldsymbol{x} + \boldsymbol{u}$ as state space vector. The available knowledge about definitions and balance equations $\boldsymbol{\zeta} = \boldsymbol{H}(\boldsymbol{\xi})$ is encapsulated in the observation equation system with fixed dimension $q \in \mathbb{N}$. It is modeled as $z = \boldsymbol{H}(\xi) + \boldsymbol{v}$ where z is a q-dimensional observation vector and $\boldsymbol{v}$ an additive noise vector independent of $\boldsymbol{u}$. If all state equations (definitions and balance equations) are linear, then $\boldsymbol{H}$ is a $(q\text{x}p)$ observation matrix. Generally, some equations are non-linear leading to $\boldsymbol{H}$: dom($\boldsymbol{\xi}$) → dom(z). Lenz and Rödel (1991) showed for linear models that, given the data ($\boldsymbol{x}$, z) and the observation model $\boldsymbol{H}$, $\boldsymbol{\xi}$ and $\boldsymbol{\zeta}$ can be estimated by a generalized least-squares (GLS) approach by $\hat{\boldsymbol{\xi}} = \boldsymbol{x} + \boldsymbol{K}(z - \boldsymbol{H}\,\boldsymbol{x})$ and $\hat{\boldsymbol{\zeta}} = \boldsymbol{H}\boldsymbol{x}$ where $\boldsymbol{K} = \boldsymbol{PH}'(\boldsymbol{HPH}' + \boldsymbol{R})^{-1}$ and the covariance matrices of the errors $\boldsymbol{u}$, $\boldsymbol{v}$ are given by $\boldsymbol{P} = \Sigma_{uu}$ and $\boldsymbol{R} = \Sigma_{vv}$. The estimators are best (UMVUE) for a quadratic loss function if $\boldsymbol{u}$, $\boldsymbol{v}$ are jointly Gaussian distributed.

In the following we shall relax the assumption of a joint Gaussian distribution and, moreover, assume cross-correlation between the variables according to some prior information. This implies to substitute GLS estimation by MCMC simulation, cf. Chib (2004) and Köppen and Lenz (2005). In this sense the study can be viewed as a study of robustness with respect to non-normality and dependencies between the variables. First, we introduce a simple model, and then we present the simulation approach and close with various scenarios showing the main effects of deviations between GLS estimates and estimates based upon our MCMC simulation.

2 Business Indicators Model

It is sufficient for our purposes to consider a simplified business indicators model M based on two equations and five variables. We have the structural equation system

Sales = Profit + Cost

Return-on-Investment (ROI) = Profit / Capital.

Evidently, there are two endogenous and three exogenous random variables. The only assumption about M we need in the following is that each equation fulfills the separability condition. This means that each equation of M is uniquely resolvable for each variable showing up on its right hand side (RHS). In order to simplify the notation we write $x \sim N(\mu, \sigma^2)$ instead of $x = \mu + u$ with $u \sim N(0, \sigma^2)$. For instance, costs ~ $N(80, 8^2)$. The mean is either estimated from the observed values or is known. The variance is assumed to be known too. The distributions considered include the normal (Gaussian), a skewed multivariate normal, exponential, gamma and the Dirichlet distribution. The correlation coefficients between pairs of variables can vary from ±0.7, ±0.6, ±0.4 to 0.0.

3 Estimation by MCMC Simulation Technique

Each of the random variables x, z is described by its density function. We use the Metropolis-Hastings algorithm, cf. Hastings (1970), for the MCMC simulation of the random variables which are transformed according to M. In the first step we consider each of the p variables and assign that subset of equations to it, where it shows up either as a LHS or RHS variable. In the later case the corresponding equations are to be solved for the given variable to make it a LHS variable. For example, resolving for profit in M we get profit = sales - costs and profit = ROI * capital. The next step is to start MCMC sampling of all RHS variables. Note that the distribution of each LHS variable is either fully specified or unknown. In the last case it will be estimated from the corresponding equations where it shows up as a LHS or RHS variable. If a random variable shows up in $1 < k \leq q$ equations the k simulations must be "fused". Therefore k pairs of lower ($\underline{q_\kappa}$) and upper ($\overline{q}_\kappa$) α/2–quantiles for that variable are computed. Let us define $\underline{q_{max}} = max\{\underline{q_1}, \underline{q_2}, ..., \underline{q_k}\}$ and $\overline{q_{\min}} = \min\{\overline{q}_1, \overline{q}_2, ..., \overline{q}_k\}$. Then we have

Definition 1: A data set of an equation system M is called M-inconsistent (contradictive) if at least for one variable it is true that $\overline{q_{min}} \leq \underline{q_{max}}$.

In Fig.1 we illustrate M-inconsistency for the case of a variable x, say, sampled as x_1 and x_2 from two equations of M. Evidently, in the upper case the overlap $\boldsymbol{I}$ is empty, i.e. the data set is of bad quality while in the lower case the overlap $\boldsymbol{I}_q$ is non empty and the data are (weak) consistent. The final step is to project the joint distribution on the subspace spanned by $x_1 - x_2 = 0$ getting the density $f_{x_1,x_2}(x,x)$ for all $x \notin I_q = \left[\underline{q_{\max}}, \overline{q_{\min}}\right]$. The algorithm SamPro – sampling and projection – summarizes the procedure.

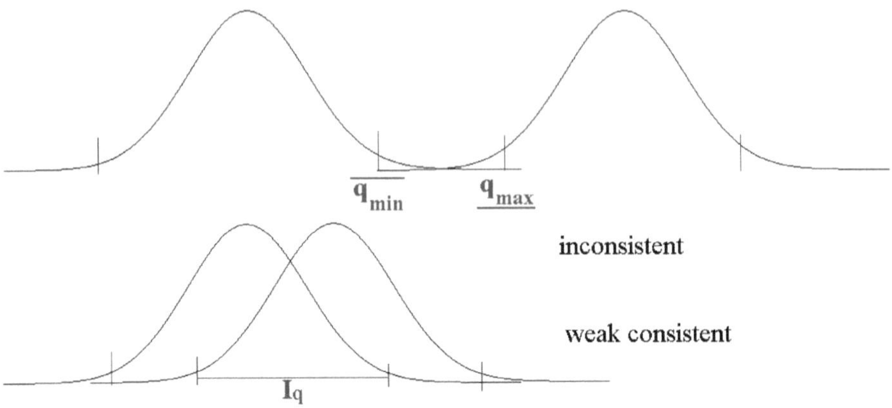

Figure 1: M-inconsistency and M-consistency

SamPro-Algorithm

input: a stochastic equation system M, one observation per variable (missing values allowed), error probability α

output: estimates for all variables, i.e. densities, means and standard deviations

begin

resolve (set LHS[1] ≡ RHS) for all variables in all equations

sample from the joint density function of all RHS variables for LHS

estimate all LHS variables

estimate α/2, 1-α/2 - quantiles $\underline{q_{max}}, \overline{q_{min}}$ for each variable, i=1,2,…,p.

if $\overline{q_{min}} > \underline{q_{max}}$ then "M -inconsistency found" and stop

else **compute** the distribution $\hat{f}_{xz}$ restricted to the subspace $x\text{-}z = 0$

end

4 Scenarios and Robustness Analysis

Let us start with generating scenarios given the two-equation model M, i.e. sales = profit + cost as a linear equation and ROI = profit / capital as a nonlinear relation.

---

[1] $z = x_1 + x_2$ then z is LHS and the x_1 and x_2 are RHS.

In all of our experiments with up to 2.5 million replications each, we assume that a realization of the random variables profit, cost and capital is at hand. Moreover, their types of distributions are varying. This implies that at least the mean of the various distributions can be determined. Furthermore, prior information is available about the standard deviation or variance of the measurement errors. The other two variables, i.e. sales and ROI, are handled in experimental group A as variables with missing values (null values) and later in group B as (correctly or noisy) observed values. If missing values of variables exist, they must be estimated, i.e. imputed.

The MCMC simulation results are compared with GLS estimation approach using the software package QUANTOR, originated as PRTI by Schmid (1979). In experimental group A the first three experiments analyze the effect of skewed distributions compared with Gaussian distributions, if all variables are not correlated. The next two experiments separately investigate the effects of cross-correlation. Finally, the interaction between non-normality and correlation is of concern.

In experimental group B the data set is complete and Gaussian distributions are assumed, that means no missing values exist. The effects of negative, zero and positive correlation are studied for two cases: The measurement of the variables fulfill ("M -inconsistent variables") or do not fulfill the balance equation system.

Experimental Group A: Effects of non-normality and / or correlation in the case of missing values

4.1 Scenario 1: Normality; no correlation

Specification of the distributions:

Profit $\sim N(20,2^2)$; Cost $\sim N(80, 8^2)$; Capital $\sim N(60, 6^2)$

Missing values:

Sales, ROI unknown

Result: The MCMC simulation and the GLS estimation result for sales as part of a linear relation are compatible with respect to mean and standard deviation (sd). The same is true for the imputation of ROI as a nonlinear relationship. Note, that the Gaussian hypothesis about the distribution of all variables is valid.

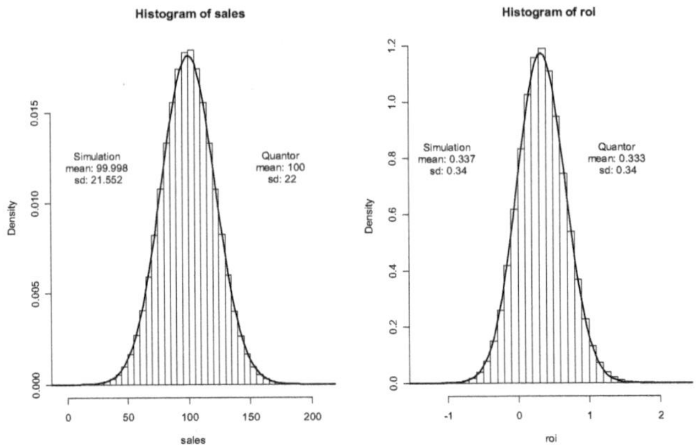

Figure 2: Normal distributed variables with no correlation

4.2 Scenario 2: Effect of Skewness; no correlation

Specification of the distributions:

Profit ~ Exp(1/20) vs. N(20, 20^2); Cost ~ Gamma(8,0.1) vs. N(80, 28^2); Capital ~ Gamma(15, 0.25) vs. N(60, 8.6^2)

Missing values:

Sales, ROI unknown

Result: In the linear case the mean and the standard deviation are similar, as the experiments for the variable sales show. In the nonlinear case the mean and the standard deviation (sd) are overestimated by about 15%.

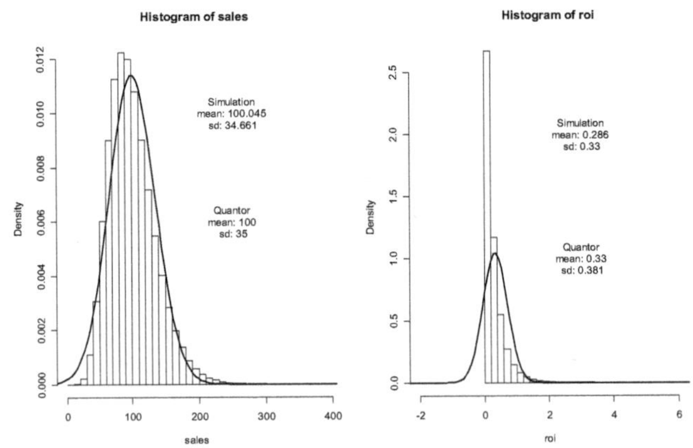

Figure 3: Effect of skewed distributions of all (observed) variables; no correlation

4.3 Scenario 3: Normality and negative cross-correlation

Specification of the distributions:

Profit ~ N(20, 2^2); Cost ~ N(80, 8^2); Capital ~ N(60, 6^2)

Missing values:

Sales, ROI unknown

Correlation used for simulation:

ρ(profit, cost) = -0.7; ρ(profit, capital) = -0.7

Result: While the means are nearly equal, the standard deviations differ between +18% and –16%.

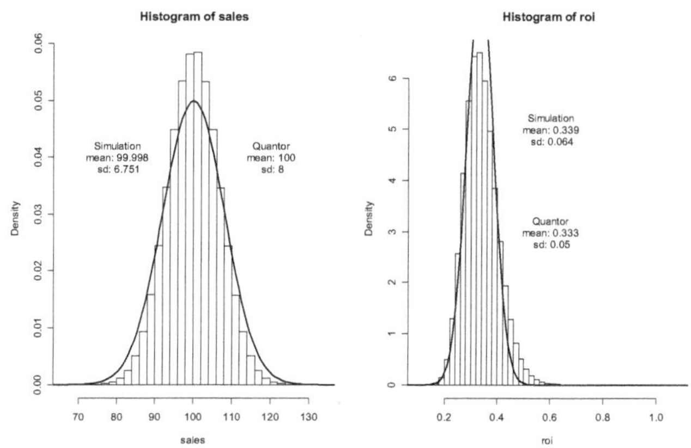

Figure 4: Normality and negative correlation (ρ = -0.7)

4.4 Scenario 4: Normality and positive cross-correlation

Specification of the distributions:

Profit ~ N(20, 2^2); Cost ~ N(80, 8^2); Capital ~ N(60, 6^2)

Missing values:

Sales, ROI unknown

Correlation used for simulation:

ρ(profit, cost) = 0.7; ρ(profit, capital) = 0.7
The positive sign of the correlation coefficients is contra intuitive from a manager's point of view. Nevertheless, it is used here more formally as an opposite case to negative correlation in scenario 3.

Result: While the means have nearly the same values, the percentage of differences changes sign: -19% for sales vs. +16% for ROI.

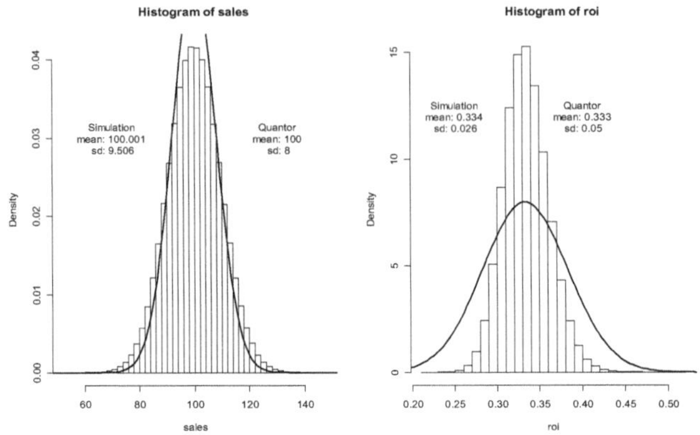

Figure 5: Normality and positive correlation (ρ = 0.7)

4.5 Scenario 5: Skewness and positive cross-correlation

Specification of the distributions:

$$
\begin{aligned}
&(Profit, Costs, Capital) \\
&\sim MSN\left(\xi=(2,8,6), \Omega=\begin{pmatrix} 1 & 0.6 & 0.7 \\ 0.6 & 1 & 0.2 \\ 0.7 & 0.2 & 1 \end{pmatrix}, \alpha=(2500,80,300)\right) \; vs.
\end{aligned}
$$

$$
\begin{aligned}
Profit &\sim N(20, 4.3^2) \\
Costs &\sim N(80, 8.3^2) \\
Capital &\sim N(60, 7.3^2)
\end{aligned}
$$

Missing values:

Sales, ROI unknown

Correlation used for simulation: This correlation is imposed by multivariate skewed normal distribution (MSN) (see Azzalini and Valle (1996) with the above given parameters. Note that the parameterisation is adopted from Azzalini and Capitanio (1999).

ρ(profit, cost) = 0.4; ρ(profit, capital) = 0.5

Result: While the means have nearly the same value for both variables, this is not true for the standard deviation (sd). For sales we get sd = 10.87 by simulation and sd' = 9.4 under the Gaussian assumption. The corresponding values for ROI are sd = 0.063 vs. sd' = 0.069.

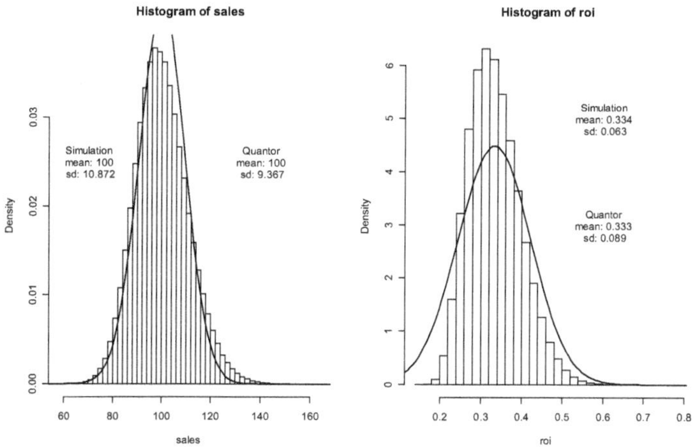

Figure 6: Skewness and positive cross-correlation

4.6 Scenario 6: Skewness and negative cross-correlation

Specification of the distributions:

(Profit, Cost, Capital) ~ Dir(30, 40, 8) vs.

Profit ~ $N(20, 2.8^2)$; Cost ~ $N(80, 8.8^2)$; Capital ~ $N(60, 20^2)$

Missing values:

Sales, ROI unknown

Correlation used for simulation imposed by Dirichlet (Dir) distribution:

ρ(profit, cost) = -0.8; ρ(profit, capital) = -0.3

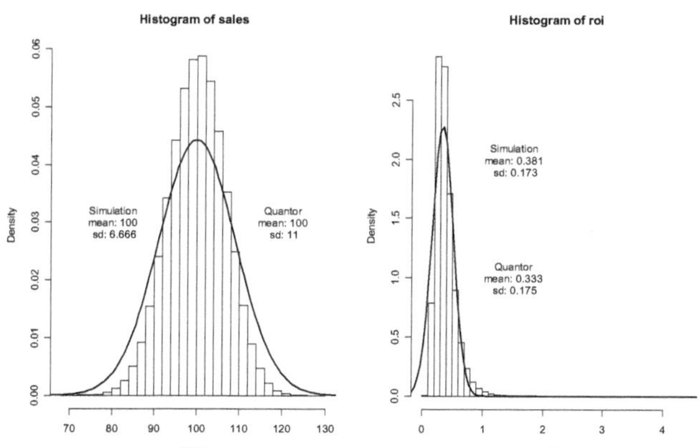

Figure 7: Non-Normality and negative cross-correlation

Result: While the means of the variable 'sales' are identical, this is not true for ROI. The GLS approach over-estimated the simulated (exact) values about 12%. Quite opposite, the standard deviation of sales is over-estimated by GLS by about 65% while sd of ROI is about the same.

Experimental Group B: Normality, cross-correlation, no missing values

Scenario 1: Normality, M - consistent observations of sales and ROI, and cross-correlation

Specification of the distributions:

Profit $\sim N(20,2^2)$; Cost $\sim N(80, 8^2)$; Capital $\sim N(60, 6^2)$; Sales $\sim N(100, 10^2)$; ROI $\sim N(0.333, 0.333^2)$

Note that as above the distribution of profit is threefold determined by the prior $N(20,2^2)$, by profit = sales – cost and by profit = ROI * capital.

Missing values: no

Correlation Matrix specified as: $\rho \in \{-0.4. 0.0, 0.4\}$. Economic reasoning leads to different signs of the correlation coefficient ρ. This expert knowledge might vary from company or business sector and represents in this example only one possible specification out of many. The lower and upper bounds of ρ are necessary to ensure a positive definite correlation matrix ***R***:

$$\boldsymbol{R} = \begin{bmatrix} & \text{profit} & \text{cost} & \text{capital} & \text{sales} & \text{roi} \\ \text{profit} & 1 & \rho & 0 & -\rho & -\rho \\ \text{cost} & \rho & 1 & 0 & -\rho & \rho \\ \text{capital} & 0 & -\rho & 1 & 0 & \rho \\ \text{sales} & -\rho & -\rho & 0 & 1 & -\rho \\ \text{roi} & -\rho & \rho & \rho & -\rho & 1 \end{bmatrix}$$

Results: The variance of costs, capital, sales and ROI is proportional to ρ. The variance of the estimated profit is non monotonic in ρ and has its maximum at $\rho = 0.2$. The means of all variables are more or less constant.

Variable	Mean	Sd	Mean	Sd
	Prior		Posterior	
Profit	20.00	2	19.87	1.58
Costs	80.00	8	79.95	6.29
capital	60.00	6	59.81	4.89
sales	100.00	10	99.96	6.37
ROI	0.33	0.33	0.33	0.03

Table 1: Means and Standard Deviations (Sd) of all observed and simulated variables, Gaussian distributions, no correlation, no missing values, M-consistent observations.

We close scenario 1 by presenting three three-dimensional scatter plots showing the simulated values of the variable profit determined from the prior distribution and the two RHS of the model equations for $\rho = -0.4, 0.0, 0.4$.

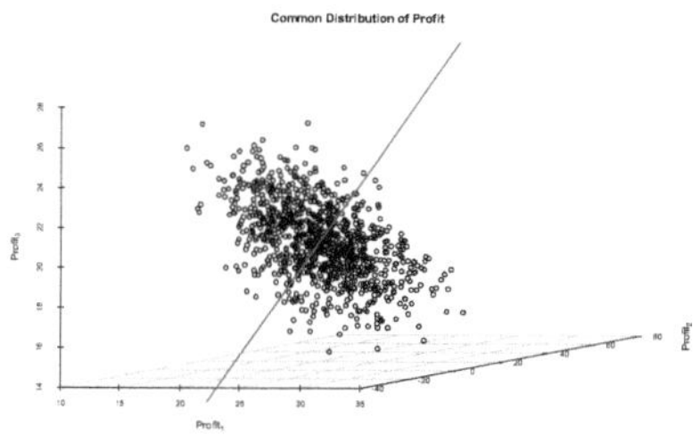

Fig. 9a: Scatter plot of simulated profit values for $\rho = -0.4$

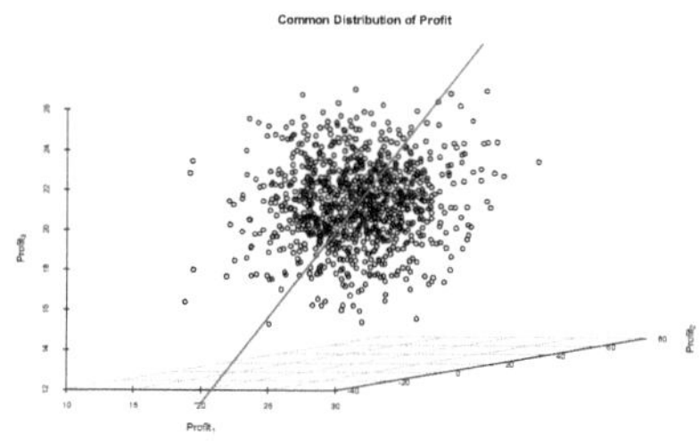

Fig. 9b: Scatter plot of simulated profit values for $\rho = 0.0$

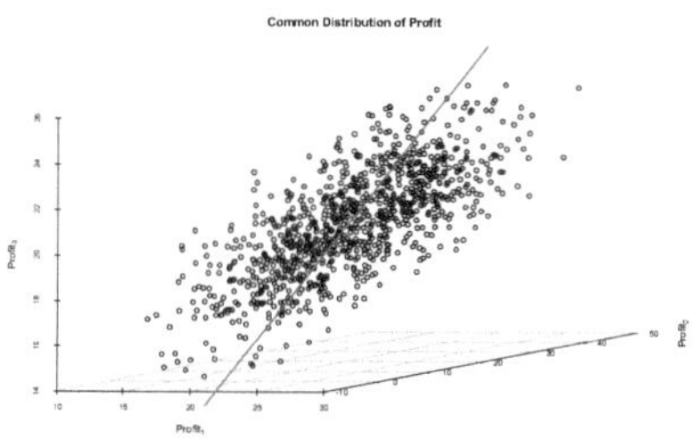

Fig. 9c: Scatter plot of simulated profit values for $\rho = +0.4$

Scenario 2: Normality, M-inconsistent observation of sales and ROI, and cross-correlations as in scenario 1

Specification of the distributions:

Profit $\sim N(30,3^2)$; Cost $\sim N(80, 8^2)$; Capital $\sim N(60, 6^2)$; Sales $\sim N(100, 10^2)$; ROI $\sim N(0.333, 0.333^2)$

Note: As in scenario 1 the distribution of profit is threefold determined by the prior $N(30,3^2)$, and by the equations profit = sales – costs and profit = ROI * capital. The mean and standard deviation of profit are increased from $N(20,2^2)$ to $N(30,3^2)$. This implies M-inconsistency of the observed values (means) of profit = sales - costs and profit = ROI * capital.

Missing values: no

Correlation Matrix as above with $\rho \in \{-0.4. 0.0, 0.4\}$

.

Results: The variance of costs, capital, sales and ROI is proportional to ρ. The variance of profit is non monotonic and gets a maximum at ρ = 0.2. The mean of profit is monotonically increasing, the means of the remaining variables are more or less constant. The case ρ = 0.4 leads to incoherency of profit, cf. Fig 9c, thus implying the M-incoherency of the whole equation system with the data set. Note that the observed value of each variable is equal to its corresponding (estimated) mean. To ensure that the first moments fulfill the equation system in a case of weak consistency, it might be necessary to iterate the SamPro algorithm. But after a few iterations this is achieved. In Tab. 2 the results of the 5th iteration are given. The first moments fulfill the equation system up to a small error.

Variable	Mean	Sd	Mean	Sd
	Prior		Posterior	
profit	30	3	24.93	0.85
costs	80	8	78.51	1.84
capital	60	6	67.03	2.33
sales	100	10	103.46	1.84
ROI	0.333	0.333	0.372	0.013

Table 2: Means and Standard Deviations of all observed and simulated variables, Gaussian distributions, no correlation, no missing values

Finally, we present three scatter plots in Fig. 9a-c for the variable profit with ρ ∈ {-0.4. 0.0, 0.4}. Note the effect of a "too large" mean of profit, i.e. $N(30,3^2)$, on the overlap of the point cloud and the (linear) subspace spanned by simulated values of profit, $profit_1$ and $profit_2$.

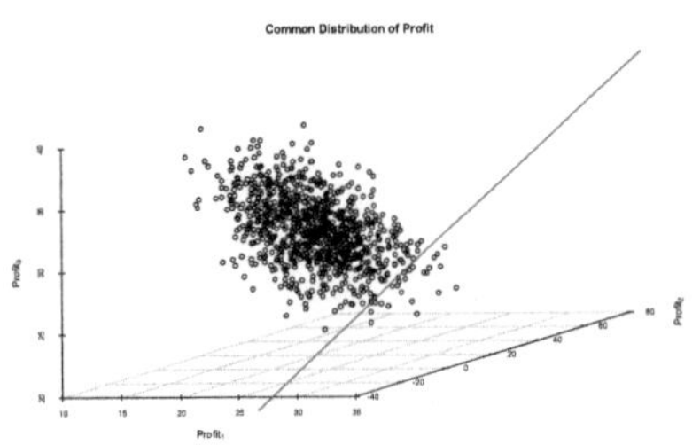

Fig. 10a: Scatter plot of simulated profit values for ρ = - 0.4

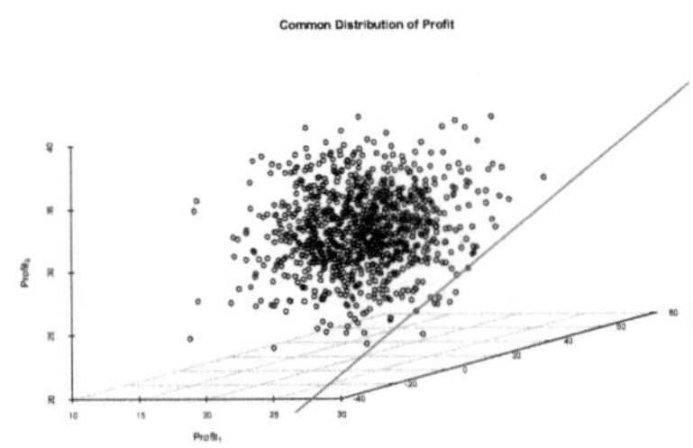

Fig. 10b: Scatter plot of simulated profit values for ρ = 0.0

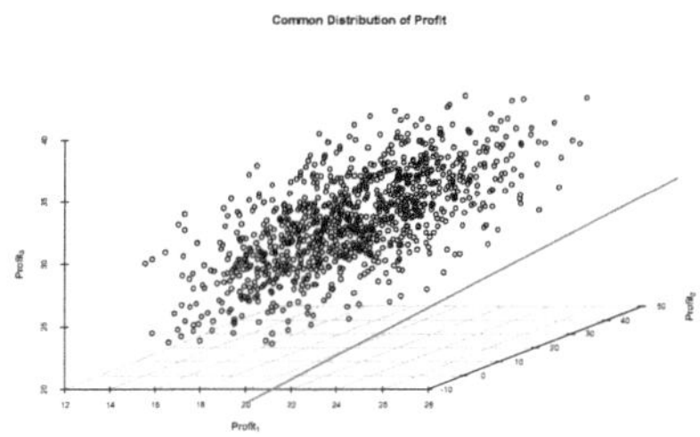

Fig. 10c: Scatter plot of simulated profit for $\rho = +0.4$

It is worthwhile mentioning that the simulated values of an M-consistent variable should lie on the straight line as in Fig. 9c, i.e. should fulfil all balance equations. As mentioned above the empty intersection of variable profit is caused by a "too large" (estimated) mean of the related distribution.

5 Conclusion

We can summarize our study of a non-Gaussian non linear equation model as follows:

1. In the uncorrelated case, (the means of) the simulated quantities are about the same as the GLS estimates.
2. Skewness of distributions mostly has only a small effect on the estimates.
3. Positive cross-correlations of the variables can lead to severe problems: The equation system may become M-inconsistent with respect to a given data set, i.e. the overlap of the sets of simulated values of at least one variable determined from all equations, where it is part of, may become empty. Under a Gaussian regime with infinite domains and under the independence assumption this effect cannot happen.

Using a GLS approach is relative to MCMC simulation computational cost-effective. But skewness and correlation may lead to quite different estimates and the introduction of robust estimators, like median, improves all estimates. In the case of M-inconsistency it may be necessary to iterate the simulation algorithm several times for satisfying a given balance equation system. Of course, any iteration increases the computational efforts. Furthermore, note that if a given data set is contradictive to the corresponding equation system M-inconsistency is revealed by our MCMC simulation algorithm quite in contrast to the GLS approach used by QUANTOR which assumes a Gaussian regime.

6 References

John Aitchison. The Statistical Analysis of Compositional Data. Kluwer, 1986.

Adelchi Azzalini and Antonella Capitanio. Statistical Applications of the Multivariate Skew Normal Distribution, Journal of the Royal Statistical Society. Series B, 61, 579-602, 1999.

Adelchi Azzalini and Alessandra Dalla Valle. The Multivariate Skew-Normal Distribution, Biometrika, 83, 715-726, 1996.

Carlo Batini and Monica Scannapieco. Data Quality Concepts, Methodologies and Techniques, Springer, 2006.

Siddhartha Chib. Handbook of Computational Statistics - Concepts and Methods, chapter Markov Chain Monte Carlo Technology, pages 71–102. Springer, 2004.

I. P. Fellegi and D. Holt. A Systematic Approach to Automatic Edit and Imputation, JASA, 71, 17-35, 1976.

W. Keith Hastings. Monte Carlo sampling methods using markov chains and their applications. Biometrika, 57:97–109, 1970.

Veit Köppen and Hans-J. Lenz. Simulation of non-linear stochastic equation systems. In A.N. Pepelyshev, S.M. Ermakov, V.B. Melas, eds., Proceeding of the Fifth Workshop on Simulation, pages 373–378, St. Petersburg, Russia, July 2005. NII Chemistry Saint Petersburg University Publishers.

Hans-J. Lenz and Roland M. Müller. On the solution of fuzzy equation systems. In G. Della Riccia, H-J. Lenz, and R. Kruse, eds., Computational Intelligence in Data Mining, CISM Courses and Lectures. Springer, New York, 2000.

Hans-J. Lenz and Egmar Rödel. Statistical quality control of data. In Peter Gritzmann, Rainer Hettich, Reiner Horst, and Ekkehard Sachs, editors, 16th Symposium on Operations Research, pages 341–346. Physica Verlag, Heidelberg, 1991.

Gunar E. Liepins and V.R.R. Uppuluri. Data Quality Control Theory and Pragmatics, Marcel Dekker, 1991.

Beat Schmid, (1979). Bilanzmodelle. Simulationsverfahren zur Verarbeitung unscharfer Teilinformationen, ORL-Bericht No. 40, ORL Institut, ETH Zürich, 1979.

Adian F. M. Smith and Alan E. Gelfand. Bayesian statistics without tears: A sampling-resampling perspective. The American Statistician, 46(2):84–88, may 1992.

G.Barrie Wetherill and Marion E. Gerson. Computer Aids to Data Quality Control, The Statisticians, 36, 598-592, 1987.

The Sensitivity of Common Capability Indices to Departures from Normality

Fred Spiring

Department of Statistics, The University of Manitoba,Winnipeg, Manitoba, CANADA R3T 2N2
spiring@cc.umanitoba.ca, fspiring@mts.net

Summary. The process capability index *Cpw* provides a general representation for a wide variety of process capability indices including *Cp, Cpk, Cpm* and *Cpmk*. In this manuscript we will develop a procedure to investigate the sensitivity of *Cpw* to departures from normality and discuss the impact on inferences drawn for a variety of regions and weights. The focus will be on the widely used indices and in particular those indices whose inference focuses on the ability of the process to be clustered around the target. The robustness of $\hat{Cpw}$ to distributional assumptions and the resulting impact on the inferences will be compared.

1 Measuring Process Capability

Juran (1979) suggested that Japanese companies initiated the use of process capability indices by relating process variation to customer requirements in the form of the ratio

$$Cp = \frac{\mathrm{USL} - \mathrm{LSL}}{6\sigma}$$

where the difference between the upper specification limit (USL) and the lower specification limit (LSL) provides a measure of allowable process spread (i.e., customer requirements) and 6σ a measure of actual process spread (i.e., process performance).

Processes with small variability, but poor proximity to the target, sparked the derivation of several indices that incorporate targets into their assessment. The most common of these measures assume the target (T) to be the midpoint of the specification limits and include

$$Cpu = \frac{\mathrm{USL} - \mu}{3\sigma}, \quad Cpl = \frac{\mu - \mathrm{LSL}}{3\sigma},$$

$$Cpk = \min(Cpl,\ Cpu),$$

H.-J. Lenz et al. (eds.), *Frontiers in Statistical Quality Control 9*,
DOI 10.1007/978-3-7908-2380-6_18,

$$Cpm = \frac{\text{USL} - \text{LSL}}{6\sqrt{\sigma^2 + (\mu - \text{T})^2}}$$

and

$$Cpk = (1 \text{ - k})\ Cp$$

where $\text{k} = \dfrac{2|T-\mu|}{USL-LSL}$, $0 \le \text{k} \le 1$ and LSL < μ < USL. The two definitions of *Cpk* are presented interchangeably and are equivalent when $0 \le \text{k} \le 1$.

The versions of these measures that do not assume T to be the midpoint of the specifications include

$$Cpu^{*} = \frac{\text{USL} - \text{T}}{3\sigma}\left(1 - \frac{|\text{T} - \mu|}{\text{USL} - \text{T}}\right),\ Cpl^{*} = \frac{\text{T} - \text{LSL}}{3\sigma}\left(1 - \frac{|\text{T} - \mu|}{\text{T} - \text{LSL}}\right),$$

$$Cpk^{*} = \min(Cpl^{*}, Cpu^{*})$$

and

$$Cpm^{*} = \frac{\min[\text{USL} - \text{T}, \text{T} - \text{LSL}]}{3\sqrt{\sigma^2 + (\mu - \text{T})^2}}.$$

A hybrid of these measures, *Cpmk*, is defined to be

$$Cpmk = \frac{\min[\text{USL} - \mu, \mu - \text{LSL}]}{3\sqrt{\sigma^2 + (\mu - \text{T})^2}}.$$

By definition, *Cp, Cpl, Cpu, Cpk, Cpm* and *Cpmk* are unitless, thereby fostering comparisons among and within processes regardless of the underlying mechanics of the product or service being monitored. In all cases, as process performance improves, either through reductions in variation and/or moving closer to the target, these indices increase in magnitude. As process performance improves relative to customer requirements, customer satisfaction increases as the process has greater ability to be near the target. In all cases larger index values indicate a more capable process. Using the indices to assess the ability of the process to meet customer expectations (with larger values of the indices indicating higher customer satisfaction and lower values, poorer customer satisfaction) is consistent with the concept of "fitness for use". Individually *Cpu, Cpu*, Cpl* and *Cpl** consider only unilateral tolerances (i.e., USL or LSL respectively) in assessing process capability.

A unified index (Spiring 1997), Cpw, of the form

$$Cpw = \frac{USL - LSL}{6\sqrt{\sigma^2 + w(\mu - T)^2}}$$

can be used to represent a wide spectrum of capability indices. Allowing *w* to take on various values permits *Cpw* to assume equivalent computational algorithms for a variety of indices including *Cp, Cpm, Cpk* and *Cpmk*. Setting $w = 0$, results in

$$Cpw = \frac{USL - LSL}{6\sqrt{\sigma^2}} = Cp,$$

while $w = 1$,
$$Cpw = \frac{USL - LSL}{6\sqrt{\sigma^2 + (\mu - T)^2}} = Cpm.$$

Defining $d = \frac{USL - LSL}{2}$, $\mu - \frac{USL + LSL}{2}$ and $p = \frac{|\mu - T|}{\sigma}$ then

$$w = \begin{cases} \left(\frac{d^2}{(d-|a|)^2} - 1\right)\left(\frac{1}{p^2}\right), & 0 < p \\ 0 & \text{elsewhere} \end{cases}$$

allows $Cpw = Cpk$, while

$$w = \begin{cases} \left(\frac{d}{d-|a|}\right)^2\left(\frac{1}{p^2} + 1\right) - \frac{1}{p^2}, & 0 < p \\ 0 & \textit{elsewhere} \end{cases}$$

results in $Cpw = Cpmk$ and

$$w = \begin{cases} \left(\frac{d}{d-u|a|}\right)^2\left(\frac{1}{p^2} + v\right) - \frac{1}{p^2}, & 0 < p \\ 0, & \textit{elsewhere} \end{cases}$$

where u and v are weights as defined in Vännman (1995) result in $Cpw=Cp(u, v)$.

2 Interpreting Process Capability Measures

Process capability measures have traditionally been used to provide insights into the number of non-conforming product (i.e., yield). Practitioners cite a *Cp* value of one as representing 2700 parts per million (ppm) non-conforming, while 1.33 represents 63 ppm; 1.66 corresponds to .6 ppm; and 2 indicates < .1 ppm. *Cpk* has similar connotations, with a *Cpk* of 1.33 representing a maximum of 63 ppm non-conforming. Practitioners use the value of the process capability index and its associated number non-conforming to identify capable processes. A process with a *Cp* greater than or equal to one has traditionally been deemed capable. While a *Cp* of less than one indicates that the process is producing more than 2700 ppm non-conforming and used as an indication that the process is not capable of meeting customer requirements.

Inherent in any discussion of yield as a measure of process capability, is the assumption that product produced just inside the specification limit is of equal quality to that produced at the target. This is equivalent to assuming what Tribus and Szonyi (1989) refer to as the square-well loss function (Figure 1) for the quality variable. By design *Cp, Cpl, Cpu, Cpk* and *Cpmk* are used to identify changes in the amount of product beyond the specification limits and therefore consistent with the square-well loss function. Any change in their magnitude, certeris paribus, is due entirely to changes in the distance between the specification limits and the process mean.

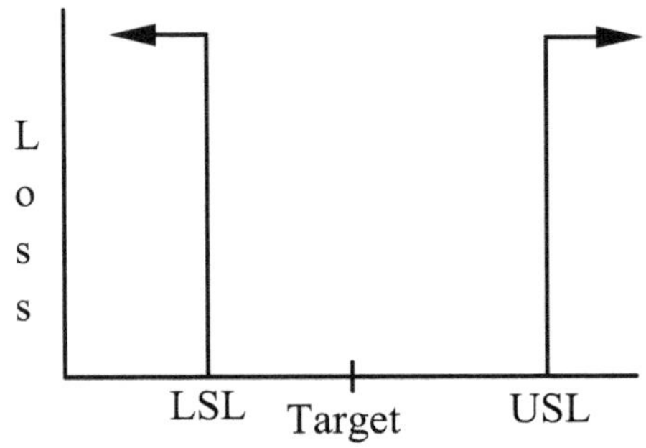

Figure 1. Square-Well Loss Function

Taguchi (1986) used the quadratic loss function to motivate the idea that a product imparts no loss, only if that product is produced at its target. He maintains that even small deviations from the target result in a loss of quality and that as the product increasingly deviates from its target there are larger and larger losses in quality. This approach to quality and quality assessment is different from the traditional approach, where no loss in quality is assumed until the product deviates beyond its upper or lower specification limit (i.e., square-well loss function). Taguchi's philosophy highlights the need to have small variability around the target. In this context the most capable process will be one that produces all of its product at the target, with the next best being the process with the smallest variability around the target.

The motivation for *Cpm* does not arise from examining yield, but from looking at the ability of the process to be in the neighborhood of the target. This motivation has little to do with the number of non-conforming although upper bounds on the number of non-conforming can be determined for numerical values of *Cpm* (Spiring 1991). Several authors (Boyles 1991; Johnson 1992; Spiring 1997) have discussed the relationship between *Cpm* and the quadratic loss function (see Figure 2) and its affinity with the philosophies that support a loss in quality for any departure from the target.

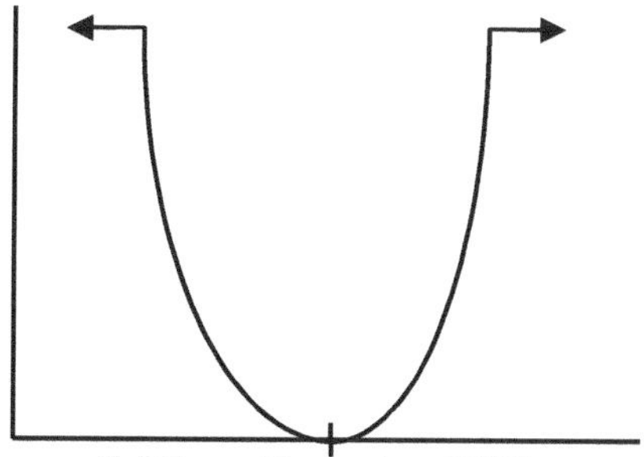

Figure 2. Modified Quadratic Loss Function

Cpk, *Cpm* and *Cpmk* have different functional forms, are represented by different loss functions and have different relationships with *Cp* as the process drifts from the target (Spiring 2007). Hence although *Cpm* and *Cpk* are lumped together as second generation measures they are very different in their development and assessment of process capability. As a hybrid of these second generation assessments of process capability, *Cpmk* behaves similarly to *Cpk* in its assessment of process capability.

3 Effects of Non-normality

If the process measurements do not arise from a normal distribution none of the indices discussed provide valid measures of yield. Each index uses a function of σ as a measure of actual process spread in its determination of process capability. But as several authors (Hoaglin et al. 1983; Mosteller and Tukey 1977; Tukey 1970; Huber 1977) have pointed out, although the standard deviation has become synonymous with the term dispersion, its physical meaning need not be the same for different families of distributions, or for that matter, within a family of distributions. Therefore the actual process spread (a function of 6σ) does not provide a consistent meaning over various distributions. To illustrate, suppose that precisely 99.73% of the process measurements fall within the specification limits (i.e., a yield of 2700 parts per million). The values of *Cp* (as well as *Cpl, Cpu, Cpk, Cpm* and *Cpmk*) are 0.5766, 0.7954, 1.0000, 1.2210 and 1.4030 respectively when the measurements arise from a uniform, triangular, normal, logistic and double exponential distribution. As long as 6σ carries a yield interpretation when assessing process capability, none of the indices should be used if the underlying process distribution is not normal.

If we assume process capability assessments to be studies of the ability of the process to produce product around the target, then *Cpm* and *Cpmk* provide practitioners with an assessment of capability regardless of the distribution associated with the measurements. Clustering around the target, rather than a measure of non-conforming releases the physical meaning attached to 6σ. The denominator of *Cpm* and *Cpmk* then provide a measure of the clustering around the target and compares this with customer tolerance.

Eliminating the physical ppm meaning and investigating Euclidean distances around the target allows *Cpm* and *Cpmk* to compare the capability of various processes, or processes over time, regardless of the underlying distribution. The underlying distribution will impact the inferences that can be made from samples gathered from the population (Leung and Spiring 2007), however the population parameter is no longer distributionally sensitive. With this warning, we will provide a technique that can be used to examine a wide range of indices in order to better understand the effect of non-normality on the behaviour of the estimated process capability indices.

We will investigate the sensitivity of the various indices to departures from normality using a modified Gayen (1949) approach where the pdf associated with $\sum_{i=1}^{n}(x_i - T)^2$ is transformed to reflect the pdf associated with $\hat{C}pw$. Analogous to Chan et al. (1988), the third and fourth moments are then varied to examine the impact of moderate departures from normality on the derived distribution of $\hat{C}pw$.

Defining $\sigma' = \sqrt{\sigma^2 + w(\mu - T)^2}$, and letting $x_1, x_2, \ldots, x_n$ denote a random sample of size n, the maximum likelihood estimators for both μ and σ^2 result in the estimator

$$\hat{C}pw = \frac{USL - LSL}{6\sqrt{\sum_{i=1}^{n}\frac{(x_i - \bar{x})^2}{n} + w(\bar{x} - T)^2}} = \frac{USL - LSL}{6\sqrt{\frac{\sum_{i=1}^{n}(x_i - \bar{x})^2 + nw(\bar{x} - T)^2}{n}}}.$$

The probability density function of $\hat{C}pw$ becomes

$$g_{\hat{C}pw}(t) = h_1\left(\frac{a^2}{t^2}\right)\left|-\frac{2a^2}{t^3}\right| \qquad 0 < t < \infty$$

where $a = \mu - \frac{USL + LSL}{2}$ and $h_1(y) = \int_0^{\frac{y}{nw}} h_{Y,Z}(y,z)\, dz$ (see Appendix).

The third (λ_3) and fourth (λ_4) moments are then varied to examine the impact of moderate departures from normality for the characteristic under investigation on the sampling distribution of $\hat{C}pw$. For example, the approximate probability density function of $\hat{C}pw$ when n = 5, T = 0, USL-LSL = 4, w = 1 (i.e. *Cpw* = *Cpm*) is

$$g_{\hat{C}pw}(t) = e^{-\frac{0.52}{9t^2}} \left\{ 0.133 + 0.002375\lambda_4 \left(-10.74 + \frac{20}{9t^2}\right)\left(-3.26 + \frac{20}{9t^2}\right) + 0.00044\lambda_3^2 \left(-16.94 + \frac{20}{9t^2}\right)\left(-7.62 + \frac{20}{9t^2}\right)\left(-2.441 + \frac{20}{9t^2}\right)\right\}\left(\frac{20}{9t^2}\right)^{3/2}$$

While for n=5, T=0, USL-LSL=4, w=0 (i.e., Cpw=Cp) results in

$$g_{\hat{C}pw}(t) = 800e^{-\frac{1.1111}{t^2}} \left\{ 0.25 + 0.00026\left(-15.52 + \frac{20}{9t^2}\right)\left(-6.61 + \frac{20}{9t^2}\right)\left(-1.87 + \frac{20}{9t^2}\right)\lambda_3^2 + 0.0042 * \left(-9.46 + \frac{20}{9t^2}\right)\left(-2.536 + \frac{20}{9t^2}\right)\lambda_4 \right\} \Big/ 81t^5$$

The ability to determine the pdfs associated with various parameters (Cpw weight, sample size, specification limits and targets) allows us to examine the impact of nonzero values of the third and fourth moments on the pdf of $\hat{C}pw$. The plots of Figure 3 depict the distorted pdf of x and the resulting pdf of $\hat{C}pw$ for selected values of λ_3 and λ_4, provided that the combinations of λ_3 and λ_4 produce a valid non-negative probability density distribution (see Barton and Dennis 1952). Tables 1 through 4 allow comparisons among p-values associated with the resulting pdf of $\hat{C}pw$ for selected values of λ_3 and λ_4.

The results summarized in Figure 3 and Tables 1 - 4 suggest moderate distortions in the underlying distribution of the process measurements have a substantial impact on the shape of the pdf associated with $\hat{C}pw$ and the resulting tail probabilities. In those cases where the practitioner is monitoring process capability on a regular basis inferences will be affected. For those process measurements exhibiting normal distribution characteristics but with non-zero estimates of λ_3 and λ_4, corrections should be made either to the action limits or to the specified level of α associated with the decision making process or capability chart (see section 4 Example: Perforation Breaking Strength).

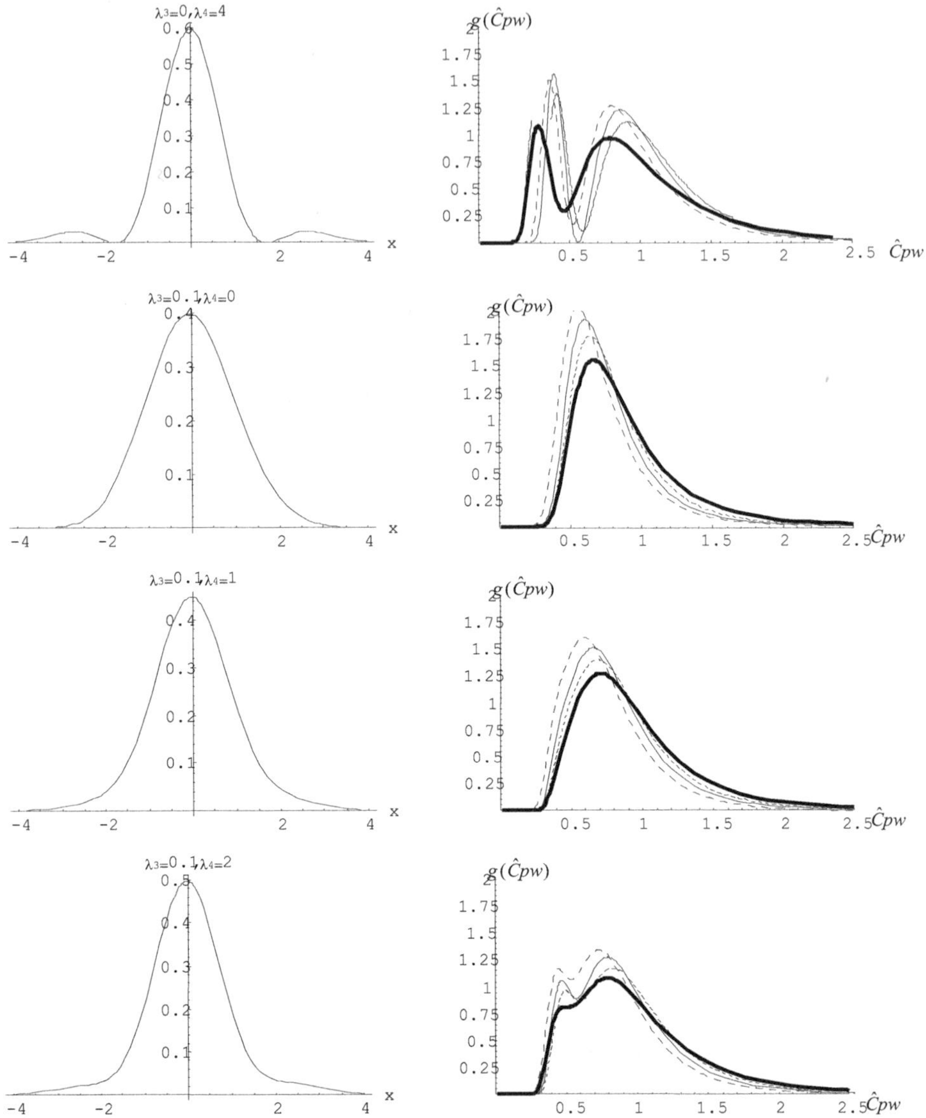

Figure 3. Selected pdfs associated with $\hat{C}pw$ for w = 0, .5, 1 and 2

Legend

$g(\hat{C}pw)$ with $w = 0.0$	thick solid line
$g(\hat{C}pw)$ with $w = 0.5$	dotted line
$g(\hat{C}pw)$ with $w = 1.0$	thin solid line
$g(\hat{C}pw)$ with $w = 2.0$	dashed line

Table 1. P[$\hat{C}pw$ > c = 1.5| w = 0, n = 5, T = 0, a =4/3, | λ_3 |, λ_4]

λ_4 / $\lvert\lambda_3\rvert$	0	.5	1.0	2.0	3.0	4.0
.0	.088338	.100771	.113204	.138070	.162935	.18780
.1	.088235	.100668	.113101	.137967	.162833	.18770
.2	*	.100359	.112792	.137658	.162524	*
.3	*	.099845	.112278	.137144	.162010	*
.4	*	.099125	.111558	.136424	.161290	*
.5	*	*	.110632	.135498	.160364	*
.6	*	*	*	.134367	.159233	*

Table 2. P[$\hat{C}pw$ > c = 1.5| w = 0.5, n = 5, T = 0, a =4/3, | λ_3 |, λ_4]

λ_4 / $\lvert\lambda_3\rvert$	0	.5	1.0	2.0	3.0	4.0
.0	.026686	.033079	.039471	.052255	.065039	.077823
.1	.026499	.032891	.039283	.052067	.064851	.077636
.2	*	.032319	.038711	.051496	.064280	*
.3	*	.031363	.037755	.050539	.063323	*
.4	*	.030022	.036414	.049198	.061983	*
.5	*	*	.034689	.047473	.060257	*
.6	*	*	*	.045364	.058148	*

Table 3. P[$\hat{C}pw$ > c = 1.5| w = 1, n = 5, T = 0, a =4/3, | λ_3 |, λ_4]

λ_4 / $\lvert\lambda_3\rvert$	0	.5	1.0	2.0	3.0	4.0
.0	.036445	.044891	.053338	.070232	.087125	.10402
.1	.036204	.044651	.053098	.069991	.086885	.10378
.2	*	.043930	.052377	.069270	.086164	*
.3	*	.042728	.051175	.068068	.084962	*
.4	*	.041045	.049492	.066386	.083279	*
.5	*	*	.047329	.064222	.081116	*
.6	*	*	*	.061578	.078472	*

Table 4. P[$\hat{C}pw$ > c = 1.5| w = 2, n = 5, T = 0, a =4/3, | λ_3 |, λ_4]

$\lvert\lambda_3\rvert$ \ λ_4	0	.5	1.0	2.0	3.0	4.0
.0	.026686	.033079	.039471	.052255	.065039	.077823
.1	.026499	.032891	.039283	.052067	.064851	.077636
.2	*	.032319	.038711	.051496	.064280	*
.3	*	.031363	.037755	.050539	.063323	*
.4	*	.030022	.036414	.049198	.061983	*
.5	*	*	.034689	.047473	.060257	*
.6	*	*	*	.045364	.058148	*

* denotes combinations where g($\hat{Cpw}$) is not positive definite

A summary of the results from Tables 1 through 4 (Table 5) suggests that $\hat{Cpm}$ is more robust to similar departures from normality than $\hat{Cp}$, while the index associated with a fixed weight of 0.5 is more robust than either $\hat{Cpm}$ and $\hat{Cp}$.

Table 5. Summary of Probability Results

w	Minimum	Maximum	Normal	Δ
1	.03620	.10402	.03645	.06782
0	.08824	.18780	.08838	.09957
.5	.02650	.07782	.02669	.05132
1.5	.03024	.08815	.03045	.05791

Similar to $\hat{Cpm}$, $\hat{Cpw}$ is not distributionally robust. However, if the practitioner can identify the amount of distortion from normality in terms of λ_3 and λ_4, corrections can be made that will provide the practitioner with viable decision rules or action limits. It appears that λ_4 has a substantially greater impact than that of λ_3 on inferences made. In order to maintain the same confidence in the decision that the process capability has significantly improved (increased) in the presence of nonzero values of λ_4, larger values of $\hat{Cpw}$ are required.

Ceteris paribus, and ignoring process yield, investigating the impact of w on g($\hat{Cpw}$) and the resulting decision process can be investigated. For example for $w = 0$ and $w = 1$, direct comparisons between the robustness of *Cp* and *Cpm* and the resulting inferences can be investigated. Further for $w = 1$ and specific values of w depending upon values of a, d and p comparisons among *Cpm, Cpk* and *Cmpk* can be investigated.

4 Example: Perforation Breaking Strengths

The adjusted breaking strengths of a perforation between non-folded lottery tickets was identified as a key process variable associated with the mechanical vending of lottery tickets (Spiring (1997)). An initial study of the process suggested that adjusted breaking strength (i.e., the strength required to initiate and completely separate a lottery ticket from its neighbor along the perforation) had non-zero values of λ_3 and λ_4. The first fifty observations from production were used to assess the normality of the breaking strengths. Both the histogram and the normal probability plot (Figure 4) suggest that breaking strengths deviate moderately from a normal distribution. The customer specifications and numerical results for the first fifty observations were as follows

N	50
USL	2.0
LSL	-2.0
Target	0
μ	-0.135
σ	0.83
λ_3	0.10
λ_4	0.99
Cpm	0.79

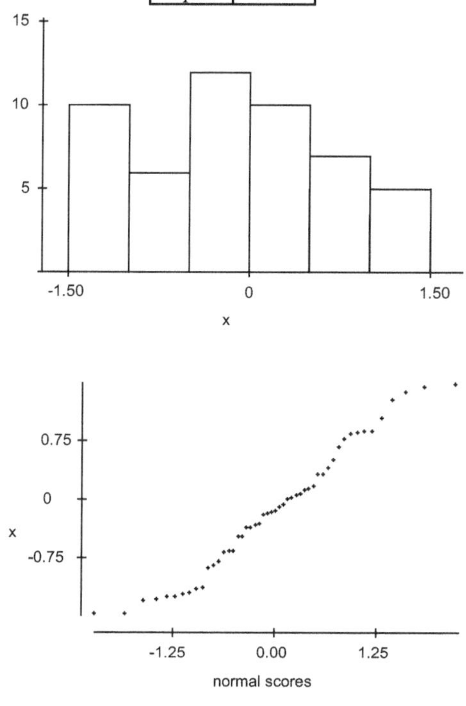

Figure 4. Histogram and probability plot of Breaking Strengths

Subsequent to the initial capability assessment (i.e., the first fifty observations), the process was routinely monitored. $\overline{X}$, s and $\hat{Cpm}$ based on samples of size 5 were calculated daily, with the first subgroup of size five being

x_1	0.63
x_2	-1.04
x_3	0.37
x_4	0.99
x_5	-0.48
$\overline{X}$	0.09
s	0.835
$\hat{Cpm}$	0.71

To investigate *Cpm* (assuming that the observations behave similar to the first fifty), the upper and lower critical values (i.e., adjusted action limits) were determined (for n = 5, $\lambda_3 = 0.1$ and $\lambda_4 = 1.0$) to be $C_L = .2690$ and $C_U = 3.004$, where C_U and C_L are such that Pr($\hat{Cpm} > C_U$) = Pr($\hat{Cpm} < C_L$) = 0.00135. Since $\hat{Cpm} = 0.71$ for the first subgroup, falls within the interval [.2690, 3.004], the practitioner concluded that the process capability, assessed using the index *Cpm,* had not changed from the initial capability study (α=0.0027). Figure 5 contains the results from the first 10 days including each subgroup $\hat{Cpm}$, the average $\hat{Cpm}$ for the ten subgroups (i.e., $\overline{Cpm}$), C_U and C_L assuming $\lambda_3=\lambda_4=0$ and the adjusted C_U and C_L for $\lambda_3 = 0.1$, $\lambda_4 = 1.0$. Adding the adjusted action limits to the $\hat{Cpm}$ run chart results in similar conclusions for the remaining subgroups.

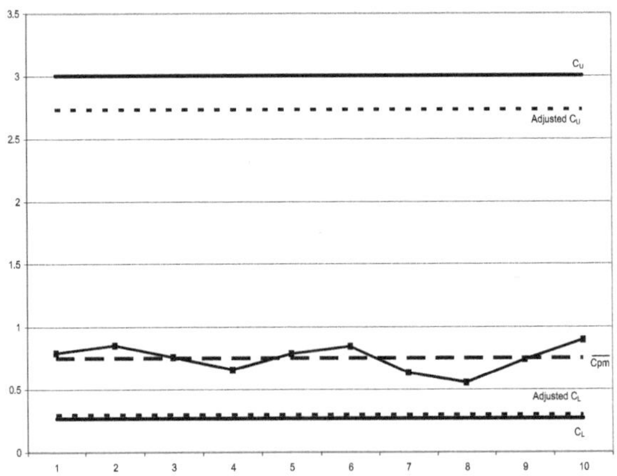

Figure 5. *Cpm* chart of Breaking Strengths by subgroup with Average $\hat{Cpm}$ ($\overline{Cpm}$), Adjusted (dashed lines) and Non-Adjusted (solid lines) Action Limits

5 Summary and Discussion

Robustness studies for those process capability indices whose magnitudes are translated into parts per million non-conforming are meaningless as the parameters are sensitive to departures from normality. Regardless of how robust an estimator maybe, its associated yield is distributionally sensitive and hence any robustness claims carry little meaning. Similarly, developing actual and approximate confidence intervals for these capability indices where the process characteristics arise from non-normal distributions is an academic pursuit with no application.

For those capability indices that attempt to assess the ability of the process to cluster around the target, the robustness of the estimator and associated inferences is a valid concern. The robustness of an estimator of *Cpw* in the face of quantifiable departures from normality can be determined and adjustments made to the critical values associated with assessing changes in the process capability. We have confirmed that moderate distortions in the underlying distribution of the process measurements have a substantial impact on the shape of the pdf associated with $\hat{Cpw}$ and the resulting tail probabilities.

Through the development of g($\hat{Cpm}$) adjustments for the impact of skewness and kurtosis on the tail probabilities and the associated inferences are possible. Further, action limit adjustments for various process values including T, USL and LSL as well as various sampling characteristics including sample size, λ_3 and λ_4 are now possible over a wide range of capability assessors. The development of similar corrective actions for $\hat{Cpmk}$ as well as other hybrid capability indices that allow the target to be other than the midpoint of the specification limits, while assessing clustering around the target, follow directly.

6 References

Barton, D. E. and Dennis, K. E. (1952) The Conditions under which Gram-Charlier and Edgeworth Curves are Positive Definite and Unimodal. Biometrika 39: 425-427.

Boyles, R.A. (1991) The Taguchi Capability Index. Journal of Quality Technology, 23: 17-26.

Chan, L. K., Cheng, S. W. and Spiring, F. A. (1988) Robustness of the Process Capability index, Cp, to Departure from Normality. Statistical Theory and Data Analysis II. North-Holland / Elsevier, Amsterdam, New York: pp. 223-239.

Gayen, A.K. (1949) The Distribution of 'Student's' t in Random Samples of Any Size Drawn from Non-Normal Universes. Biometrika 36: 353-369

Hoaglin, D.C., Mosteller, F., Tukey, J.W. (1983) Understanding Robust and Exploratory Data Analysis. Wiley, New York.

Huber, P.J. (1977) Robust Statistical Procedures. Society for Industrial and Applied Mathematics, Philadelphia.
Johnson, T. (1992) The Relationship of Cpm to Squared Error Loss. Journal of Quality Technology, 24: 211-215.
Juran, J.M. (1979) Quality Control Handbook. McGraw-Hill, New York.
Leung, B. P. K and Spiring, F. A. (2007) Adjusted Action Limits for Cpm based on Departures from Normality. International Journal of Production Economics, 107: 237-249.
Mosteller, F. and Tukey, J.W. (1977) Data Analysis and Regression. Addison-Wesley, Reading, MA.
Spiring, F. A. (1991) A New Measure of Process Capability. Quality Progress, 24: 57-61.
Spiring, F. A. (1995) Process Capability: A Total Quality Management Tool. Total Quality Management, 6: 21-33.
Spiring, F. A. (1997) A Unifying Approach to Process Capability Indices. Journal of Quality Technology, 29: 49-58.
Spiring, F. A. (2007) A Comparison of Process Capability Indices. Encyclopedia of Statistics in Quality and Reliability. Wiley, New York.
Taguchi, G. (1986) Introduction to Quality Engineering: Designing Quality into Products and Processes. Kraus, White Plains, New York.
Tribus, M. and Szonyi, G. (1989) An Alternate view of the Taguchi Approach. Quality Progress, 22: 46-52.
Tukey, J.W. (1970) A Survey of Sampling from Contaminated Distributions In Contributions to Probability and Statistics, I. Olkin, S. Ghurye, W. Hoeffding, W. Madow and H. Mann (Editors). Stanford University Press: pp. 448-485.
Vännman, K. (1995) A Unified Approach to Capability Indices", Statistica Sinica, Vol. 5, pp. 805-820.

7 Appendix

Consider a random sample of n measurements, $x_1, x_2, \ldots, x_n$, of a characteristic taken from a process where X ~ N(0, 1) but contaminated such that there are non zero values of λ_3 and λ_4.

Rewriting $$\hat{C}pw = \frac{USL - LSL}{6\sqrt{\sum_{i=1}^{n}\frac{\left(x_i - \bar{x}\right)^2}{n} + w\left(\bar{x} - T\right)^2}} = \frac{USL - LSL}{6\sqrt{\frac{\sum_{i=1}^{n}\left(x_i - \bar{x}\right)^2 + nw\left(\bar{x} - T\right)^2}{n}}},$$

and defining s_1, s_2 as $\begin{cases} s_1 = \sum_{i=1}^{n} x_i \\ s_2 = \sum_{i=1}^{n} (x_i - \bar{x})^2 \end{cases}$ and y, z as $\begin{cases} y = nw\left(\frac{s_1}{n} - T\right)^2 + s_2 \\ z = \left(\frac{s_1}{n} - T\right)^2 \end{cases}$

with inverse $\begin{cases} s_1 = n(\pm\sqrt{z} + T) \\ s_2 = y - nwz \end{cases}$

and Jacobian $J = \left|\frac{d(s_1, s_2)}{d(y,z)}\right| = \begin{vmatrix} 0 & \pm\frac{n}{2\sqrt{z}} \\ 1 & -nw \end{vmatrix} = \mp\frac{n}{2\sqrt{z}}$

allows us to use a technique developed in Gayen (1949) for the joint probability density function of S_1 and S_2

$$g_{S_1,S_2}(s_1,s_2) = W(n-1)\left\{1 + \frac{n\lambda 3}{3!}\left[H_3\left(\frac{s_1}{n}\right) + 3\frac{s_2}{n}H_1\left(\frac{s_1}{n}\right)\right]\right.$$

$$+\frac{n\lambda_4}{4!}\left[H_4\left(\frac{s_1}{n}\right) + 6\frac{s_2}{n}H_2\left(\frac{s_1}{n}\right) + \frac{3(n-1)}{n+1}\left(\frac{s_2}{n}\right)^2\right]$$

$$+\frac{n\lambda_3^2}{72}\left[n\left(\frac{s_1}{n}\right)^6 - 3(2n+3)\left(\frac{s_1}{n}\right)^4 + 9(n+4)\left(\frac{s_1}{n}\right)^2\right.$$

$$\left. -15 + 6\frac{s_2}{n}\left[n\left(\frac{s_1}{n}\right)^4 - 3(n+3)\left(\frac{s_1}{n}\right)^2 + 6\right]\right.$$

$$\left.\left. + \frac{9}{n+1}\left(\frac{s_2}{n}\right)^2\left[n(n+1)\left(\frac{s_1}{n}\right)^2 - 3(n-1)\right] + \frac{6n(n-2)}{(n+3)(n+1)}\left(\frac{s_2}{n}\right)^3\right]\right\}$$

where $H_v\left(\frac{s_1}{n}\right)$ is the Hermite polynomial of degree v in $\left(\frac{s_1}{n}\right)$, λ_3 and λ_4 are measures of skewness and kurtosis, respectively and

$$W(n_o) = \left(\frac{e^{-\frac{s_1^2}{2n}}}{\sqrt{2\pi n}}\right)\left(\frac{S_2^{\frac{1}{2}(n_o-2)} e^{-\frac{1}{2}S_2}}{2^{\frac{n_o}{2}}\Gamma\left(\frac{1}{2}n_o\right)}\right).$$

Using the above transformations, the Edgeworth expansion for the joint probability density function for $Y = nw\left(\frac{S_1}{n} - T\right)^2 + S_2$ and $Z = \left(\frac{S_1}{n} - T\right)^2$ becomes

$$h_{Y,Z}(y,z)=g_{S_1,S_2}(s_1,s_2)|J|=g\left[n\left(-\sqrt{z}+T\right),y-nwz\right]\left|\frac{n}{2\sqrt{z}}\right|+g\left[n\left(\sqrt{z}+T\right),y-nwz\right]\left|\frac{-n}{2\sqrt{z}}\right|$$

$$=W_1(n-1)\left\{1+\frac{n\lambda_3}{3!}\left[H_3\left(-\sqrt{z}+T\right)+3\left(\frac{y}{n}-wz\right)H_1\left(-\sqrt{z}+T\right)\right]\right.$$

$$+\frac{n\lambda_4}{4!}\left[H_4\left(-\sqrt{z}+T\right)+6\left(\frac{y}{n}-wz\right)H_2\left(-\sqrt{z}+T\right)+\frac{3(n-1)}{n+1}\left(\frac{y}{n}-wz\right)^2\right]$$

$$+\frac{n\lambda_3^2}{72}\left[n\left(-\sqrt{z}+T\right)^6-3(2n+3)\left(-\sqrt{z}+T\right)^4+9(n+4)\left(-\sqrt{z}+T\right)^2\right.$$

$$\left.-15+6\left(\frac{y}{n}-wz\right)\right]\left[n\left(-\sqrt{z}+T\right)^4-3(n+3)\left(-\sqrt{z}+T\right)^2+6\right]$$

$$\left.+\frac{9}{n+1}\left(\frac{y}{n}-wz\right)^2\left[n(n+1)\left(-\sqrt{z}+T\right)^2-3(n-1)\right]+\frac{6n(n-2)}{(n+3)(n+1)}\left(\frac{y}{n}-wz\right)^3\right]\right\}\frac{n}{2\sqrt{z}}+$$

$$W_2(n-1)\left\{1+\frac{n\lambda_3}{3!}\left[H_3\left(\sqrt{z}+T\right)+3\left(\frac{y}{n}-wz\right)H_1\left(\sqrt{z}+T\right)\right]\right.$$

$$+\frac{n\lambda_4}{4!}\left[H_4\left(\sqrt{z}+T\right)+6\left(\frac{y}{n}-wz\right)H_2\left(\sqrt{z}+T\right)+\frac{3(n-1)}{n+1}\left(\frac{y}{n}-wz\right)^2\right]$$

$$+\frac{n\lambda_3^2}{72}\left[n\left(\sqrt{z}+T\right)^6-3(2n+3)\left(\sqrt{z}+T\right)^4+9(n+4)\left(\sqrt{z}+T\right)^2\right.$$

$$\left.-15+6\left(\frac{y}{n}-wz\right)\right]\left[n\left(\sqrt{z}+T\right)^4-3(n+3)\left(\sqrt{z}+T\right)^2+6\right]$$

$$\left.+\frac{9}{n+1}\left(\frac{y}{n}-wz\right)^2\left[n(n+1)\left(\sqrt{z}+T\right)^2-3(n-1)\right]+\frac{6n(n-2)}{(n+3)(n+1)}\left(\frac{y}{n}-wz\right)^3\right]\right\}\frac{n}{2\sqrt{z}}$$

where

$$W_1(n-1)=\left(\frac{e^{-\frac{s_1^2}{2n}}}{\sqrt{2\pi n}}\right)\left(\frac{s_2^{\frac{n-3}{2}}e^{-\frac{s_2}{2}}}{2^{\frac{1}{2}(n-1)}\Gamma\left(\frac{n-1}{2}\right)}\right)$$

$$=\frac{e^{-\frac{n^2\left(-\sqrt{z}+T\right)^2}{2n}}(y-nwz)^{\frac{n-3}{2}}e^{-\left(\frac{y-nwz}{2}\right)}}{\sqrt{2\pi n}\,2^{\frac{1}{2}(n-1)}\Gamma\left(\frac{n-1}{2}\right)}$$

$$=\frac{e^{-\frac{1}{2}\left[(1-w)nz+y-2n\sqrt{z}T+nT^2\right]}(y-nwz)^{\frac{n-3}{2}}}{\sqrt{2\pi n}\,2^{\frac{1}{2}(n-1)}\Gamma\left(\frac{n-1}{2}\right)}.$$

Similarly,

$$W_2(n-1)=\frac{e^{-\frac{1}{2}\left[(1-w)nz+y+2n\sqrt{z}T+nT^2\right]}(y-nwz)^{\frac{n-3}{2}}}{\sqrt{2\pi n}\,2^{\frac{1}{2}(n-1)}\Gamma\left(\frac{n-1}{2}\right)}.$$

The marginal probability density function of Y can then be obtained by considering

$$h_1(y)=\int_0^{\frac{y}{nw}}h_{Y,Z}(y,z)\,dz$$

for $\hat{C}pw=\frac{USL-LSL}{6\sqrt{\frac{y}{n}}}=\frac{a}{\sqrt{y}}$ where $a=\frac{\sqrt{n}(USL-LSL)}{6}$.

Letting $y=\frac{a^2}{\hat{C}pw^2}$ with $J=\frac{\partial y}{\partial Cpw}=-\frac{2a^2}{\hat{C}pw^3}$ results in the pdf

$$g_{\hat{C}pw}(t)=h_1\left(\frac{a^2}{t^2}\right)\left|-\frac{2a^2}{t^3}\right| \qquad 0<t<\infty.$$

A Note on the Estimation of Restricted Scale Parameters of Gamma Distributions

Yuan-Tsung Chang

Dept. of Social Information, Faculty of Studies on Contemporary Society, Mejiro University, 4-31-1 Shinjuku-ku, Tokyo 161-8539 Japan
cho@mejiro.ac.jp

Summary. In this paper an admissible estimator of the scale parameter of a Gamma distribution is derived, which is bounded above. Next, the simultaneous estimation of p scale parameters of Gamma distributions are considered that are bounded above. Moreover, simultaneous estimation of two ordered scale parameters of two Gamma distributions is investigated in terms of the mean square error (MSE). Finally, some simulation results are given for illustrating the improvement obtained with the new estimators, when compared with the usual estimators.

1 Introduction

The Gamma distribution plays an important role in reliability theory, as it serves often as a model for lifetime times and waiting times, leading to the necessity of estimating the parameters of a Gamma distribution. Usually, the estimation problem is solved assuming that there are no restrictions with respect to the values of the parameter. However, in reality the parameters are always restricted and very often prior information about the restriction is available. In this case, the usual methods of estimation turn out to be inappropriate. Therefore, we consider for the scale parameter some types of constraints, for example, bounded above or order restrictions between two scale parameters, and derive estimators, which take into account the restrictions. The suggested estimators perform better than the usual estimators.

Problems concerning statistical inference for restricted parameters are of considerable interest. The earlier works related to statistical inference under restrictions are reviewed by Barlow et al (1972), Robertson and Wright (1988). The more recent literature about restricted statistical inference after 1988 is reviewed in two monographs, Silvapulle and Sen (2004), Eeden (2006). Many authors have considered inference on parameters in case of a is simple order restriction or tree order restriction on parameters and have suggested estimators, which improve the performance of unbiased estimators, especially for the normal distribution.

H.-J. Lenz et al. (eds.), *Frontiers in Statistical Quality Control 9*,
DOI 10.1007/978-3-7908-2380-6_19,

Katz (1961) has shown the admissibility or minimaxity of estimators when an upper or lower bound restriction is given in the case of the expectation parameter in exponential family under squared error loss. Casella and Strawderman (1981) proposed a minimax estimator for a normal mean that is restricted to a bounded interval. Berry (1993) considered admissible minimax estimators of the location parameter for the exponential distribution, when upper bound, lower bound or two-sided restrictions are given. Eeden (1995) suggested an admissible estimator for the scale parameter of a Gamma distribution with given lower bound under scale invariant squared error loss. When there is an order restriction between two scale parameters of two Gamma distributions, Chang and Shinozaki (2002) have considered the estimation of a linear function of two ordered scale parameters of two Gamma distributions. They derived a necessary and sufficient condition for the ratio of two coefficients, which guarantees that the maximum likelihood estimator (MLE) and the modified MLE dominate the crude unbiased estimator (UE) and the admissible estimator based solely on one sample under quadratic loss, respectively. Marchand and Strawderman (2005) improved the minimum risk equivariant estimator of a scale parameter under a lower bound restriction.

Although Katz(1961) has given a proof of the admissibility for an estimator of the expectation parameter in the exponential family, when there is a lower bound(upper bound) restriction, Eeden(1995) has pointed out that there is an error in Katz's proof and showed that Katz's condition is not satisfied for the Gamma scale parameter with lower bound restriction.

On the other hand, Berger(1980) developed a general technique for simultaneous estimation of Gamma scale parameters to improve the standard estimators under several types of loss functions that include the sum of squared error loss and scale invariant squared error loss.

In this paper we investigate the estimation of restricted scale parameters of Gamma distributions. We discuss the admissibility of an estimator of an upper bounded scale parameter of the Gamma distribution under squared error loss. We also discuss the simultaneous estimation of upper bounded and lower bounded Gamma scale parameters. In Section 2, we discuss the admissible estimation of an scale parameter, which is bounded above, of the Gamma distribution under squared error loss. In Section 3, we consider the simultaneous estimation of Gamma scale parameters with upper bound and lower bound restrictions under sum of squared error loss and sum of scale invariant squared error loss. In Section 4, we consider the simultaneous estimation of two ordered scale parameters of two Gamma distributions. In Section 5, some simulation results are given for illustrating the performance of the risk function of simultaneous estimation of p upper bounded scale parameters of Gamma distributions and simultaneous estimation of two ordered Gamma scale parameters. Finally in Section 6 we give some conclusions.

2 An Admissible Estimator for an Upper-Bounded Gamma Scale Parameter

Let X be a $Gamma(\alpha, \theta)$ random variable with density function

$$f_\theta(x) = \frac{1}{\Gamma(\alpha)} \frac{x^{\alpha-1}}{\theta^\alpha} e^{-x/\theta}, \qquad 0 < x < \infty \tag{2.1}$$

where α is known and θ is unknown, but satisfies the relation $0 < \theta \leq a$, where a is a positive constant.

Consider the following density function of a prior distribution

$$g_n(\theta) = \frac{c_n}{n\theta^2} e^{-1/n\theta}, \qquad 0 < \theta \leq a \tag{2.2}$$

where $c_n = e^{1/na}$. Then, Bayes estimator of θ is given by:

$$\delta_n(x) = E[\theta|x] = \frac{\displaystyle\int_0^a \theta^{-(\alpha+1)} e^{-(x+1/n)/\theta} d\theta}{\displaystyle\int_0^a \theta^{-(\alpha+2)} e^{-(x+1/n)/\theta} d\theta}. \tag{2.3}$$

For simplicity let $a = 1$, for a moment. By the variable transformation $(x + 1/n)/\theta = t$, we have:

$$\delta_n(x) = \left(x + \frac{1}{n}\right) \frac{\displaystyle\int_{x+1/n}^\infty t^{\alpha-1} e^{-t} dt}{\displaystyle\int_{x+1/n}^\infty t^\alpha e^{-t} dt}.$$

Further by applying a partial integration to the numerator, we have

$$\int_{x+1/n}^\infty t^{\alpha-1} e^{-t} dt = -\frac{1}{\alpha}\left(x + \frac{1}{n}\right)^\alpha e^{-(x+1/n)} + \frac{1}{\alpha}\int_{x+1/n}^\infty t^\alpha e^{-t} dt.$$

Then

$$\begin{aligned}\delta_n(x) &= \frac{1}{\alpha}\left(x + \frac{1}{n}\right)\left(1 - \frac{(x+1/n)^\alpha e^{-(x+1/n)}}{\displaystyle\int_{x+1/n}^\infty t^\alpha e^{-t} dt}\right) \\ &= \frac{1}{\alpha}\left(x + \frac{1}{n}\right)(1 - \phi_n(x)),\end{aligned}$$

where

$$\phi_n(x) = \frac{(x+1/n)^\alpha e^{-(x+1/n)}}{\displaystyle\int_{x+1/n}^\infty t^\alpha e^{-t} dt} = \frac{(x+1/n)^\alpha e^{-(x+1/n)}}{\displaystyle\int_x^\infty \left(t + \frac{1}{n}\right)^\alpha e^{-(t+1/n)} dt}.$$

The function $\phi_n(x)$ has the following properties:
Since θ is restricted by $0 < \theta \leq 1$ it follows $0 < \delta_n(x) = E(\theta|x) \leq 1$, which implies $0 < \phi_n(x) \leq 1$. It is also easy to check that

$$\phi_n'(x) = \phi_n(x)\left(\frac{\alpha}{x+1/n} - (1-\phi_n(x))\right) \geq 0,$$

and, thus, $\phi_n(x)$ is a non-decreasing function of x.

The Bayes estimator $\delta_n(x)$ of θ in (2.3) can also be expressed as:

$$\delta_n(x) = \frac{\int_0^a \theta^{-(\alpha+1)} e^{-(x+1/n)/\theta} d\theta}{\int_0^a \theta^{-(\alpha+2)} e^{-(x+1/n)/\theta} d\theta} = \frac{x+1/n}{\alpha} - \frac{a^{-\alpha} e^{-(x+1/n)/a}}{\alpha \int_0^a \theta^{-(\alpha+2)} e^{-(x+1/n)/\theta} d\theta}$$

with the limit for $n \to \infty$ given by:

$$\delta(x) = \lim_{n\to\infty} \delta_n(n) = \frac{x}{\alpha} - \frac{a^{-\alpha} e^{-x/a}}{\alpha \int_0^a \theta^{-(\alpha+2)} e^{-x/\theta} d\theta} \tag{2.4}$$

This limit can also be derived by Katz's (1961) general formula as follows: Let $\omega = -1/\theta$ then the density of x is

$$f_\omega(x) = \frac{x^{\alpha-1}(-\omega)^\alpha e^{x\omega}}{\Gamma(\alpha)}, \qquad x > 0, \tag{2.5}$$

where ω is restricted to $-\infty < \omega \leq -1/a$. Let $\beta(\omega) = (-\omega)^\alpha$, and $\mu(x) = x^{\alpha-1}/\Gamma(\alpha)$ be σ–finite measure on $(0,\infty)$. Then by Katz's general formula (for details see Theorem 2.1) we obtain the estimator of $E_\omega(x) = \varphi(\omega) = -\beta'(\omega)/\beta(\omega) = \alpha/(-\omega)$ as follows:

$$\widehat{\frac{\alpha}{-\omega}} = x - \frac{e^{-x/a} a^{-\alpha}}{\int_{-\infty}^{-1/a} (-\omega)^\alpha e^{x\omega} d\omega}.$$

Moreover, the estimator of $-\omega^{-1}$ is given by:

$$\widehat{-\omega^{-1}} = \frac{x}{\alpha} - \frac{e^{-x/a} a^{-\alpha}}{\alpha \int_{-\infty}^{-1/a} (-\omega)^\alpha e^{x\omega} d\omega}. \tag{2.6}$$

Since, $\omega = -1/\theta$, we have:

$$\int_{-\infty}^{-1/a} (-\omega)^\alpha e^{x\omega} d\omega = \int_0^a \theta^{-(\alpha+2)} e^{-x/\theta} d\theta,$$

Thus, we obtain the estimator (2.4), which is an admissible estimator of θ with restriction $0 < \theta \leq a$.

By Theorem 2.1. we show that (2.6) is an admissible estimator of $-1/\omega$:

Theorem 2.1 Suppose that the density function of X is given by (2.5) and that ω is restricted to $-\infty < \omega \leq -1/a$. Then the estimator

$$x - \frac{e^{-x/a} a^{-\alpha}}{\int_{-\infty}^{-1/a} (-\omega)^{\alpha} e^{x\omega} d\omega}.$$

is an admissible estimator of $\varphi(\omega) = E_{\omega}(X)$.

Proof. The prior distribution of ω is given by:

$$g_n(\omega) = \frac{c_n}{n} e^{\omega/n}, \qquad -\infty < \omega \leq -1/a, \tag{2.7}$$

where $c_n = e^{1/na}$, and the corresponding Bayes estimator of $\varphi(\omega) = \alpha(-\omega)^{-1}$, $\delta_n(x)$, is given by:

$$\delta_n(x) = E[\varphi(\omega)|x] = \frac{\int_{-\infty}^{-1/a} \alpha(-\omega)^{\alpha-1} e^{(x+1/n)\omega} d\omega}{\int_{-\infty}^{-1/a} (-\omega)^{\alpha} e^{(x+1/n)\omega} d\omega}. \tag{2.8}$$

Applying integration by parts to the numerator of (2.8) yields:

$$\int_{-\infty}^{-1/a} \alpha(-\omega)^{\alpha-1} e^{(x+1/n)\omega} d\omega$$
$$= -a^{-\alpha} e^{-(x+1/n)/a} + \left(x + \frac{1}{n}\right) \int_{-\infty}^{-1/a} (-\omega)^{\alpha} e^{(x+1/n)\omega} d\omega.$$

It follows:

$$\delta_n(x) = x + \frac{1}{n} - \frac{g_{x,n}(1/a)}{G(x+1/n)} \geq 0,$$

where

$$g_{x,n}(1/a) = a^{-\alpha} e^{-(x+1/n)/a}$$

and

$$G(x+1/n) = \int_{-\infty}^{-1/a} (-\omega)^{\alpha} e^{(x+1/n)\omega} d\omega.$$

It should be noted that $G(x+1/n)$ is an increasing function of n with $n = 1, 2, \cdots$, and

$$\frac{g_{x,n}(1/a)}{G(x+1/n)} \leq x + \frac{1}{n}. \tag{2.9}$$

The risk of $\delta_n(x)$ under squared error loss is given by:

$$\begin{aligned}\rho(\delta_n,\omega) &= E_\omega\left[x+\frac{1}{n}-\frac{g_{x,n}(1/a)}{G(x+1/n)}-\varphi(\omega)\right]^2\\ &= \varphi'(\omega)+2\varphi(\omega)E_\omega\left[\frac{g_{x,n}(1/a)}{G(x+1/n)}\right]-2E_\omega\left[\left(x+\frac{1}{n}\right)\frac{g_{x,n}(1/a)}{G(x+1/n)}\right]\\ &\quad +E_\omega\left[\frac{g^2_{x,n}(1/a)}{G^2(x+1/n)}\right]+\frac{1}{n^2}. \qquad (2.10)\end{aligned}$$

Using (2.9) it is easy to verify that all the terms of (2.10) are finite.

In order to calculate the Bayes risk of $\delta_n(x)$ we need to calculate the expected value of each term of (2.10) with respect to the prior distribution of ω. Since $\varphi(\omega)=-\beta'(\omega)/\beta(\omega)$, we have:

$$\begin{aligned}E\left[\varphi(\omega)E_\omega\left[\frac{g_{x,n}(1/a)}{G(x+1/n)}\right]\right] &= \int_{-\infty}^{-1/a}\varphi(\omega)E_\omega\left[\frac{g_{x,n}(1/a)}{G(x+1/n)}\right]\frac{c_n}{n}e^{\omega/n}d\omega\\ &= -\frac{c_n}{n}\int_{-\infty}^{-1/a}\int_0^\infty \beta'(\omega)\frac{g_{x,n}(1/a)}{G(x+1/n)}e^{(x+1/n)\omega}d\mu(x)d\omega.\end{aligned}$$

By changing the order of integration and partially integrating with respect to ω first, we get:

$$\begin{aligned}&\int_{-\infty}^{-1/a}\int_0^\infty \beta'(\omega)\frac{g_{x,n}(1/a)}{G(x+1/n)}e^{(x+1/n)\omega}d\mu(x)d\omega\\ &= \int_0^\infty \frac{g^2_{x,n}(1/a)}{G(x+1/n)}d\mu(x)-\int_0^\infty\left(x+\frac{1}{n}\right)g_{x,n}(1/a)d\mu(x).\end{aligned}$$

Similarly, we obtain:

$$E\left[E_\omega\left[\left(x+\frac{1}{n}\right)\frac{g_{x,n}(1/a)}{G(x+1/n)}\right]\right]=\frac{c_n}{n}\int_0^\infty\left(x+\frac{1}{n}\right)g_{x,n}(1/a)d\mu(x),$$

and

$$E\left[E_\omega\left[\frac{g^2_{x,n}(1/a)}{G^2(x+1/n)}\right]\right]=\frac{c_n}{n}\int_0^\infty\frac{g^2_{x,n}(1/a)}{G(x+1/n)}d\mu(x).$$

Thus, we get following expression for the Bayes risk of $\delta_n(x)$:

$$r_n(\delta_n)=E[\varphi'(\omega)]-\frac{c_n}{n}\int_0^\infty\frac{g^2_{x,n}(1/a)}{G(x+1/n)}d\mu(x)+\frac{1}{n^2}. \qquad (2.11)$$

It is easy to check that all terms are finite.

The limit of $\delta_n(x)$ is

$$\delta(x) = \lim_{n\to\infty} \delta_n(x) = x - \frac{g_x(1/a)}{G(x)}, \tag{2.12}$$

where $g_x(1/a) = a^{-\alpha}e^{-x/a}$ and $G(x) = \int_{-\infty}^{-1/a} (-\omega)^{\alpha} e^{x\omega} d\omega.$

The risk of $\delta(x)$ under squared error loss is given by:

$$\begin{aligned}\rho(\delta,\omega) &= E_\omega\left[x - \frac{g_x(1/a)}{G(x)} - \varphi(\omega)\right]^2 \\ &= \varphi'(\omega) + 2\varphi(\omega)E_\omega\left[\frac{g_x(1/a)}{G(x)}\right] - 2E_\omega\left[\frac{xg_x(1/a)}{G(x)}\right] + E_\omega\left[\frac{g_x^2(1/a)}{G^2(x)}\right].\end{aligned}$$

It easy to verify that all these terms is finite. Moreover, we can show that:

$$E\left[\varphi(\omega)E_\omega\left[\frac{g_x(1/a)}{G(x)}\right]\right] = -\frac{c_n}{n}\int_{-\infty}^{-1/a}\int_0^\infty \beta'(\omega)\frac{g_x(1/a)}{G(x)}e^{(x+1/n)\omega}d\mu(x)d\omega$$

$$= -\frac{c_n}{n}\int_0^\infty \frac{g_{x,n}(1/a)g_x(1/a)}{G(x)}d\mu(x) + \frac{c_n}{n}\int_0^\infty\left(x+\frac{1}{n}\right)\frac{g_x(1/a)G(x+1/n)}{G(x)}d\mu(x).$$

Similarly, we have:

$$E\left[E_\omega\left[\frac{xg_x(1/a)}{G(x)}\right]\right] = \frac{c_n}{n}\int_0^\infty \frac{xg_x(1/a)G(x+1/n)}{G(x)}d\mu(x)$$

and

$$E\left[E_\omega\left[\frac{g_x^2(1/a)}{G^2(x)}\right]\right] = \frac{c_n}{n}\int_0^\infty \frac{g_x^2(1/a)G(x+1/n)}{G^2(x)}d\mu(x).$$

Since $G(x+1/n) \le G(x)$ and $\int_0^\infty g_x(1/a)d\mu(x) = 1$ the following Bayes risk of $\delta(x)$ is obtained:

$$\begin{aligned}r_n(\delta) &= E[\varphi'(\omega)] - 2\frac{c_n}{n}\int_0^\infty \frac{g_{x,n}(1/a)g_x(1/a)}{G(x)}d\mu(x) \\ &+ 2\frac{c_n}{n^2}\int_0^\infty \frac{G(x+1/n)g_x(1/a)}{G(x)}d\mu(x) + \frac{c_n}{n}\int_0^\infty \frac{G(x+1/n)g_x^2(1/a)}{G^2(x)}d\mu(x) \\ &\le E[\varphi'(\omega)] - 2\frac{c_n}{n}\int_0^\infty \frac{g_{x,n}(1/a)g_x(1/a)}{G(x)}d\mu(x) + 2\frac{c_n}{n^2} \\ &+ \frac{c_n}{n}\int_0^\infty \frac{g_x^2(1/a)}{G(x)}d\mu(x),\end{aligned} \tag{2.13}$$

Next, we will show that $\delta(x)$ is an admissible estimator of $\varphi(\omega)$ for $-\infty < \omega \le -1/a$.

Suppose that δ is inadmissible. Then, there exists an estimator δ' such that $\rho(\delta',\omega) \le \rho(\delta,\omega)$ for all ω with $-\infty < \omega \le -1/a$, and strict inequality for some ω. Because the risk is a continuous function of ω, we can find an $\epsilon > 0$ and an interval $(\omega_1, \omega_2) \subset (-\infty, -1/a)$ such that

$$\rho(\delta, \omega) - \rho(\delta', \omega) > \epsilon \qquad \text{for all} \qquad \omega \in (\omega_1, \omega_2),$$

implying:

$$n(r_n(\delta) - r_n(\delta')) > \epsilon e^{1/na} \int_{\omega_1}^{\omega_2} e^{\omega/n} d\omega = \epsilon e^{1/na} n \left(e^{\omega_2/n} - e^{\omega_1/n} \right),$$

Then it follows:

$$\lim_{n\to\infty} n(r_n(\delta) - r_n(\delta')) > \epsilon \lim_{n\to\infty} n \left(e^{\omega_2/n} - e^{\omega_1/n} \right) = \epsilon \lim_{n\to\infty} \frac{e^{\omega_2/n} - e^{\omega_1/n}}{1/n}$$
$$= \epsilon \lim_{n\to\infty} \frac{-\frac{\omega_2}{n^2} e^{\omega_2/n} + \frac{\omega_1}{n^2} e^{\omega_1/n}}{-1/n^2} = \epsilon(\omega_2 - \omega_1) > 0.$$

Now, consider the quantity:

$$n(r_n(\delta) - r_n(\delta'))/n(r_n(\delta) - r_n(\delta_n)). \tag{2.14}$$

From (2.11) and (2.13) we obtain the following upper bound for the difference $n(r_n(\delta) - r_n(\delta_n))$:

$$\begin{aligned} & n(r_n(\delta) - r_n(\delta_n)) \\ & \le c_n \int_0^\infty \frac{g_x^2(1/a)}{G(x)} d\mu(x) - 2c_n \int_0^\infty \frac{g_x(1/a) g_{x,n}(1/a)}{G(x)} d\mu(x) \\ & + c_n \int_0^\infty \frac{g_{x,n}^2(1/a)}{G(x+1/n)} d\mu(x) + 2\frac{c_n}{n} - \frac{1}{n}. \end{aligned} \tag{2.15}$$

From (2.9) it is easy to check that the second and the third term of (2.15) are finite. Therefore, we can change the order of limit and integration with respect to the second and the third term of (2.15). We obtain:

$$\begin{aligned} & \lim_{n\to\infty} c_n \int_0^\infty \frac{g_x^2(1/a)}{G(x)} d\mu(x) - 2 \lim_{n\to\infty} c_n \int_0^\infty \frac{g_x(1/a) g_{x,n}(1/a)}{G(x)} d\mu(x) \\ & + \lim_{n\to\infty} c_n \int_0^\infty \frac{g_{x,n}^2(1/a)}{G(x+1/n)} d\mu(x) \\ = & \int_0^\infty \frac{g_x^2(1/a)}{G(x)} d\mu(x) - 2 \int_0^\infty \lim_{n\to\infty} \frac{g_x(1/a) g_{x,n}(1/a)}{G(x)} d\mu(x) \\ & + \int_0^\infty \lim_{n\to\infty} \frac{g_{x,n}^2(1/a)}{G(x+1/n)} d\mu(x) = 0. \end{aligned}$$

Hence:

$$\lim_{n\to\infty} n(r_n(\delta) - r_n(\delta_n)) = 0.$$

It follows that for $n \to \infty$ the quantity (2.14) tends to infinity, too:

$$\lim_{n\to\infty} \frac{n(r_n(\delta) - r_n(\delta'))}{n(r_n(\delta) - r_n(\delta_n))} \to \infty,$$

Thus, for sufficiently large n we have shown that:

$$r_n(\delta') < r_n(\delta_n).$$

This contradicts the fact that δ_n is the Bayes estimator of $\varphi(\omega)$ with respect to the prior distribution (2.1), for $n = 1, 2, \ldots$. It follows that $\delta(x)$ as given by (2.12) is an admissible estimator of $\varphi(\omega)$ for $-\infty < \omega \leq -1/a$. This implies that (2.6) is an admissible estimator of θ for $0 < \theta \leq a$.

3 Simultaneous Estimation of Gamma Scale Parameters with an Upper-bound or an Lower-bound Restriction

First, we consider simultaneous estimation of Gamma scale parameters when each scale parameter has an upper-bound restriction under sum of squared-error loss, and suggest an improvement of the admissible estimator based solely on each sample.

Let $\mathbf{X} = (X_1, \cdots, X_p)$ be a p-variate random variable, where, $X_i, i = 1, \ldots p$, are independent $Gamma(\alpha_i, \theta_i)$ random variables, each having a density function given by (2.1), where α_i is known and the unknown scale parameter is restricted by $0 < \theta_i \leq a$. The admissible estimator of the scale parameter θ_i is given by (2.4):

$$\delta_i(\mathbf{x}) = \frac{x_i}{\alpha_i} - \frac{a^{-\alpha_i} e^{-x_i/a}}{\alpha_i \displaystyle\int_0^a \theta^{-(\alpha_i+2)} e^{-x_i/\theta} d\theta} = \frac{x_i - t(x_i)}{\alpha_i}, \qquad i = 1, \cdots, p.$$

For simultaneous estimation of $\theta = (\theta_1, \cdots, \theta_p)$ under sum of squared-error loss, Chang (1982) has derived a uniform improvement of $\delta_i(\mathbf{x})$ given by:

$$\hat{\delta}_i(\mathbf{x}) = \frac{x_i - t(x_i)}{\alpha_i} + \frac{c\alpha_i(\alpha_i - 1)x_i^{-1}}{D} + \frac{2c\alpha_i^3(\alpha_i + 1)^2 x_i^{-3}}{D^2} \tag{3.1}$$

where $D = \sum_{j=1}^p \alpha_j^2(\alpha_j + 1)^2 x_j^{-2}$. As can be seen, $\hat{\delta}_i(\mathbf{x})$ extends $\delta_i(\mathbf{x})$ not in direction of $\frac{x_i - t(x_i)}{\alpha_i}$. Here, we consider another estimator which extends $\delta_i(\mathbf{x})$ in the same direction and which is given by:

$$\hat{\delta}_i^s(\mathbf{x}) = \frac{x_i - t(x_i)}{\alpha_i} + \frac{c\alpha_i(\alpha_i - 1)(x_i - t(x_i))^{-1}}{D^*} + \frac{2c\alpha_i^3(\alpha_i + 1)^2(x_i - t(x_i))^{-3}}{D^{*2}} \tag{3.2}$$

where $D^* = \sum_{j=1}^{p} \alpha_j^2(\alpha_j+1)^2(x_j - t(x_j))^{-2}$.

In Theorem 3.2 it is shown that $\hat{\delta}_i^s(\mathbf{x})$ improves $\delta_i(\mathbf{x})$ uniformly.

To evaluate the difference of the MSE of the two estimators, we need the following lemma, which is proved by integration by parts (Berger (1980)).

Lemma 3.1. Suppose that $X \sim^{ind} Gamma(\alpha, \theta)$, if $\alpha > 1$ and $g(x)$ is absolutely continuous on $(0, \infty)$ with $g'(x) = \frac{dg(x)}{dx}$ satisfying

$(i) E[|g'(X)|] < \infty$ *and* $E[|g(X)|] < \infty$

$(ii) \lim_{x \to 0} g(x)x^{\alpha-1}e^{-x/\theta} = \lim_{x \to \infty} g(x)x^{\alpha-1}e^{-x/\theta} = 0,$ *for* $\theta > 0$.

then

$$E[X^{-1}g(X)] = \frac{1}{\alpha-1}\left\{E\left[\frac{g(X)}{\theta}\right] - E\left[g'(X)\right]\right\}.$$

Theorem 3.2. Suppose that $X_i \sim^{ind} Gamma(\alpha_i, \theta_i)$, and that there is an upper-bound restriction for θ_i, $0 < \theta_i \le a, i = 1, \cdots, p$. Let $\hat{\delta}^s(\mathbf{x}) = (\hat{\delta}_1^s(\mathbf{x}), \cdots, \hat{\delta}_p^s(\mathbf{x}))$ and $\delta(\mathbf{x}) = (\delta_1(\mathbf{x}), \cdots, \delta_p(\mathbf{x}))$. Then the estimator $\hat{\delta}^s(\mathbf{x})$ for simultaneous estimation of $(\theta_1, \cdots, \theta_p)$ has a uniformly smaller risk than $\delta(\mathbf{x})$, if $\alpha_i > 1, p \ge 3$ and $0 < c < 2(p-2)$.

Proof. The difference of the risk of $\hat{\delta}^s(\mathbf{x})$ and of $\delta(\mathbf{x})$ is given by:

$$\begin{aligned}
\Delta R &= R(\delta(\mathbf{x}), \theta) - R(\hat{\delta}^s(\mathbf{x}), \theta) \\
&= -\sum_{i=1}^{p}\left\{E\left[\frac{2c\alpha_i(\alpha_i-1)(x_i-t(x_i))^{-1}}{D^*}\left(\frac{x_i-t(x_i)}{\alpha_i}-\theta_i\right)\right.\right. \\
&\quad +\frac{4c\alpha_i^3(\alpha_i+1)^2(x_i-t(x_i))^{-3}}{D^{*2}}\left(\frac{x_i-t(x_i)}{\alpha_i}-\theta_i\right) \\
&\quad +\frac{4c^2\alpha_i^4(\alpha_i-1)(\alpha_i+1)^2(x_i-t(x_i))^{-4}}{D^{*3}} \\
&\quad \left.\left.+\frac{c^2\alpha_i^2(\alpha_i-1)^2(x_i-t(x_i))^{-2}}{D^{*2}}+\frac{4c^2\alpha_i^6(\alpha_i+1)^4(x_i-t(x_i))^{-6}}{D^{*4}}\right]\right\}.
\end{aligned}$$

Since $x_i - t(x_i) \ge 0$ and by applying Lemma 3.1, we get:

$$\begin{aligned}
&E\left[\frac{(x_i-t(x_i))^{-1}\theta_i}{D^*}\right] \ge E\left[\frac{x_i^{-1}\theta_i}{D^*}\right] \\
&= E\left[\frac{1}{(\alpha_i-1)D^*} - \frac{2\alpha_i^2(\alpha_i+1)^2(x_i-t(x_i))^{-3}\theta_i}{(\alpha_i-1)D^{*2}} + \frac{2\theta_i\alpha_i^2(\alpha_i+1)^2 t'(x_i)}{(\alpha_i-1)D^{*2}}\right] \\
&\ge E\left[\frac{1}{(\alpha_i-1)D^*} - \frac{2\alpha_i^2(\alpha_i+1)^2(x_i-t(x_i))^{-3}\theta_i}{(\alpha_i-1)D^{*2}}\right],
\end{aligned}$$

as $\theta_i \geq 0$ and $t'(x) \geq 0$ holds. It follows:

$$\Delta R \geq \sum_{i=1}^{p} \left\{ E\left[\frac{2c}{D^*} - \frac{4c\alpha_i^2(\alpha_i+1)^2(x_i-t(x_i))^{-2}}{D^{*2}}\right] - E\left[\frac{4c^2\alpha_i^4(\alpha_i+1)^2(\alpha_i-1)(x_i-t(x_i))^{-4}}{D^{*3}}\right] - E\left[\frac{c^2\alpha_i^2(\alpha_i-1)^2(x_i-t(x_i))^{-2}}{D^{*2}} - \frac{4c^2\alpha_i^6(\alpha_i+1)^4(x_i-t(x_i))^{-6}}{D^{*4}}\right] \right\}.$$

From the definition of D^* we obtain:

$$\frac{\alpha_i^2(\alpha_i+1)^2(x_i-t(x_i))^{-2}}{D^*} \leq 1, \quad \frac{\alpha_i^4(\alpha_i+1)^2(x_i-t(x_i))^{-4}}{D^{*3}} \leq \frac{\alpha_i^2(x_i-t(x_i))^{-2}}{D^{*2}},$$

and

$$\frac{\alpha_i^6(\alpha_i+1)^4(x_i-t(x_i))^{-6}}{D^{*4}} \leq \frac{\alpha_i^2(x_i-t(x_i))^{-2}}{D^{*2}}.$$

Therefore:

$$\Delta R \geq E\left[\frac{2pc-4c}{D^*} - \frac{c^2}{D^{*2}}\sum_{i=1}^{p}(4\alpha_i^2(\alpha_i-1)+\alpha_i^2(\alpha_i-1)^2+4\alpha_i^2)(x_i-t(x_i))^{-2}\right]$$
$$= E\left[\frac{2pc-4c-c^2}{D^*}\right] > 0,$$

if $p \geq 3$ and $0 < c < 2(p-2)$.

Next we consider simultaneous estimation of p-variate Gamma scale parameters, when each scale parameter has a lower-bound restriction, $\theta_i \geq a(=1)$, under sum of scale invariant squared error loss, $\sum_{i=1}^{p}(d_i/\theta_i-1)^2$.

For scale-invariant squared error loss, Eeden(1995) gave the following admissible estimator of θ_i:

$$\xi_i(\mathbf{x}) = \frac{x_i}{\alpha_i+1} + \frac{g_{\alpha_i+2}(x_i)}{(\alpha_i+1)G_{\alpha_i+2}(x_i)} = \frac{x_i}{\alpha_i+1} + t^*(x_i),$$

where

$$g_\beta(x) = x^\beta e^{-x}, \qquad G_\beta(x) = \int_0^x t^{\beta-1}e^{-t}dt.$$

In Theorem 3.3 we show that for simultaneous estimation of $\theta = (\theta_1, \cdots, \theta_p)$, the estimator $\xi_i^s(\mathbf{x})$ given below has uniformly smaller risk than $\xi_i(\mathbf{x})$ for $\xi_i(\mathbf{x}), i = 1, \cdots, p$, if $b > 0$.

$$\xi_i^s(\mathbf{x}) = \frac{x_i}{\alpha_i+1}\left(1 - \frac{c(\alpha_i+1)\log x_i}{b+\sum_{j=1}^{p}\alpha_j(\alpha_j+1)(\log x_j)^2}\right) + t^*(x_i) \qquad if\ x_i \geq 1$$

Theorem 3.3. Suppose that $X_i \sim^{ind} Gamma(\alpha_i, \theta_i)$, and that there is a lower-bound restriction for θ_i, $\theta_i \geq 1, i = 1, \cdots, p$. Let $\xi^s(\mathbf{x}) = (\xi_1^s(\mathbf{x}), \cdots, \xi_p^s(\mathbf{x}))$ and $\xi(\mathbf{x}) = (\xi_1(\mathbf{x}), \cdots, \xi_p(\mathbf{x}))$. Then in simultaneous estimation of $(\theta_1, \cdots, \theta_p)$ under sum of scale-invariant squared-error loss, $\xi^s(\mathbf{x})$ has uniformly smaller risk than $\xi(\mathbf{x})$, if

$$0 < c \leq \frac{2\{p - 2 - \sum(\alpha_i + 1)^{-1} - 3b^{-\frac{1}{2}}[\sum \alpha_i/(\alpha_i + 1)]^{\frac{1}{2}}\}}{1 + (2p/b) + (16/[27bp]) + b^{-\frac{1}{2}}[4p + \sum\{\alpha_i(\alpha_i + 1)\}^{-1}]^{\frac{1}{2}}}.$$

Proof. The proof is the same as that of Theorem 3.2 and we can check the conditions in Berger (1980). Therefore, we omit the details here.

It should be noted (see Berger (1980)), that by choosing b large enough c meets the above requirement provided:

$$p > 2 + \sum(\alpha_i + 1)^{-1},$$

which will be satisfied, if $\alpha_i \geq 3$ for $p \geq 3$.

4 Simultaneous Estimation of Two Ordered Scale Parameters

In this section, we deal with the simultaneous estimation of (θ_1, θ_2), when there is an order restriction of type $\theta_1 \leq \theta_2$. We propose two estimators that improve the modified MLE of two ordered scale parameters. The modified MLE of two ordered scale parameters was suggested by Chang and Shinozaki (2002) as follows:

$$\begin{aligned}\hat{\theta}_i(\mathbf{x}) &= \frac{x_i}{\alpha_i + 1} + (-1)^i \frac{((\alpha_2 + 1)x_1 - (\alpha_1 + 1)x_2)^+}{(\alpha_i + 1)(\alpha_1 + \alpha_2 + 2)} \\ &= \frac{x_i}{\alpha_i + 1} + (-1)^i t_i(\mathbf{x}), \quad i = 1, 2,\end{aligned}$$

where $t_i(\mathbf{x}) = ((\alpha_2 + 1)x_1 - (\alpha_1 + 1)x_2)^+/((\alpha_i + 1)(\alpha_1 + \alpha_2 + 2))$.

Chang & Shinozaki(2002) showed that $\hat{\theta}_1$ improves $x_1/(\alpha_1 + 1)$, if $\alpha_1 \leq \alpha_2$ and $\alpha_2 \geq 1$, and that $\hat{\theta}_2$ improves $x_2/(\alpha_2 + 1)$, when there is an order restriction for θ_i, where $x_i/(\alpha_i + 1)$ is an admissible estimator of θ_i based solely on x_i. The competitor of $\hat{\theta}_i$ is:

$$\hat{\theta}_i^s(\mathbf{x}) = \frac{x_i}{\alpha_i + 1} + (-1)^i t_i(\mathbf{x}) + \frac{c(\alpha_i^2 - 1)x_i^{-1}}{T} + \frac{2c(\alpha_i + 1)^5 x_i^{-3}}{T^2}, \quad i = 1, 2, \tag{4.1}$$

where $T = \sum_{j=1}^2 (\alpha_j + 1)^4 x_j^{-2}$. In Theorem 4.1 it is shown that $(\hat{\theta}_1^s, \hat{\theta}_2^s)$ improves $(\hat{\theta}_1, \hat{\theta}_2)$ for simultaneous estimation of (θ_1, θ_2) under sum of squared error loss.

Theorem 4.1. Suppose that $X_i \sim^{ind} Gamma(\alpha_i, \theta_i), i = 1, 2$, and that there is the order restriction $\theta_1 \leq \theta_2$. Let $\hat{\theta}^s(\mathbf{x}) = (\hat{\theta}_1^s(\mathbf{x}), \hat{\theta}_2^s(\mathbf{x}))$ and $\hat{\theta}(\mathbf{x}) = (\hat{\theta}_1(\mathbf{x}), \hat{\theta}_2(\mathbf{x}))$. Then, the estimator $\hat{\theta}^s(\mathbf{x})$ for simultaneous estimation of (θ_1, θ_2) under sum of squared error loss has a uniformly smaller risk, than $\hat{\theta}(\mathbf{x})$, if $0 \leq c \leq 2$.

Proof. The difference of the risks of $\hat{\theta}^s(\mathbf{x})$ and $\hat{\theta}(\mathbf{x})$ is given by:

$$\begin{aligned}
\Delta R &= R(\hat{\theta}(\mathbf{x}), \theta) - R(\hat{\theta}^s(\mathbf{x}), \theta) \\
&= -\sum_{i=1}^{2} \left\{ E\left[\frac{2c(\alpha_i - 1)}{T} - \frac{2c(\alpha_i^2 - 1)x_i^{-1}\theta_i}{T} + \frac{(-1)^i 2c(\alpha_i^2 - 1)x_i^{-1}t_i(\mathbf{x})}{T} \right] \right. \\
&\quad + E\left[\frac{4c(\alpha_i + 1)^5 x_i^{-3}}{T^2} \left(\frac{x_i}{\alpha_i + 1} + (-1)^i t_i(\mathbf{x}) - \theta_\mathrm{i} \right) \right] \\
&\quad \left. + E\left[\frac{c^2(\alpha_i^2 - 1)^2 x_i^{-2}}{T^2} + \frac{4c^2(\alpha_i^2 - 1)(\alpha_i + 1)^5 x_i^{-4}}{T^3} + \frac{4c^2(\alpha_i + 1)^{10} x_i^{-6}}{T^4} \right] \right\}.
\end{aligned} \tag{4.2}$$

From Lemma 3.1 we obtain, if $\alpha_i \geq 1$:

$$E\left[\frac{x_i^{-1}\theta_i}{T}\right] = E\left[\frac{1}{(\alpha_i - 1)T} - \frac{2(\alpha_i + 1)^4 x_i^{-3}\theta_i}{(\alpha_i - 1)T^2}\right],$$

$$\begin{aligned}
E\left[\frac{x_1^{-1}t_1(\mathbf{x})}{T}\right] &= \frac{1}{(\alpha_1 - 1)\theta_1} E\left[\frac{t_1(\mathbf{x})}{T}\right] - \frac{2(\alpha_1 + 1)^4}{\alpha_1 - 1} E\left[\frac{x_1^{-3}t_1(\mathbf{x})}{T^2}\right] \\
&\quad - \frac{(\alpha_2 + 1)}{(\alpha_1^2 - 1)(\alpha_1 + \alpha_2 + 2)} E\left[\frac{I_{(\alpha_2+1)x_1 \geq (\alpha_1+1)x_2}}{T}\right]
\end{aligned} \tag{4.3}$$

and

$$\begin{aligned}
E\left[\frac{x_2^{-1}t_2(\mathbf{x})}{T}\right] &= \frac{1}{(\alpha_2 - 1)\theta_2} E\left[\frac{t_2(\mathbf{x})}{T}\right] - \frac{2(\alpha_2 + 1)^4}{\alpha_2 - 1} E\left[\frac{x_2^{-3}t_2(\mathbf{x})}{T^2}\right] \\
&\quad + \frac{(\alpha_1 + 1)}{(\alpha_2^2 - 1)(\alpha_1 + \alpha_2 + 2)} E\left[\frac{I_{(\alpha_2+1)x_1 \geq (\alpha_1+1)x_2}}{T}\right].
\end{aligned} \tag{4.4}$$

where $I_{(\alpha_2+1)x_1 \geq (\alpha_1+1)x_2}$ denotes the indicator function of the set $\{(x_1, x_2) | (\alpha_2 + 1)x_1 \geq (\alpha_1 + 1)x_2\}$. For simplicity we denote $I_{(\alpha_2+1)x_1 \geq (\alpha_1+1)x_2}$ by I, hereafter.

From (4.3) and (4.4) we obtain:

$$\sum_{i=1}^{2}\left\{E\left[\frac{(-1)^i(\alpha_i^2-1)x_i^{-1}t_i(\mathbf{x})}{T}\right]\right\}$$
$$= E\left[\frac{I}{T}-\left(\frac{1}{\theta_1}-\frac{1}{\theta_2}\right)\frac{((\alpha_2+1)x_1-(\alpha_1+1)x_2)^+}{(\alpha_1+\alpha_2+2)T}\right]$$
$$-2E\left[\sum_{i=1}^{2}\frac{(-1)^i(\alpha_i+1)^5x_i^{-3}t_i(\mathbf{x})}{T^2}\right]\leq E\left[\frac{I}{T}-2\sum_{i=1}^{2}\frac{(-1)^i(\alpha_i+1)^5x_i^{-3}t_i(\mathbf{x})}{T^2}\right], \tag{4.5}$$

since $\theta_1 \leq \theta_2$. From the definition of T we get:

$$\frac{(\alpha_i+1)^5x_i^{-4}}{T^3}\leq\frac{(\alpha_i+1)x_i^{-2}}{T^2},\quad \frac{(\alpha_i+1)^{10}x_i^{-6}}{T^4}\leq\frac{(\alpha_i+1)^2x_i^{-2}}{T^2}.$$

Then the difference of risk (4.2) becomes:

$$\Delta R \geq E\left[\frac{4c}{T}-\frac{2cI}{T}-c^2\sum_{i=1}^{2}\frac{(4(\alpha_i^2-1)(\alpha_i+1)+(\alpha_i^2-1)^2+4(\alpha_i+1)^2)x_i^{-2}}{T^2}\right]$$
$$\geq E\left[\frac{2c}{T}-\frac{c^2}{T}\right]\geq 0 \qquad \textit{if } \ 0\leq c\leq 2,$$

since it is easy to check that $4(\alpha_i^2-1)(\alpha_i+1)+(\alpha_i^2-1)^2+4(\alpha_i+1)^2=(\alpha_i+1)^4$.

Next we consider the estimator which extends the modified MLE, when $(\alpha_1+1)x_2 \leq (\alpha_2+1)x_1$:

$$\tilde{\theta}_i^s(\mathbf{x}) = \begin{cases} \dfrac{x_i}{\alpha_i+1}, & \text{if } (\alpha_1+1)x_2 > (\alpha_2+1)x_1 \\ \dfrac{x_i}{\alpha_i+1} + (-1)^i t_i(\mathbf{x}) + h(\mathbf{x}), & \text{if } (\alpha_1+1)x_2 \leq (\alpha_2+1)x_1 \end{cases}$$

where $h(\mathbf{x}) = \frac{c(\alpha_i^2-1)x_i^{-1}}{T} + \frac{2c(\alpha_i+1)^5x_i^{-3}}{T^2}$. In the following theorem we show that $(\tilde{\theta}_1^s, \tilde{\theta}_2^s)$ improves $(\hat{\theta}_1, \hat{\theta}_2)$ under sum of squared error loss.

Theorem 4.2. Suppose that $X_i \sim^{ind} Gamma(\alpha_i, \theta_i), i = 1, 2,$ and that there is the order restriction $\theta_1 \leq \theta_2$. Let $\tilde{\theta}^s(\mathbf{x}) = (\tilde{\theta}_1^s(\mathbf{x}), \tilde{\theta}_2^s(\mathbf{x}))$ and $\hat{\theta}(\mathbf{x}) = (\hat{\theta}_1(\mathbf{x}), \hat{\theta}_2(\mathbf{x}))$. Then $\tilde{\theta}^s(\mathbf{x})$ has a uniformly smaller risk for simultaneous estimation of (θ_1, θ_2) under sum of squared error loss, $\tilde{\theta}^s(\mathbf{x})$ than $\hat{\theta}(\mathbf{x})$, if $0 \leq c \leq 2$.

Proof. The difference of the risks of $\tilde{\theta}^s(\mathbf{x})$ and of $\hat{\theta}(\mathbf{x})$ is given by:

$$\begin{aligned}
\Delta R &= R(\hat{\theta}(\mathbf{x}), \theta) - R(\tilde{\theta}^s(\mathbf{x}), \theta) \\
&= -\sum_{i=1}^{2}\Bigg\{E\left[\left(\frac{2c(\alpha_i-1)}{T} - \frac{2c(\alpha_i^2-1)x_i^{-1}\theta_i}{T} + \frac{(-1)^i 2c(\alpha_i^2-1)x_i^{-1}t_i(\mathbf{x})}{T}\right)I\right] \\
&+ E\left[\frac{4c(\alpha_i+1)^5 x_i^{-3}}{T^2}\left(\frac{x_i}{\alpha_i+1} + (-1)^i t_i(\mathbf{x}) - \theta_i\right)I\right] \\
&+ E\left[\left(\frac{c^2(\alpha_i^2-1)^2 x_i^{-2}}{T^2} + \frac{4c^2(\alpha_i^2-1)(\alpha_i+1)^5 x_i^{-4}}{T^3} + \frac{4c^2(\alpha_i+1)^{10}x_i^{-6}}{T^4}\right)I\right]\Bigg\}.
\end{aligned} \tag{4.6}$$

If $\alpha_i \geq 1$, we have from Lemma 3.1:

$$E\left[\left(\frac{x_1^{-1}\theta_1}{T}\right)I\right] = -A_1 + E\left[\left(\frac{1}{(\alpha_1-1)T} - \frac{2(\alpha_1+1)^4 x_1^{-3}\theta_1}{(\alpha_1-1)T^2}\right)I\right], \tag{4.7}$$

and

$$E\left[\left(\frac{x_2^{-1}\theta_2}{T}\right)I\right] = A_2 + E\left[\left(\frac{1}{(\alpha_2-1)T} - \frac{2(\alpha_2+1)^4 x_2^{-3}\theta_2}{(\alpha_2-1)T^2}\right)I\right], \tag{4.8}$$

where

$$A_1 = \frac{C_1\theta_1^{-\alpha_1+1}\theta_2^{-\alpha_2}\int_0^\infty x_2^{\alpha_1+\alpha_2} e^{-\left(\frac{\theta_1(\alpha_2+1)+\theta_2(\alpha_1+1)}{\theta_1\theta_2(\alpha_2+1)}\right)x_2}dx_2}{\Gamma(\alpha_1)\Gamma(\alpha_2)((\alpha_1+1)^2+(\alpha_2+1)^2)}, \tag{4.9}$$

$$A_2 = \frac{C_2\theta_1^{-\alpha_1}\theta_2^{-\alpha_2+1}\int_0^\infty x_1^{\alpha_1+\alpha_2} e^{-\left(\frac{\theta_2(\alpha_1+1)+\theta_1(\alpha_2+1)}{\theta_1\theta_2(\alpha_1+1)}\right)x_1}dx_1}{\Gamma(\alpha_1)\Gamma(\alpha_2)((\alpha_1+1)^2+(\alpha_2+1)^2)} \tag{4.10}$$

and

$$C_1 = \frac{(\alpha_1+1)^{\alpha_1-1}}{(\alpha_1-1)(\alpha_2+1)^{\alpha_1+1}},\ C_2 = \frac{(\alpha_2+1)^{\alpha_2-1}}{(\alpha_2-1)(\alpha_1+1)^{\alpha_2+1}}. \tag{4.11}$$

Note that:

$$\frac{\int_0^\infty x_2^{\alpha_1+\alpha_2} e^{-\left(\frac{\theta_1(\alpha_2+1)+\theta_2(\alpha_1+1)}{\theta_1\theta_2(\alpha_2+1)}\right)x_2}dx_2}{\Gamma(\alpha_1+\alpha_2+1)} = \left(\frac{\theta_1\theta_2(\alpha_2+1)}{\theta_2(\alpha_1+1)+\theta_1(\alpha_2+1)}\right)^{\alpha_1+\alpha_2+1} \tag{4.12}$$

and

$$\frac{\int_0^\infty x_1^{\alpha_1+\alpha_2} e^{-\left(\frac{\theta_2(\alpha_1+1)+\theta_1(\alpha_2+1)}{\theta_1\theta_2(\alpha_1+1)}\right)x_1}dx_1}{\Gamma(\alpha_1+\alpha_2+1)} = \left(\frac{\theta_1\theta_2(\alpha_1+1)}{\theta_2(\alpha_1+1)+\theta_1(\alpha_2+1)}\right)^{\alpha_1+\alpha_2+1} \tag{4.13}$$

With (4.9)-(4.13), it is easy to check that:

$$(\alpha_2^2-1)C_2\int_0^\infty x_1^{\alpha_1+\alpha_2}e^{-\left(\frac{\theta_2(\alpha_1+1)+\theta_1(\alpha_2+1)}{\theta_1\theta_2(\alpha_1+1)}\right)x_1}dx_1$$
$$=(\alpha_1^2-1)C_1\int_0^\infty x_2^{\alpha_1+\alpha_2}e^{-\left(\frac{\theta_1(\alpha_2+1)+\theta_2(\alpha_1+1)}{\theta_1\theta_2(\alpha_2+1)}\right)x_2}dx_2$$

and that $(\alpha_2^2-1)A_2-(\alpha_1^2-1)A_1\geq 0$, since $\theta_2\geq\theta_1$.

Next we evaluate $E[((-1)^i x_i^{-1}t_i(\mathbf{x})/T)I]$ using the same method as given by (4.3)-(4.5) in the proof of Theorem 4.1. We obtain:

$$\sum_{i=1}^{2}\left\{E\left[\left(\frac{(-1)^i(\alpha_i^2-1)x_i^{-1}t_i(\mathbf{x})}{T}\right)I\right]\right\}$$
$$\leq E\left[\left(\frac{1}{T}-2\sum_{i=1}^{2}\frac{(-1)^i(\alpha_i+1)^5x_i^{-3}t_i(\mathbf{x})}{T^2}\right)I\right]$$

and, hence, the difference of risk becomes:

$$\Delta R\geq E\left[\left(\frac{2c}{T}-c^2\sum_{i=1}^{2}\frac{(4(\alpha_i^2-1)(\alpha_i+1)+(\alpha_i^2-1)^2+4(\alpha_i+1)^2)x_i^{-2}}{T^2}\right)I\right]$$
$$=E\left[\left(\frac{2c}{T}-\frac{c^2}{T}\right)I\right]\geq 0,\qquad if\ \ 0\leq c\leq 2.$$

5 Simulation results

In this section we give some simulation results to illustrate the extent of the improvement of the risks. The simulations are made using R and each simulation run contains 10000 replications. Simulation results on simultaneous estimation of $p(=10)$ upper bounded Gamma scale parameters, and simultaneous estimation of two ordered Gamma scale parameters are given in Section 5.1 and Section 5.2.

5.1 Simulation Results on Simultaneous Estimation of $p(=10)$ Upper Bounded Gamma Scale Parameters

We consider the simulation of simultaneous estimation of $p(=10)$ upper bounded Gamma scale parameters under sum of squared error loss. The upper bound is set to be 1 and $c=2(p-2)$. The comparison involves the crude unbiased estimator (UE), the suggested estimators $\hat{\delta}(x),\hat{\delta}^s(x)$ and the admissible estimator $\delta(x)$. Compared are the relative risks $R(\theta,UE)/R(\theta,\delta)$, $R(\theta,\hat{\delta})/R(\theta,\delta)$ and $R(\theta,\hat{\delta}^s)/R(\theta,\delta)$, respectively. For the simulation study the following cases with respect to θ are considered:

i) $\theta = (1, \ldots, 1)$
ii) $\theta = (0.1, \ldots, 0.1)$
iii) $\theta = (1, 1/2, 1/3, 1/4, 1/5, 1/6, 1/7, 1/8, 1/9, 1/10)$
iv) $\theta = (1, 1, 1, 1/5, 1/5, 1/5, 1/10, 1/10, 1/10, 1/10)$

For the cases i)-iii), we consider the cases $\alpha = (2, \ldots, 2)$, $(3, \ldots, 3)$, $(4, \ldots, 4)$, $(10, \ldots, 10)$, (2, 2, 2, 10, 10, 10, 20, 20, 20, 20) and (20, 20, 20, 10, 10, 10, 2, 2, 2, 2). For case iv), we consider the cases α=(2,2, 2, 10, 10, 10, 20, 20, 20, 20), (20, 20, 20, 10, 10, 10, 2, 2, 2, 2), (100,100,100,50, 50, 50, 2, 2, 2, 2) (2, 2, 2, 50, 50, 50, 20, 20, 20, 20) and (2, 2, 2, 50, 50, 50, 70, 70, 70, 70). The simulation results are displayed in Table 1.

Table 1. Relative risks of UE , $\hat{\delta}$ and $\hat{\delta}^s$ and the admissible estimator δ

Case	α	$\frac{R(\theta,UE)}{R(\theta,\delta)}$	$\frac{R(\theta,\hat{\delta})}{R(\theta,\delta)}$	$\frac{R(\theta,\hat{\delta}^s)}{R(\theta,\delta)}$
i)	(2, 2, 2, 2, 2, 2, 2, 2, 2, 2)	1.8923	0.6404	0.7001
	(3, 3, 3, 3, 3, 3, 3, 3, 3, 3)	1.6878	0.5019	0.5800
	(4, 4, 4, 4, 4, 4, 4, 4, 4, 4)	1.5726	0.4343	0.5226
	(10,10,10,10,10,10,10,10,10,10)	1.3327	0.3993	0.4906
	(2, 2, 2,10,10,10,20,20,20,20)	1.7000	0.7766	0.8028
	(20,20,20,10,10,10, 2, 2, 2, 2)	1.7329	0.8004	0.8226
ii)	(2, 2, 2, 2, 2, 2, 2, 2, 2, 2)	1.2171	0.8613	0.8636
	(3, 3, 3, 3, 3, 3, 3, 3, 3, 3)	1.0728	0.8657	0.8670
	(4, 4, 4, 4, 4, 4, 4, 4, 4, 4)	1.0253	0.8844	0.8850
	(10,10,10,10,10,10,10,10,10,10)	1.0000	0.9538	0.9538
	(2, 2, 2,10,10,10,20,20,20,20)	1.1528	0.9876	0.9877
	(20,20,20,10,10,10, 2, 2, 2, 2)	1.1703	0.9866	0.9867
iii)	(2, 2, 2, 2, 2, 2, 2, 2, 2, 2)	1.9676	0.9512	0.9503
	(3, 3, 3, 3, 3, 3, 3, 3, 3, 3)	1.7396	0.9518	0.9502
	(4, 4, 4, 4, 4, 4, 4, 4, 4, 4)	1.5743	0.9582	0.9563
	(10,10,10,10,10,10,10,10,10,10)	1.2973	0.9803	0.9786
	(2, 2, 2,10,10,10,20,20,20,20)	1.9965	0.9988	0.9987
	(20,20,20,10,10,10, 2, 2, 2, 2)	1.1982	0.9264	0.9216
iv)	(2, 2, 2, 10, 10, 10, 20, 20, 20, 20)	1.8959	0.9996	0.9994
	(20, 20, 20, 10, 10, 10, 2, 2, 2, 2)	1.2205	0.8938	0.8846
	(100,100,100,50, 50, 50, 2, 2, 2, 2)	1.1381	0.9321	0.9290
	(2, 2, 2, 50, 50, 50, 20, 20, 20, 20)	1.9064	0.9998	0.9997
	(2, 2, 2, 50, 50, 50, 70, 70, 70, 70)	1.9081	1.0000	1.0000

Note that, when $\theta = (1, \ldots, 1)$, the improvement of the estimators $\hat{\delta}$ and $\hat{\delta}^s$ is considerable, especially, for $\alpha = (10, \ldots, 10)$. In contrast, the improvement is small if $\theta = (0.1, \ldots, 0.1)$ or θ=(1,1/2,1/3,1/4,1/5,1/6,1/7,1/8,1/9, 1/10). However, if the upper bound is fixed at 0.1 for the case ii),i.e., $\theta = (0.1, \ldots, 0.1)$, we get the same improvement as in case i). In case iv), the improvements by $\hat{\delta}$ and $\hat{\delta}^s$ are small for any α. However, if we group θ as $(1,1,1)$, $(1/5,1/5,1/5)$ and $(1/10,1/10,1/10,1/10)$ and fix the upper bounds to be $1, 1/5, 1/10$ for each group, then we can get the same improvement as in case i).

5.2 Simulation results for simultaneous estimation of two ordered Gamma scale parameters

Let $X_i \sim Gamma(\alpha_i, \theta_i), i = 1,2$. The simulation refers to the simultaneous estimation under sum of squared error loss of the two Gamma scale parameters (θ_1, θ_2), where $\theta_1 \leq \theta_2$. The simulation involves the admissible estimator $\hat{\theta}^{ad}(\mathbf{x}) = (x_1/(\alpha_1+1), x_2/(\alpha_2+1))$ of θ_i, which is based solely on x_i, and the suggested estimators, $\hat{\theta}^s(\mathbf{x}), \tilde{\theta}^{\mathbf{s}}(\mathbf{x})$ and a modified MLE $\hat{\theta}(\mathbf{x})$. The comparison is made by the following relative risks $R(\theta,\hat{\theta}^{ad})/R(\theta,\hat{\theta})$, $R(\theta,\hat{\theta}^s)/R(\theta,\hat{\theta})$ and $R(\theta,\tilde{\theta}^s)/R(\theta,\hat{\theta})$.

For the simulation, we set $\theta_1 = 1$, $\theta_2 = 1,(1),10$ and $c = 2$. We made simulations for 12 cases with respect to (α_1, α_2): (1,1), (1,3), (1,5), (1,7), (1,9), (3,3), (3,5), (3,7), (3,9), (10,10), (10,20), (10,30). The simulation results are shown in Figure 1.

Note that relative risk of $\hat{\theta}^s$ gets larger when α_1 is small, except for $\alpha_1 = 1, \alpha_2 = 1$. For larger α_1 and α_2 the relative risk of $\hat{\theta}^s$ gets smaller. The relative risk of $\tilde{\theta}^s$ gets smaller for all cases, except for $\alpha_1 = 1, \alpha_2 = 1$. The risk of $\tilde{\theta}^s$ is worse than that of $\hat{\theta}^s$ for all cases. When $\theta_2 = 1$, the relative risk of the admissible estimator is not good, especially, when α_1 is large.

6 Concluding and remarks

In simultaneous estimation of p-variate upper bounded scale parameters of Gamma distributions, we suggest the estimator $\hat{\delta}_i^s$, (3.2), which extends the individually admissible estimator of an upper bounded scale parameters in the same direction. From Table 1. we see that the risks of $\hat{\delta}_i$ and $\hat{\delta}_i^s$ are almost the same. In simultaneous estimation of two ordered scale parameters of two Gamma distributions, $\Gamma(\alpha_i, \theta_i), i = 1,2$, the suggested two estimators for (θ_1, θ_2), $(\hat{\theta}_1^s, \hat{\theta}_2^s)$ and $(\tilde{\theta}_1^s, \tilde{\theta}_2^s)$, satisfy the order restriction if $\alpha_2 \geq \alpha_1$.

Acknowledgements

The author would like to thank Prof. Nobuo Shinozaki, Keio University for many discussions, suggestions and for checking the manuscript.

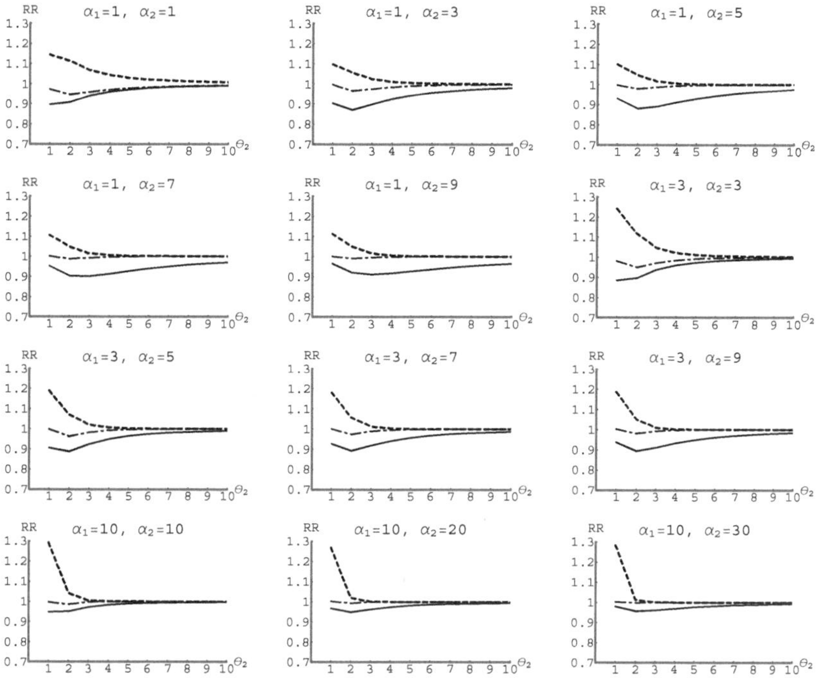

Fig. 1. Relative risks of each estimator: dashed line for $\tilde{\theta}^s$, solid line for $\hat{\theta}^s$, and dotted line for admissible estimator based solely on x_i.

References

1. Barlow, R. E., Bartholomew, D. J., Bremner, J. M. and Brunk, H. D. (1972). Statistical Inference under Order Restrictions, Wiley, New York.
2. Berger, J. (1980). Improving on inadmissible estimators in continuous exponential families with applications to simultaneous estimation of gamma scale parameters. *Ann. Statist.*, 8, 545-571.
3. Berry, J.C. (1993). Minimax estimation of a restricted exponential location parameter. *Statist. Decisions*, 11, 307-316.
4. Casella, G. and Strawderman, W.E (1981). Estimating a bounded normal mean. *Ann. Statist.*, 9, 870-878.
5. Chang, Y.-T. (1982). Stein-type estimators for parameters in truncated spaces. *Keio Science and Technology Reports*, 34, 83-95.
6. Chang, Y.-T. and Shinozaki, N. (2002). A comparison of restricted and unrestricted estimators in estimating linear functions of ordered scale parameters of two gamma distributions. *Ann. Inst. Statist. Math.* Vol. 54, No.4, 848-860.
7. Katz,M.W. (1961). Admissible and minimax estimates of parameters in truncated spaces. *Ann. Math. Statist.*, 32, 136-142.
8. Marchand, È. and Strawderman, W.E. (2005) On improving on the minimum risk equivariant estimator of a scale parameter under a lower-bound constraint. *Journal of Statist. planning and inference*, 134. 90-101.

9. Robertson, T., Wright, F. T. and Dykstra, R. L.(1988). Order Restricted Statistical Inference, Wiley, New York.
10. Shinozaki, N. and Chang, Y.-T. (1999). A comparison of maximum likelihood and best unbiased estimators in the estimation of linear combinations of positive normal means. *Statistics & Decisions*, 17, 125-136.
11. Silvapulle, M. J. and Sen, P. K. (2004). Constrained Statistical Inference, Wiley, New Jersey.
12. Van Eeden,C. (1995). Minimax estimation of a lower-bounded scale parameter of a gamma distribution for scale-invariant squared-error loss. *Canad.J.Statist.*, 23, 245-256.
13. Van Eeden, C (2006). Restricted parameter space estimation problems. Lecture notes in Statistics 188, Springer.

Attractive Quality and Must-be Quality from the Viewpoint of Environmental Lifestyle in Japan

Tsuyoshi Kametani, Ken Nishina, and Kuniaki Suzuki

Nagoya Institute of Technology, Gokiso-cho, Showa-ku, Nagoya 466-8555, Japan

Summary. In this paper, the evaluation of environmental quality based on environmental lifestyle is discussed using a questionnaire survey, which is conducted by applying the following three approaches:.

1. Kano's approach of subjective quality evaluation: Accordingly, the quality elements can be categorized in "attractive quality", "one-dimensional quality","must-be quality","indifferent quality" and "reverse quality". Kano's proposal is applied to express the variability of subjective quality.
2. Akiba and Enkawa's proposal concerning the transistion steps of categorized quality elements: The hypothesis is that quality evaluation by customers changes as follows: "indifferent" quality → "attractive" quality → "one-dimensional" quality → "must-be" quality. In this paper, the transition is referred to as "the maturity of quality".
3. The third approach is the theory concerning the environmental lifestyle by Nishio. She classified female subjects to four types, "green", "ego - eco", "pre-eco" and "non-eco" in a survey from the viewpoints of the degree of environmental consciousness and of the practical action for the protection of environment.

Our questionnaire survey was performed in November 2005. The questionnaire is composed of two parts. One is for classifying the 277 female respondents to the four types of the environmental lifestyle following Nishio, the other is for categorizing the quality elements of refrigerator and shampoo, which are the environment-friendly products following Kano's theory.

From the results of our survey, we have established a relationship between "the maturity of quality" concerning environment and the environmental lifestyle. This information might be relevant for planning of products.

1 Introduction

It has been nearly 20 years since people from the marketing area in Japan started say-ing that they no longer understood what kind of products would sell. It was a time of change, from an age in which a product was judged to be good primarily from "looking" at its physical properties, to one in which a product would not sell

H.-J. Lenz et al. (eds.), *Frontiers in Statistical Quality Control 9*,
DOI 10.1007/978-3-7908-2380-6_20, © Springer-Verlag Berlin Heidelberg 2010

unless customers could "feel" a story or vision behind the product. This trend initially became pervasive as "brands," but from the eve of the 21st century a new element was rapidly spreading: the "environmental friendliness" of products.

As environmental pollution problems such as destruction of the ozone layer, global warming, and environmental endocrine disrupters, which affected the earth itself, came to the fore, the main object of environmental concern shifted from factories causing local pollution to individual consumers around the world. It took little time after that for the emergence of "green consumers." When deciding to make a purchase, "women" more than "they" are said place emphasis on the environmental friendliness of the product, but behind that purchase and consumption is a close relation to the lifestyle of them/women. This is not the usual purchasing behavior in which an individual seeks his or her own pleasure, convenience, or economic benefit, but rather something in which a person feels an image of him- or herself as someone who is contributing to society or considering the future of his or her children. This kind of consumption behavior, which corresponds to a person's values and lifestyle, is very important to the person's sense of self as seen in his or her actions. One clothes oneself with "environmentally friendly" behavior and purchases as one would clothe oneself with a favorite brand.

As shown above, the purchase decision-making of consumers is thought to be influenced by their lifestyle with respect to the environment.

This article takes up the theory of attractive quality and must-be quality of Kano et al. [1] to explain consumer purchase decision-making from the viewpoint of product attributes. In addition, it looks at Nishio's "typification of environmental awareness" [2] to classify consumers according to environment-related lifestyle.

In this article we attempt to use consumers' environmental lifestyles to explain variability in assessments of environmentally friendly product attributes, and from this to obtain a deeper level of information for product planning.

2 Attractive Quality and Must-be Quality

Kano et al.'s [1] theory of attractive quality and must-be quality is very appropriate for explaining the purchase decision-making of consumers from the view-point of product attributes.

With respect to the product elements making up product quality, Kano et al. [1] investigated the relationship between the objective aspect indicated by fulfillment of physical criteria and the subjective aspect of sense of satisfaction with product quality. From the two-way relationship shown in Fig. 1, with physical fulfillment on the horizontal axis and user satisfaction on the vertical axis, they proposed the classification of quality elements shown in Table 1.

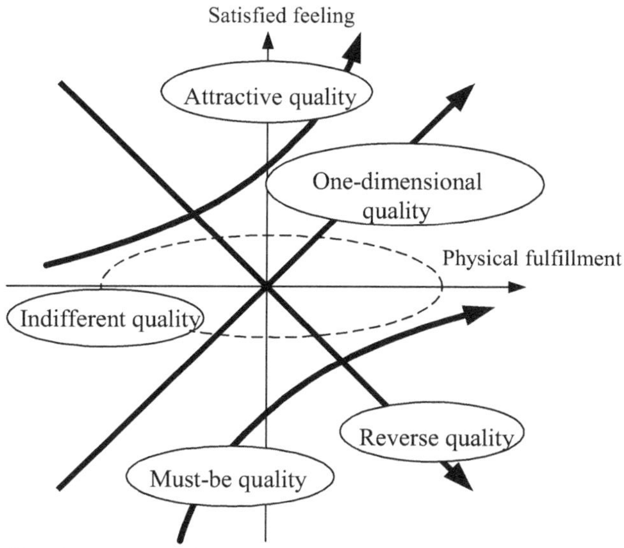

Fig. 1 Quality elements seen from the relation between physical aspects and consumer satisfaction [1]

Table 1. Definitions of quality elements of Kano et al. [1]

Classification	Content
Attractive Quality Element	Quality elements that when fulfilled provide satisfaction but when not fulfilled are acceptable.
One-Dimensional Quality Element	Quality elements that result in satisfaction when fulfilled and in dissatisfaction when not fulfilled.
Must-Be Quality Element	Quality elements that are absolutely expected (taken for granted when fulfilled), but result in dissatisfaction when not fulfilled.
Indifferent Quality Element	Quality elements that neither result in satisfaction nor dissatisfaction, regardless of whether they are fulfilled or not.
Reverse Quality Element	Quality elements that result in dissatisfaction when fulfilled and satisfaction when not fulfilled.

According to Kano et al., consumers' purchase decision-making is affected by the relationship between the physical aspects of quality elements and the subjective aspects of consumers themselves. For this reason, various studies have been conducted from the perspective of improving quality in the product planning stage, so-called "planning quality."

Akiba and Enkawa [3] showed from a study of televisions that, as the market matured, the quality elements that consumers emphasized, when evaluating a product, shifted from basic functions to durability and maintainability, then from durability and maintainability to operability, and finally to ease of placement, personal preference, and optional functions. Moreover, in the pattern of shifting emphasis in consumers' product evaluation, starting from basic functions and moving up to optional functions, the quality elements that customers had at one time emphasized become "must-be quality" elements. At this point the quality elements emphasized by customers become "one-dimensional quality," and at a further step beyond that "attractive quality." The quality elements beyond that are "indifferent quality."

Enkawa [4] looked at evaluation factors for the service industry, and found that among the quality classification categories of attractive quality, must-be quality and the others, there is an order with the direction indifferent → attractive → one-dimensional → must-be, and that such an order moves with time from the quality element of indifference until reaching must-be. Regularity in this order was then proposed in this transition of quality classifications that is caused by market maturity.

According to the claims of Akiba et al. [3] and Enkawa [4], the elements that consumers emphasize in evaluating products, meaning the changes in quality classifications, follow a time axis of market maturity. Then, from the level of market maturity, the quality classification that is assessed by consumers moves in the order

indifferent quality → attractive quality → one-dimensional quality
→ must-be quality.

3 Classifying consumer type by ecology

Nishio [2] proposed a "typification of ecological awareness" by classifying consumers by type according to their level of interest and behavior with respect to environmental problems. She categorized consumers into four types indicated on the two axes of "degree of ecological involvement" and "degree of implementation of ecological behavior." In the present study these four types are called environmental lifestyle. Here, "degree of ecological involvement" is a scale expressing an individual's insistence, thoughts, and interest in carrying out environmental preservation and corresponding behaviors.

Each of the following ten items was scored 1 point for "Completely disagree," 2 points for "Disagree somewhat," 3 points for "Cannot say either way," 4 points for "Agree somewhat," and 5 points for "Completely agree." However, questions 1 through 3 were scored in reverse. The total score for the ten items was taken to be the "degree of ecological involvement." The questions were as follows.

1. Environmental problems are serious, but there are more serious problems in my life.
2. Although I would like to preserve the environment, I do not want to give up my current comfort and convenience.
3. It is difficult for me to think of environmental problems as my own problems.
4. Of the various problems faced by society, I am interested above all in envi-ronmental problems.
5. I try to communicate the importance of environmental problems to the people around me, and get them to adopt environmentally-friendly behaviors.
6. I feel like my individuality is reflected by the types of behaviors I adopt to preserve the environment.
7. When selecting a product, I consider its effect on the environment or what has been done to make it environmentally-friendly.
8. I am very knowledgeable about environmental problems.
9. I regularly watch television programs or read articles about environmental problems.
10. Environmental problems should be addressed immediately.

"Degree of implementation of ecological behavior" expresses the degree of action one takes with concern for ecology. "Degree of implementation of ecological behavior" was measured based on subject responses as to whether or not she adopted 20 ecological behaviors. The "degree of implementation of ecological behavior" was scored by giving one point for each positive answer a respondent gives, and totaling the score. The questions were the following.

- I unplug electrical appliances that I am not using.
- I give used milk cartons for recycling.
- I do not pour oil down the drain.
- When I go shopping I carry a basket with me (I do not take the store's shopping bags).
- I limit automobile use and ride a bicycle or public transportation.
- I gather full loads of laundry before washing.
- I recycle or reuse plastic bottles.
- I participate in environmental volunteer activities.
- I am careful to turn off lights.
- I try to limit trash by not buying disposable items or more than I need.
- I dispose of kitchen garbage at home by using a garbage processor or composting.
- I do not use synthetic detergents as dishwashing or laundry soap.
- I send clothes, I no longer need, to a recycling shop or flea market.
- I try whenever possible to repair broken items and use them.

- I set my air conditioner at 28°C or higher, and my heater at 20°C or lower.
- I do not let my car's engine idle.
- I seek out and buy organic vegetables.
- I turn off the faucet conscientiously, and do not leave water running.
- I use the determined amount of washing powder.
- I drive a low-emission vehicle (hybrid, etc.), or am considering one as my next car.

Consumers were divided into four types according to the four quadrants of a coordinate system related to ecology, as shown in Fig. 2.

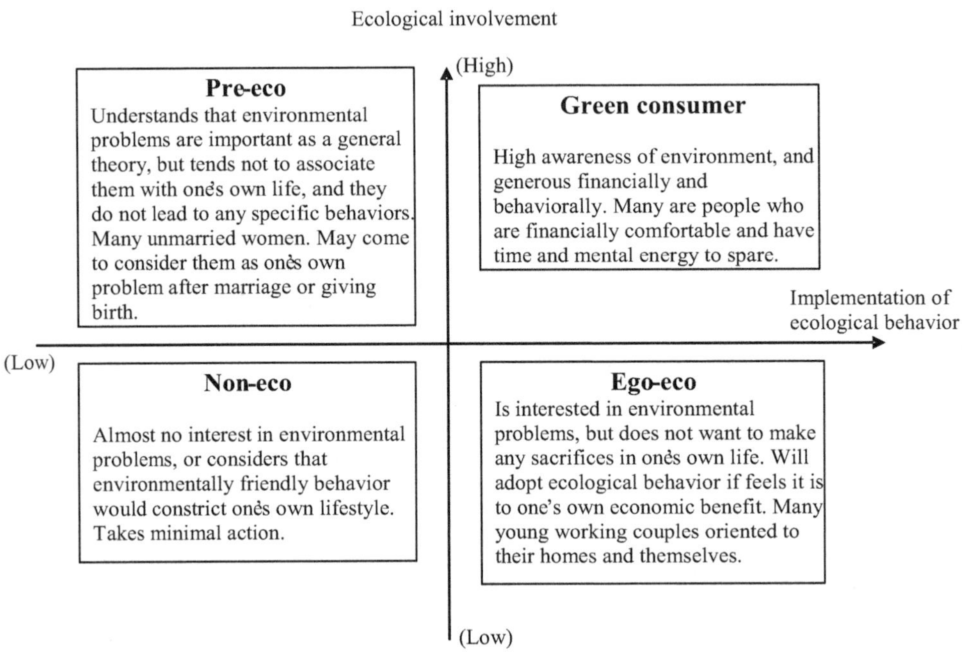

Fig. 2. Four types of environmental lifestyle

4 Questionnaire survey

4.1 Purpose of survey

The purpose of this survey was to validate our hypothesis that the variance in assessments of environmentally-friendly quality attributes can be explained by consumers' environmental lifestyles. We also considered the relationship between quality elements and the time axis of maturation of consumer environmental life-styles.

4.2 Survey Outline

All questioned persons in this study were women. This was, because it was assumed, as a general image, that purchasing environmentally friendly products, is a behaviour being characteristic for women.

The products should be selected on condition that the persons imagined an environmentally-friendly product group. Moreover, the products should be of a type that "everybody is familiar with" and that "women make the purchase (decision)". Finally, to avoid having a great number of questions on the questionnaire, the number of products was narrowed to two with different properties and price ranges. Thus, a refrigerator and shampoo was selected for the study.

Table 2. Quality element items used in questions

Quality element items	Question items
Basic function	Good performance
Usability	Easy to handle
Ease of placement	Size is just right
Preference	Good design
	Good reputation
Reliability	Trustworthy manufacturer
	Long-lasting
Environmental friendliness	Conserves resources
	Carries environmental label
	Waste materials are not harmful to the earth
Availability	Easy to obtain
Harmful effects	No adverse effects from use
Economically advantageous	Economical
Applicability	Multifunctional
Maintainability	Good after-sales service

Note: In the text, "Conserves resources", "Carries environmental label" and "Waste materials are not harmful to the earth" are abbreviated with "resources", "label" and "waste material", respectively.

A questionnaire was conducted to determine environmental lifestyle types following the proposal of Nishio. On questions related to quality elements, the persons were asked what they would feel in the case that the product satisfied or did not satisfy the respective quality elements in Table 2. On the actual questionnaire, subjects were not asked whether products satisfied or did not satisfy the quality elements, but to classify their feeling using the five categories of "like," "acceptable," "no feeling," "must-be," and "do not like" in the case that each element was good and bad.

Table 3. Demographic attributes of subjects

Total number of subjects		277
Age	20s	69
	30s	82
	40s	69
	50s	54
	60s	3
Marital status	Married	190
	Single	87
Children	Yes	179
	No	98
No. of other people in household	1	42
	2	63
	3	80
	4	34
	5	14
	6	5
	8	1
	0	23
Work	Full-time homemaker	98
	Part-time employment	58
	Full-time employment	92
	Other	29
Annual income	Less than 4 million yen	67
	4 million to less than 6 million yen	76
	6 million to less than 8 million yen	59
	8 million to less than 10 million yen	34
	10 million yen or more	33
	Unspecified	8

The survey was conducted in cooperation with the Chubu Electric Power Co., by posting a call for questionnaire respondents to the 5,000 registered members of the company's mobile phone website. Questionnaires were sent by post to 398 women aged in the 20s–50s, who had requested it. The number of responses was 359, i.e., a response rate of 90%. Table 3 shows the demographic attributes of the subjects. If a respondent failed to answer all the questions or made a clear mistake in responding, all the data from that respondent was excluded. As a result, the number of valid responses was 277.

Table 4. Determination of attractive quality items – skeptical quality items [1]

Not fulfilled Fulfilled	Like	Essential	Don't care either way	No choice but to accept	Don't like
Like	S	A	A	A	—
Essential	R	I	I	I	M
Don't care either way	R	I	I	I	M
No choice but to accept	R	I	I	I	M
Don't like	R	R	R	R	S

Note: A: Attractive evaluation; M: Must-be evaluation; O: One-dimensional evaluation; I: Indifferent evaluation; R: Reverse evaluation; S: Skeptical evaluation.

5 Survey results

5.1 Overall quality elements and quality classifications

Quality classifications were established from the responses to the questions on quality elements in Table 2, following the method of Kano [1]. Table 4 shows Kano's method of establishing quality classifications. Table 4 is a matrix of quality classifications for each response of satisfaction/dissatisfaction with quality elements.

The incidence of skeptical quality and reverse quality were extremely low in this study, and so they were excluded from the analysis.

First, the relation between quality elements and quality classifications was analyzed without stratifying subjects. The results of the analysis of shampoo are shown as an example. A correspondence analysis of quality classifications × quality elements was conducted (Fig.3), and is summarized in the two-way contingency table shown in Table 5 (see Miyakawa [5].)

Table 5. Two-way contingency table of quality classification × quality element (shampoo)

	Performance	Usability	Ease of placement	Design	Brand	Manufacturer	Resources
Must-be	117	76	43	7	25	57	46
One-dimension	129	108	93	39	91	102	54
Attractive	19	60	81	153	85	52	82
Indifferent	12	33	60	76	76	66	96
total	277	277	277	275	277	277	278

	Label	Waste material	Availability	Harmful effects	Multifunction	Economically advantageous	total
Must-be	35	45	46	44	9	35	371
One-dimension	42	64	62	154	28	139	616
Attractive	79	86	87	53	89	80	532
Indifferent	121	82	82	26	144	23	419
total	277	277	277	277	270	277	

From Fig. 3 we can see the quality elements corresponding to maturity of quality elements: indifferent quality → attractive quality → one-dimensional quality → must-be quality. All of the product attributes related to environmental friendliness, "resources," "label," "waste material," are positioned in indifferent quality. The same trend was seen for refrigerators.

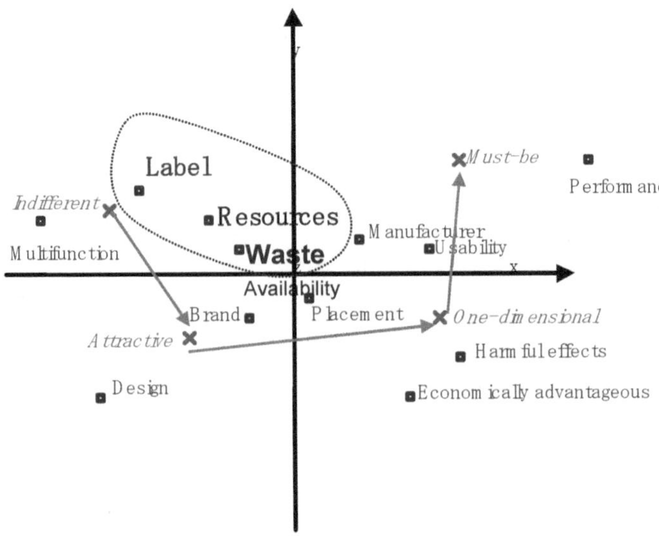

Fig. 3. Results of correspondence analysis (for shampoo)

5.2 Relation between quality elements and quality classifications with stratification by environmental lifestyle

Subjects were stratified by environmental lifestyle according to the "typifica-tion of environmental awareness" proposed by Nishio, and the results are shown in Fig. 4. The same analysis was done (data were put in a three-way contingency table) after this stratification of subjects. The results are shown in Fig. 5, but only for "waste material" to make it easier to understand on paper.

In 5.1, "resources" was positioned in indifferent quality, but a clear pattern emerges after stratification of subjects by environmental lifestyle. Here, when made to correspond to the process of quality classification maturation, we see the directionality of non-eco, ego-eco → pre-ego → green consumer. Considering the four types of environmental lifestyle shown in Fig. 2, a maturation process of that can be assumed for this similar process. Accordingly, the process of quality classification maturation corresponds with that of environmental lifestyle maturation.

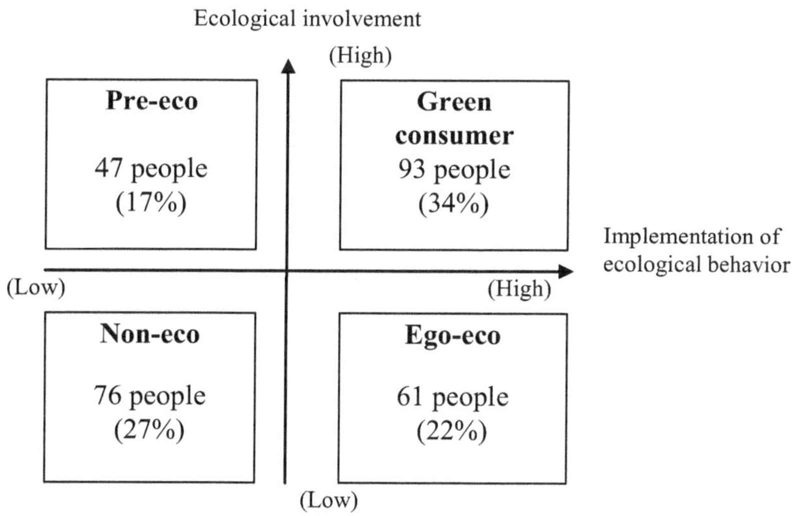

Note: Numbers in shaded boxes are numbers of people in category
(Numbers in parentheses are percentage of 277 people)

Fig. 4. Stratification by environmental lifestyle

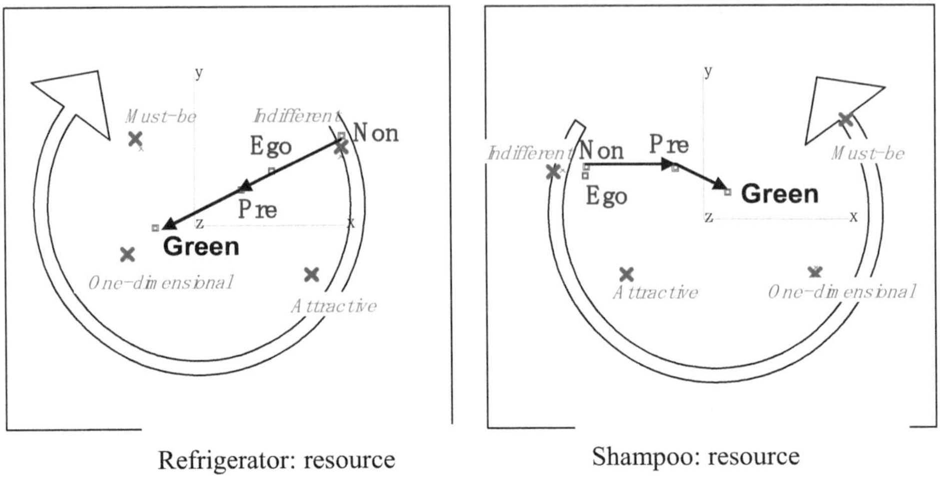

Fig.5 Results of correspondence analysis after stratification by environmental lifestyle

5.3 Discussion

Considering the results of the analysis described in 5.2, it seems that rising maturity of consumers' environmental lifestyles also leads to maturity of quality classifications. Consumer whose environmental behavior is at a low stage of maturity (non-eco), consider attributes showing environmental concern to be "indifferent quality". Conversely, consumer who has a high level of "greenness" (green consumer), evaluate environmentally-friendly product attributes as "one-dimensional quality". Consumer position on a maturity axis of environmental lifestyle can be interpreted as contributing to the variance in quality classification.

From the above one may consider the following:

(1) The hypothesis that lifestyle characteristics of individual consumers are factors that can contribute to the variance in quality classifications holds.

(2) With respect to (1), it was confirmed that there is variance in quality classifications on the maturity axis for environmental lifestyle.

(3) A relationship of correspondence is formed between the processes of environmental lifestyle maturation and quality classification maturation.

(4) The variance in quality classifications related to environmentally-friendly attributes should be kept in mind, when dealing with environmentally friendly products in product planning, or when it is assumed that green consumers exist in the target market.

(5) If (4) becomes further generalized, it may be possible to make a further step and obtain information on customers future needs by focusing on the variance in quality elements that become "indifferent quality" as environmentally-friendly attributes.

6 Conclusion

The variability in assessments of environmentally-friendly quality attributes was explained by the environmental lifestyle of consumers. This variability was also explained by the process of quality classification maturation. This suggests that the variability in quality elements that becomes indifferent quality overall may provide a deeper level of information for products planning.

Ackowledgement

The authers would like to thank Mr. Masaya Kato for his helpful support of the questionnaire survey.

References

1. Kano, N., Seraku, N., Takahashi, F., and Tsuji, S. (1996), "Attractive Quality and Must-Be Quality", *The Best Quality*, IAQ Book Series Vol. 7, ASQC Quality Press, 165 - 186.
2. Nishio, C. (1997), "Reaction of the consumer for the environmental problem and Ecological Marketing", *A research report of Academy of Marketing Course*, Japan Productivity Center for Socio Economic Development (in Japanese).
3. Akiba, M., Enkawa, T. (1986) *Evaluation of Durable Goods as Consumer's Aspect*, NIKKAN KOGYO SHIMBUN,LTD (in Japanese).
4. Enkawa, T. (1992) "Customer Evaluation Perspectives and Quality Planning Based on Product Evaluation Factors", *Hinshitsu (Quality)*, Vol. 22, No. 1, 37 - 45 (in Japanese).
5. Miyakawa, M., (1990) "Analysis of Contingency Table by Example", Hinshitsu (Quality), Vol. 20, No. 3, 16- 22 (in Japanese).

On Identifying Dispersion Effects in Unreplicated Fractional Factorial Experiments

Seiichi Yasui[1], Yoshikazu Ojima[2], and Tomomichi Suzuki[3]

Department of Industrial Administration, Tokyo University of Science, 2641 Yamazaki, Noda, Chiba, 278-8510, JAPAN, yasui@rs.noda.tus.ac.jp

The analysis of dispersion effects is as important as the location effect analysis in the quality improvement. Unreplicated fractional factorial experiments are useful to analyse not only location effects but also dispersion effects. The statistics introduced by Box and Meyer (1986) to identify dispersion effects are based on residuals subtracting the estimates for large location effects from observations. The statistic is a simple form, however, the property is not completely discovered. In this article, the distribution of the statistic under the null hypothesis is derived in unreplicated fractional factorial experiments using an orthogonal array. The distribution under the null hypothesis cannot be expressed uniquely. The statistic has different null distributions depending on the combination of columns allocating factors. We concluded that the distributions can be classified into three types, i.e. the F distribution, unknown distributions close to the F distribution and the constant (not stochastic variable) which is one. Finally, the power of the test for detection of a single active dispersion effect is evaluated.

1 Introduction

In quality control and improvement, it is important to control the quality characteristics by grasping the relationship between the response and the controllable factors. The control succeeds by two kinds of works. One is to adjust the mean response for the target value, the other one is to minimize the spread of response. The former have been traditionally achieved by mature scientific approaches; design of experiments is one representation of useful methods. To the latter less attention has been paid than to the analysis of the mean response, however, its importance was emphasized by Taguchi (e.g., Taguchi and Wu, 1980). Hence, the

H.-J. Lenz et al. (eds.), *Frontiers in Statistical Quality Control 9*,
DOI 10.1007/978-3-7908-2380-6_21, © Springer-Verlag Berlin Heidelberg 2010

analysis of both location and dispersion effects becomes important to the quality improvement.

By using methods such as Bartlett and Kendall (1949) and Nair and Pregibon (1988) it is possible to estimate location and dispersion effects from experiments with replications. Since it is not economical to analyze dispersion effects in such experiments with a large number of runs, there has been considerable interest in estimating location and dispersion effects from unreplicated fractional factorial experiments.

Box and Meyer (1986) is one of pioneering works for analysis of the dispersion effect from unreplicated two-level (fractional) factorial experiments, and demonstrated a method for identifying dispersion effects under the assumption of effect sparsity; namely, only a few of the effects are active. Bergman and Hynen (1997) modifies the Box and Meyer (1986)procedure in two-level factorial designs. Blomkvist, Hynen and Bergman (1997) developed the generalized procedure of Bergman and Hynen (1997) for applying to two-level fractional factorials and response surface methodology. Wang (1989) derives the maximum likelihood estimators of location and dispersion effects under the normal assumption with the log-linear dispersion model. Wang (2001) expanded Wang (1989) to unreplicated multilevel fractional factorial experiments. Liao (2000) proposed the likelihood ratio test in unreplicated two-level fractional factorial experiments. Harvey (1976) proposed a two-step procedure for location-dispersion modeling in the general regression context. Wiklander (1998) developed the unbiased estimator for two-level factorial designs without replications when the linear main effect models are assumed for the location and dispersion models. Holm and Wiklander (1999) adapted Wiklander (1998) to two-level fractional factorial designs. Wiklander and Holm (2003) discussed the theoretical background of Wiklander (1998) and Holm and Wiklander (1999). Lee (1994) studied the Bayesian estimation and compared it with the maximum likelihood estimator in unreplicated factorial designs.

It is a difficult problem to identify both location and dispersion effects from unreplicated experiments. Brenneman and Nair (2001) showed the unbiasedness of some estimators under both linear and log-linear dispersion models. McGrath and Lin (2001a) pointed out that Bergman and Hynen (1997)'s statistic is biased under the multiple dispersion effect model, and proposed an unbiased estimator. Ferrer and Romero (1993) evaluated the accuracy of the estimation for Harvey (1976), iterative weighted least square method and maximum likelihood estimation. Pan (1998), McGrath and Lin (2001b) and Schoen

(2004) discussed the impact of misidentifying the location model to infer for dispersion effects. McGrath (2003) developed a procedure to determine the minimum number of additional runs needed to remove the confounding between the dispersion effect and two unidentified location effects.

Mostly, procedures for identifying dispersion effects have been based on the residuals. Thus, the performance of the statistic to identify dispersion effects is affected by the identified location model in unreplicated fractional factorial experiments. The properties and impact of procedures to identify dispersion effects were disclosed for some multiple dispersion effect models and unidentified location effects. In this article, an important property is added by a theoretical consideration of Box and Meyer (1986)'s procedure. The important property is that the performance depends on the defining relation among the location effects which are used to calculate residuals, even if all the true active location effects are correctly identified. We show that the null distributions of Box and Meyer (1986)'s statistic can be classified into three types depending on the defining relation among the location effects (allocation pattern).

In the next section, we show the property of the statistic introduced by Box and Meyer(1986). In Section 3, the null distribution is shown. In Section 4, we show the power curves defined as the probability that the null hypothesis is rejected.

2 The property of the test statistic for dispersion effects

We investigate the test statistic to detect dispersion effects in the procedure to identify dispersion effects from unreplicated fractional factorial designs introduced by Box and Meyer (1986). First, the location effects are estimated via ordinary least squares and the set of active location effects is identified. Next, the residuals are calculated by using active location effects. Let $\mathbf{r}_+$ be the residual vector collecting the residuals for $+1$ level of the interested dispersion effect, and let $\mathbf{r}_-$ be the residual vector for -1 level of it. Then the test statistic for the dispersion effect is

$$D = \frac{\mathbf{r}'_+\mathbf{r}_+}{\mathbf{r}'_-\mathbf{r}_-}, \tag{1}$$

where $\mathbf{r}'_+$ is the transposed vector of $\mathbf{r}_+$. The set of the location effects is divided into four subsets as follows;

1. true active effects found to be active,

2. true inactive effects found to be active,
3. true active effects found to be inactive,
4. true inactive effects found to be inactive.

The first and second subsets are used for calculating residuals. The third subset generates the bias for residuals. The fourth subset are pure errors. Let $\boldsymbol{\beta}_{AA}$, $\boldsymbol{\beta}_{IA}$, $\boldsymbol{\beta}_{AI}$ be the true effects vector corresponding to the first, second, third subsets, respectively. Then the design matrix is partitioned into four parts corresponding to an overall mean and three subsets, and it is divided by two parts for the row. Hence, if the experiment designed with the $n \times n$ regular orthogonal array is carried out, the underlying statistical model is

$$\begin{aligned}\mathbf{y} &= \begin{pmatrix} \mathbf{1} & X_{AA+} & X_{IA+} & X_{AI+} \\ \mathbf{1} & X_{AA-} & X_{IA-} & X_{AI-} \end{pmatrix} \begin{pmatrix} \mu \\ \boldsymbol{\beta}_{AA} \\ \mathbf{0} \\ \boldsymbol{\beta}_{AI} \end{pmatrix} + \begin{pmatrix} \boldsymbol{\varepsilon}_{+} \\ \boldsymbol{\varepsilon}_{-} \end{pmatrix} \\ &= X_T \boldsymbol{\beta}_T + \boldsymbol{\varepsilon} \\ &\begin{pmatrix} \boldsymbol{\varepsilon}_{+} \\ \boldsymbol{\varepsilon}_{-} \end{pmatrix} \sim N\left(\begin{pmatrix} \mathbf{0} \\ \mathbf{0} \end{pmatrix}, \sigma^2 \begin{pmatrix} k^2 I_{\frac{n}{2}} & O \\ O & I_{\frac{n}{2}} \end{pmatrix} \right),\end{aligned} \tag{2}$$

where $\mathbf{1}$ is a column vector with all the $n/2$ elements equal to one, μ is an overall mean and k is the magnitude of a dispersion effect and $I_{\frac{n}{2}}$ is an identity matrix of $n/2 \times n/2$. A single dispersion model is assumed in order to disclose basic property. The X_{AA+} matrix denotes the design matrix for the $+1$ level of the single dispersion effect in the subset consisting of true active location effects found to be active.

The estimated statistical model to calculate residuals is

$$\begin{aligned}\mathbf{y} &= \begin{pmatrix} \mathbf{1} & X_{AA+} & X_{IA+} \\ \mathbf{1} & X_{AA-} & X_{IA-} \end{pmatrix} \begin{pmatrix} \mu \\ \boldsymbol{\beta}_{AA} \\ \boldsymbol{\beta}_{IA} \end{pmatrix} + \begin{pmatrix} \boldsymbol{\varepsilon}_{+} \\ \boldsymbol{\varepsilon}_{-} \end{pmatrix} \\ &= X_E \boldsymbol{\beta}_E + \boldsymbol{\varepsilon} \\ &\begin{pmatrix} \boldsymbol{\varepsilon}_{+} \\ \boldsymbol{\varepsilon}_{-} \end{pmatrix} \sim N\left(\begin{pmatrix} \mathbf{0} \\ \mathbf{0} \end{pmatrix}, \sigma^2 \begin{pmatrix} I_{\frac{n}{2}} & O \\ O & I_{\frac{n}{2}} \end{pmatrix} \right).\end{aligned} \tag{3}$$

Hence, the residual vector is

$$\begin{pmatrix} \mathbf{r}_{+} \\ \mathbf{r}_{-} \end{pmatrix} = (\mathbf{y} - \hat{\mathbf{y}}) = \left(I_n - \frac{X_E X_E'}{n} \right) X_T \boldsymbol{\beta}_T + \left(I_n - \frac{X_E X_E'}{n} \right) \boldsymbol{\varepsilon}. \tag{4}$$

Due to $X_T' X_T = nI$,

$$\mathbf{r}_+ = X_{AI+}\beta_{AI} + \left(I_{\frac{n}{2}} - \frac{1}{n}(X_+X_+'), -\frac{1}{n}(X_+X_-')\right)\begin{pmatrix}\varepsilon_+\\ \varepsilon_-\end{pmatrix} \quad (5)$$

$$\mathbf{r}_- = X_{AI-}\beta_{AI} + \left(-\frac{1}{n}(X_-X_+'), I_{\frac{n}{2}} - \frac{1}{n}(X_-X_-')\right)\begin{pmatrix}\varepsilon_+\\ \varepsilon_-\end{pmatrix} \quad (6)$$

where $X_+ = (\mathbf{1}_{\frac{n}{2}}, X_{AA+}, X_{IA+})$, $X_- = (\mathbf{1}_{\frac{n}{2}}, X_{AA-}, X_{IA-})$.

Consequently,

$$E(\mathbf{r}_+) = X_{AI+}\boldsymbol{\beta}_{AI}, \quad (7)$$

$$E(\mathbf{r}_-) = X_{AI-}\boldsymbol{\beta}_{AI}, \quad (8)$$

and

$$V(\mathbf{r}_+) = k^2\sigma^2\left(I_{\frac{n}{2}} - \frac{X_+X_+'}{n}\right)^2 + \sigma^2\left(\frac{X_+X_-'X_-X_+'}{n^2}\right), \quad (9)$$

$$V(\mathbf{r}_-) = k^2\sigma^2\left(\frac{X_-X_+'X_+X_-'}{n^2}\right) + \sigma^2\left(I_{\frac{n}{2}} - \frac{X_-X_-'}{n}\right)^2, \quad (10)$$

$\mathbf{r}_+$ and $\mathbf{r}_-$ are distributed according to the multivariate normal distribution, respectively.

Lemma 1. $\left(I_{\frac{n}{2}} - \frac{X_+X_+'}{n}\right)$ *and* $\left(I_{\frac{n}{2}} - \frac{X_-X_-'}{n}\right)$ *are idempotent matrices* $\Leftrightarrow X_+X_-' = X_-X_+' = O_{\frac{n}{2}\times\frac{n}{2}}$, $O_{\frac{n}{2}\times\frac{n}{2}}$ *is zero matrix.*

Proof. $\left(I_{\frac{n}{2}} - \frac{X_+X_+'}{n}\right)^2 = I_{\frac{n}{2}} - 2\frac{X_+X_+'}{n} + \frac{X_+X_+'X_+X_+'}{n^2}$. If $\left(I_{\frac{n}{2}} - \frac{X_+X_+'}{n}\right)$ is idempotent matrix, then $\frac{X_+X_+'X_+X_+'}{n^2} = \frac{X_+X_+'}{n}$. $\therefore$ $X_+(X_+'X_+ - nI_{\frac{n}{2}})X_+' = O_{\frac{n}{2}\times\frac{n}{2}}$. $\Leftrightarrow X_+(-X_-'X_-)X_+' = O_{\frac{n}{2}\times\frac{n}{2}}(\because X_T'X_T = nI)$.
$\therefore (X_-X_+')'(X_-X_+') = O_{\frac{n}{2}\times\frac{n}{2}}$. Hence, X_-X_+' is a zero matrix.
Analogously, X_+X_-' is a zero matrix due to the condition that $\left(I_{\frac{n}{2}} - \frac{X_-X_-'}{n}\right)$ is an idempotent matrix.

On the contrary, if $X_+X_-' = X_-X_+' = O_{\frac{n}{2}\times\frac{n}{2}}$, then $\left(I_{\frac{n}{2}} - \frac{X_+X_+'}{n}\right)$ and $\left(I_{\frac{n}{2}} - \frac{X_-X_-'}{n}\right)$ are idempotent matrices by the straightforward expansion. □

If X_-X_+' and X_-X_+' are zero matrices, $\mathbf{r}_+$ and $\mathbf{r}_-$ are mutually independently distributed according to following distributions;

$$\mathbf{r}_+ \sim N\left(X_{AI+}\beta_{AI}, k^2\sigma^2\left(I_{\frac{n}{2}} - \frac{X_+X_+'}{n}\right)\right), \quad (11)$$

$$\mathbf{r}_- \sim N\left(X_{AI-}\beta_{AI}, \sigma^2\left(I_{\frac{n}{2}} - \frac{X_-X_-'}{n}\right)\right). \quad (12)$$

Let m be the number of columns of the matrix X_+,

$$\mathrm{tr}\left(I_{\frac{n}{2}} - \frac{X_+X_+'}{n}\right) = \mathrm{tr}\left(I_{\frac{n}{2}} - \frac{X_-X_-'}{n}\right) = \frac{1}{2}(n-m). \tag{13}$$

Consequently, we obtain the following result.

Theorem 1. $X_-X_+' = X_+X_-' = O_{\frac{n}{2}\times\frac{n}{2}} \Rightarrow \frac{\mathbf{r}_+'\mathbf{r}_+}{k^2\sigma^2}$ *and* $\frac{\mathbf{r}_-'\mathbf{r}_-}{\sigma^2}$ *are mutually independently distributed according to non-central chi-square distributions with* $\frac{n-m}{2}$ *degrees of freedom, and the noncentrality parameters are* $(\boldsymbol{\beta}_{AI}'X_{AI+}'X_{AI+}\boldsymbol{\beta}_{AI})/(k^2\sigma^2)$ *and* $(\boldsymbol{\beta}_{AI}'X_{AI-}'X_{AI-}\boldsymbol{\beta}_{AI})/\sigma^2$*, respectively.*

3 Null Distributions of Test Statistics

In this section, the distributions of D under the null hypothesis $H_0 : k = 1$ are investigated. Additionally, we assume that all the true location active effects are completely identified as being active, i.e. X_{AI+} and X_{AI-} are zero matrices. The distributions are classified into some types, depending on the design matrix of the estimated statistical model (3).

In case of $X_+X_-' = X_-X_+' = O_{\frac{n}{2}\times\frac{n}{2}}$, due to Theorem 1, for any k, D is distributed according to the k^2 multiple F distribution with $(n-m)/2$ and $(n-m)/2$ degrees of freedoms. We show the allocation patterns for Taguchi's 8×8 orthogonal array in Table 1. Taguchi's 8×8 orthogonal array is the matrix changing the columns of the regular 8×8 orthogonal array expressed by Yates's standard order (e.g. see Table 3 of Ojima et. al (2004)), and we express it as L_8. The number of active location effects $(m-1)$ is shown in the first row of Table 1. If X_+X_-' is the zero matrix, the number of columns of the estimated design matrix m is an even number because the element of the matrix is $+1$ or -1. The second row and below express the allocation pattern. For instance, if the five effects are allocated to 1st, 2nd, 3rd, 4th and 5th columns and the residuals are calculated based on all of those effects, D has the k^2 multiple F distribution with (8-6)/2=1 and one degree of freedoms. Suppose the three main effects (A, B and C) are allocated to 1st, 2nd and 4th columns and the interested dispersion effect is evaluated from the 1st column. In Taguchi's orthogonal array, the 1st, 2nd and 4th columns are main effect vectors of an unreplicated three-way layout. Under this situation, the sufficient condition so that the statistic D has an F distribution is that 1st, 2nd, 3rd, 4th and 5th columns are included in the design matrix X_E. The 3rd and 5th columns denote the two-factor interaction columns, the 3rd column is the interaction between A and B, the 5th column is the interaction between A and C. This design matrix is the expanded location model introduced by Bergman and Hynen (1997). Bergman and Hynen (1997) also referred to the result that the distribution of the statistic D based on the expanded location model is an F distribution. Hence, theorem 1 will be a theoretical evidence for Bergman and Hynen (1997).

Table 1. Allocations for L_8 when the D has the k^2 multiple F distribution

$m-1$	1	3	5
allocation	1	1,2,3	1,2,3,4,5
		1,4,5	1,2,3,6,7
		1,6,7	1,4,5,6,7

In case of $X_+X'_-, X_-X'_+ \neq O_{\frac{n}{2}\times\frac{n}{2}}$, the test statistic D under the null hypothesis has the distribution shown in Table 3 or the constant 1 depending on the allocation pattern. The allocation patterns in case that D is a stochastic variable are shown in Table 2. The type of the distribution is unique for each number of allocation effects $(m-1)$ except for $m-1=2$. In case of $m=2$, D has two kinds of distributions. In case m's equal 5 and 6, the D's do not have a distribution of this type. In Table 3, the quantiles of the distributions for each m are shown. The quantiles are obtained from the empirical distribution with 50,000 observations generated by Monte Carlo method. When the allocation is another pattern except for patterns denoted in Tables 1 and 2, the test statistic D is equal to one.

4 The evaluation of the power

In this section, the power defined as the probability that the dispersion effect $k \neq 0$ is detected is evaluated. In Figure 1, the power functions for each m are shown. The power is evaluated on a grid of values $k = 0.5(0.1)5.0(0.5)10.0$ by Monte Carlo simulations with 10,000 replications except for cases where the test statistic has an F distribution.

We consider the cases that the allocations are denoted in Table 2. The powers for m being an odd number are higher than those for m being an even number except for the type II allocation of $m=3$. The power for that case is the lowest for nearly the whole evaluation range.

Table 2. Allcations for L_8 when the D is a stochastic variable in case of $X_+X'_-, X_-X'_+ \neq O_{\frac{n}{2}\times\frac{n}{2}}$

$m-1$	1	2		3	4
		Type I	Type II		
allocation	2	1,2	2,4	1,2,4	1,2,3,4
	3	1,3	2,5	1,2,5	1,2,3,5
	4	1,4	2,6	1,2,6	1,2,3,6
	5	1,5	2,7	1,2,7	1,2,3,7
	6	1,6	3,4	1,3,4	1,2,4,5
	7	1,7	3,5	1,3,5	1,2,6,7
		2,3	3,6	1,3,6	1,3,4,5
		4,5	3,7	1,3,7	1,3,6,7
		6,7	4,6	1,4,6	1,4,5,6
			4,7	1,4,7	1,4,5,7
			5,6	1,5,6	1,4,6,7
			5,7	1,5,7	1,5,6,7
				2,3,4	2,3,4,5
				2,3,5	2,3,6,7
				2,3,6	4,5,6,7
				2,3,7	
				2,4,5	
				2,6,7	
				3,4,5	
				3,6,7	
				4,5,6	
				4,5,7	
				4,6,7	
				5,6,7	

By observing the power curves for the range of k until 1000.0 (the power equals to 0.99 for the k multiplied F distribution with 1 degree of freedom), we make sure that the power curves will converge to values less than or equal to 1.0.

Consequently, the allocations that the $X_+X'_-$ is not zero matrix is not better. If m is three or five, the allocation should adopt the pattern denoted in Table 1 by adding one inactive effect.

5 Conclusion

The property of the test statistic for identifying a particular dispersion effect introduced by Box and Meyer(1986) is investigated in an

Table 3. Distributions for L_8 when the D is a stochastic variable in case of $X_+X'_-, X_-X'_+ \neq O_{\frac{n}{2}\times\frac{n}{2}}$

$m-1$	1	2		3	4
Lower Prob.		Type I	Type II		
0.05	0.19	0.12	0.26	0.16	0.08
0.10	0.29	0.21	0.39	0.28	0.16
0.15	0.38	0.29	0.50	0.39	0.24
0.20	0.46	0.37	0.59	0.49	0.34
0.25	0.54	0.45	0.68	0.58	0.44
0.30	0.63	0.55	0.76	0.68	0.55
0.35	0.72	0.65	0.83	0.77	0.67
0.40	0.81	0.75	0.90	0.86	0.78
0.45	0.90	0.87	0.96	0.94	0.90
0.50	1.00	0.99	1.00	1.00	1.00
0.55	1.10	1.13	1.04	1.06	1.11
0.60	1.23	1.31	1.11	1.15	1.27
0.65	1.39	1.52	1.20	1.29	1.50
0.70	1.59	1.80	1.32	1.47	1.81
0.75	1.84	2.17	1.48	1.71	2.25
0.80	2.18	2.66	1.69	2.05	2.92
0.85	2.67	3.43	2.02	2.59	4.04
0.90	3.50	4.81	2.61	3.57	6.38
0.95	5.45	8.08	3.92	6.01	13.53

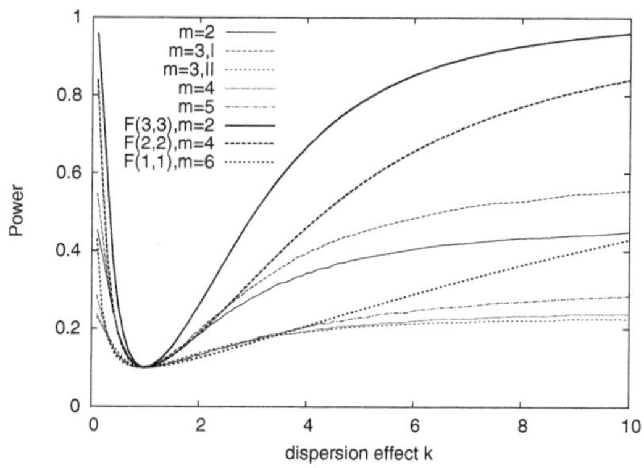

Fig. 1. The power curves the dispersion effect is detected

8×8 regular orthogonal array. The distributions of the test statistic are investigated under the null hypothesis, assuming all the true active effects are not misidentified. The distributions are classified into three types which are the F distribution, the distributions denoted in Table 3 and the constant 1(not stochastic variable). Each distribution type is associated with the allocation pattern of Tables 1 and 2. The powers defined as the probability rejecting the null hypothesis are evaluated. The power curves for $X_+X'_- \neq O$ converge to a value less than or equal to 1. Hence, the allocation should adopt the pattern denoted in Table 1, and if m is three or five, turn m into four or six by adding one inactive effect. In other words, the expanded location model introduced by Bergman and Hynen (1997) should be adopted. In practice, the active location effects are selected by testing before analyzing the dispersion effects. It is not better to analyze the dispersion effect based on only the selected active effects, because the test statistic for the dispersion effect is always one in the worst case. The factors should be allocated based on the allocation patterns for the analysis of the dispersion effect.

References

1. Bartlett, M. S. and Kendall, D. G. 1946, The Statistical Analysis of variance-Heterogeneity and the Logarithmic Transformation, Journal of the Royal Statistical Society, Ser. B, 8, 128-138.
2. Bergman, B., and Hynen, A., 1997, Dispersion Effects From Unreplicated Designs in the 2^{p-q} Series, Technometrics, 39, 191-198.
3. Blomkvist O., Hynen A. and Bergman B., 1997, A Method to Identify Dispersion Effects from unreplicated multilevel experiments, Quality and Reliability Engineering International, 13, 127-138.
4. Box, G. E. P., and Meyer, R. D., 1986, Dispersion Effects From Fractional Designs, Technometrics, 28, 19-27.
5. Ferrer, A. J. and Romero R., 1993, Small Samples Estimation of Dispersion Effects From Unreplicated Data, Communications in Statistics - Simulation and Computation, 22, 975-995.
6. Harvey, A. C., 1976, Estimating Regression Models with Multiplicative Heteroscedasticity, Econometrica, 44, 461-465.
7. Holm S. and Wiklander K., 1999, Simultaneous Estimation of Location and Dispersion in Two-Level Fractional Factorial Designs, Journal of Applied Statistics, 26, 235-242.
8. Lee, H. S., 1994, Estimates For Mean and Dispersion Effects in Unreplicated Factorial Designs, Communications in Statistics - Theory and Methods, 23, 3593-3608.

9. Liao, C. T., 2000, Identification of dispersion effects from Unreplicated 2^{n-k} Fractional Factorial Designs, Computational Statistics & Data Analysis, 33, 291-298.
10. McGrath, R. N., 2003, Separateing Location and Dispersion Effects in Unreplicated Fractional Factorial Designs, Journal of Quality Technology, 35, 306-316.
11. McGrath, R. N. and Lin, D. K., 2001a, Testing Multiple Dispersion Effects in Unreplicated Fractional Factorial Designs, Technometrics, 43, 406-414.
12. McGrath, R. N. and Lin, D. K., 2001b, Confounding of Location and Dispersion Effects in Unreplicated Fractional Factorials, Journal of Quality Technology, 33, 129-139.
13. Nair, V. N. and Pregibon, D., 1988, Analyzing Dispersion Effects from Replicated Factorial Experiments, Technometrics, 30, 247-257.
14. Ojima, Y., Suzuki T. and Yasui S., 2004 An Alternative Expression of the Fractional Factorial Designs for Two-level and Three-level Factors, Frontiers in Statistical Quality Control, Vol. 7, pp. 309-316.
15. Pan, G., 1999, The Impact of Unidentified Location Effects on Dispersion-Effects Identification From Unreplicated Fractional Factorial Designs, Technometrics, Vol. 41, pp. 313-326.
16. Schoen, E. D., 2004, Dispersion-effects Detection after Screening for Location Effects in Unreplicated Two-level Experiments, Journal of Statistical Planning and Inference, 126, 289-304.
17. Taguchi, G., and Wu, Y., 1980, An Introduction to Off-Line Quality Control, Nagoya, Japan: Central Japan Quality Control Association.
18. Wang, P. C., 1989, Tests for Dispersion Effects from Orthogonal Arrays, Computational Statistics & Data Analysis, 8, 109-117.
19. Wang, P. C., 2001, Testing Dispersion Effects From General Unreplicated Fractional Factorial Designs, Qualtiy and Reliability Engineering International, 17, 243-248.
20. Wiklander, K., 1998, A Comparison of Two Estimators of Dispersion Effects, Communications in Statistics - Theory and Methods, 27, 905-923.
21. Wiklander, K. and Holm S., 2003, Dispersion Effects in Unreplicated Factorial Designs, Applied Stochastic Models in Business and Industry, 19, 13-30.

Evaluating Adaptive Paired Comparison Experiments

Tomomichi Suzuki [1], Seiichi Yasui [2], and Yoshikazu Ojima [3]

[1,2,3] Department of Industrial Administration, Tokyo University of Science
2641 Yamazaki, Noda, Chiba, 278-8510, JAPAN
[1] E-mail: `suzuki@ia.noda.tus.ac.jp`

Summary. Paired comparison experiments are effective tools when the characteristics of the objects cannot be measured directly. In paired comparison experiments the characteristics of the objects are estimated from the result of the comparisons. The concept of paired comparison experiments was introduced by Thurstone (1927). The method by Scheffé (1952) is widely used for complete paired comparison experiments and the method by Bradley and Terry (1952) is popularly used in incomplete paired comparison experiments.

In incomplete paired comparison experiments, the design of the experiment, that is, how to form the pairs to be compared, is crucial to successful analysis. Many methods including adaptive experimental designs are proposed. The tournament systems in sports and other competitions are typical examples of such designs, but their statistical properties are not fully investigated. In this paper, we discuss how tournament systems may be evaluated and propose a new criterion. We also give examples of evaluating tournaments based on the proposed criterion.

1 Introduction

Evaluation of items of interest is usually carried out by measuring certain characteristics of the items. It is relatively easy if these characteristics can be measured directly, but relatively difficult otherwise. When measuring if difficult, sensory evaluation is often used, in which case human beings are the measurement equipment.

Paired comparison experiments are effective tools when we cannot directly measure the characteristics of the objects. It is especially useful when a person can point out the superior or preferable item when two items are presented. The tournament systems in sports and other competitions are typical examples of such designs.

H.-J. Lenz et al. (eds.), *Frontiers in Statistical Quality Control 9*,
DOI 10.1007/978-3-7908-2380-6_22, © Springer-Verlag Berlin Heidelberg 2010

Various tournament systems are adopted to rank players in competitions such as sports and games. Competitions are divided into two groups, 'race' and 'match.' In races, players compete for time or scores that can be evaluated numerically. In matches, players compete face to face for a win. Although a tournament system adopted in a competition depends on scale and period, its primary objective is to appropriately rank the participants according to their results. There are other objectives of the tournaments such as commercial aspects. This paper is motivated by the desire to obtain an accurate, representative final ranking rather than a spectacular upset. We focus on the statistical objective of ranking the players correctly.

In races, ranking the participants is rather easy once the time or the scores are obtained. On the other hand, it is generally difficult to rank the participants in competitions decided by matches. One of the reasons is that obtaining the results themselves is difficult. Popular tournament systems adopted in such cases are knockout tournaments and round robin tournaments. Many modifications are proposed for these systems but it is difficult to evaluate their effectiveness. In this paper, we discuss a method of evaluating tournament systems and propose a new criterion. We also give examples of evaluating tournaments based on the proposed criterion.

2 Paired Comparison Experiments

Paired comparisons is a methodology where the items are compared in pairs to judge which of each pair is preferred. It can be regarded as a statistical method. We briefly explain it in this section.

2.1 Thurstone's Method

The basic method for paired comparison methods was introduced by Thurstone (1927). Thurstone gave five cases and gave procedures for evaluating effects of the items in each case. In the fifth case, the proposed model is given as follows.

$$x_{ijk} = (\alpha_i - \alpha_j) + \varepsilon_{ijk} \tag{1}$$

Here, x_{ijk} denotes the evaluation (preference) of item i over item j in the kth replication when items are presented in this order. α_i and α_j stand for effects of items, and ε_{ijk} expresses the error in evaluation. It is assumed that the error is normally independently identically distributed with zero mean and a constant variance. Measurements are replicated for all the combinations of the pairs of the items, and the effects of the items are to be estimated. Therefore the decision depends on only the difference between the effects of each item. The procedure focuses on estimating the difference.

2.2 Scheffé's Method

This method is one of the most popular paired comparison methods. Scheffé's method (Scheffé 1952) is based on Thurstone's method. This method is adopted in some ISO standards such as ISO 5492 (1992). Evaluation between the items can be a numerical value rather than preference only. In Scheffé's method, the following model is assumed.

$$x_{ijk} = (\alpha_i - \alpha_j) + \gamma_{ij} + \delta_{ij} + \varepsilon_{ijk} \tag{2}$$

Here γ_{ij} expresses the combination effect ($\gamma_{ij}=-\gamma_{ji}$), and δ_{ij} expresses the order effect ($\delta_{ij} = \delta_{ji}$). Other symbols are defined the same as in Eq.(1).

2.3 Bradley and Terry's Method

This is a popular paired comparison method also known as the Bradley-Terry model. In this method (Bradley and Terry 1952; Fujino and Takeuchi 1988), the model is given below.

$$p_{ij} = \frac{\pi_i}{\pi_i + \pi_j} \tag{3}$$

In Eq.(3), p_{ij} is the probability that item i is judged superior to item j. π_i and π_j represent the effects of the items. These are the parameters that will be estimated from the results of the paired comparisons. The estimation is usually carried out by means of the method of maximum likelihood.

3 Tournament Systems

A match in a tournament can be regarded identical to comparing a pair in paired comparison experiments. A tournament system is a procedure for determining the pairings of the matches in order to rank the participants. In paired comparisons, it can be regarded as a procedure for design of experiments. In this section, popular tournament systems are introduced.

3.1 Round Robin Tournaments

A round robin tournament is a competition in which every player or team plays against every other player or team. This is one of the most popular tournament systems adopted in many competitions: soccer leagues in many countries, baseball pennant race, rugby Six Nations, and so forth. When the number of matches between the same players is one, it is called single round robin, and if two, double round robin. The latter is often used where the winning percentage between the players of the same strength is regarded as not equal to 50% such as home and away games in soccer and other sports, or white and black for chess.

One of the merits of the round robin tournaments is that all the combinations are realized so that the ranking of the players can be done for all the players. Another merit is that ranking of the players will be done relatively accurately because the number of played games is large. On the other hand, a demerit is that it takes time to finish all the games. From commercial point of view, lack of an apparent final match can also be considered as a demerit.

For the methods described in section 2, Scheffé's method and Thurstone's method require comparisons for all the possible pairs. Only round robin tournaments satisfy their requirements. So, when we adopt other tournament systems, we cannot apply the existing methods directly. Using terms in design of experiments, round robin tournaments are complete designs where comparison are made for all the possible pairs.

3.2 Knockout Tournaments

A knockout tournament is a competition in which only the winning players or teams at each stage continue to play until there is only one winner. This is also adopted in many tournaments: Wimbledon tennis, World Cup Soccer finals, judo, wrestling, and so forth.

Commercial-wise, the big merit is that there is a final. Another merit is that the number of the games is not so large. However, from the viewpoint of ranking the players, it is the least effective. In many cases, ranking the players is impossible except for a few top players. Another demerit is that the influence of the pairing is large, i.e. the outcome for a player much depends on to whom he is matched. To overcome these disadvantages, methodologies such as seeding and double elimination knockout tournaments are proposed (McGarry and Schutz 1997; Marchand 2002). Stepladder tournaments (used in bowling) and the page system (used in softball) can be considered as a variation on the knockout tournament.

From the view point of experimental design, the knockout tournaments are one type of incomplete design.

3.3 Swiss System Tournaments

A Swiss tournament is a competition in which players or teams at each round play against those with the same (or similar) results. Every player plays the same number of games and does not play against the same opponent twice. The name Swiss system is used because this type of tournament was originally adopted in chess tournaments in Switzerland in the late 1800s. The Swiss system is adopted mainly in mental sports such as chess.

A merit of the Swiss system is that it is possible to rank all the players and it can be done relatively well, if not as accurately as for round robin tournaments. From the players' point of view, they can enjoy the tournament because they are not eliminated as soon as they lose a game. A demerit of the Swiss system is the difficulties in running the tournament.

In Swiss tournaments, players play against those with similar results. Therefore, the pairing of the matches is done according to the result at the time of pairing. This pairing makes the Swiss system different from other tournaments. The pairing rules are basically described accurately [3, 8], but knowledge and experience are necessary to effectively run the tournament. Many techniques have already been proposed by Takizawa and Kakinoki (2002), Hashimoto et al. (2002), Kayama (2003), Kawai et al. (2005), Glickman (2005), and others. Commercial software exists to support the running the Swiss tournaments such as Swiss Perfect [17] and Swiss-Manager [18]. But even in this case, basic knowledge is necessary and the pairing must be made by some means.

From the view point of experimental design, Swiss system tournaments are also incomplete designs. In particular, since the pairings in the Swiss systems depend on the history of the experiment, the Swiss tournaments can be regarded as adaptive paired comparison experiments as discussed by Glickman (2005).

3.4 Hierarchical Structure of the Tournaments

As described in previous sections, there are various tournament systems. In practice, one tournament often consists of multiple tournament systems, say, preliminary rounds and final rounds in the World Cup Soccer. In other cases, round robin tournaments with different numbers of players are run successively. Tournament systems have broader range if the hierarchical structure of the tournaments is considered.

4 Tournament System Evaluation

4.1 Evaluation Criteria

The principal objective of the tournament is ranking the players, but it is not clear what is 'good' ranking is. Each tournament has its own objectives: to decide the only winner, to decide who receive medals, to decide who goes through the preliminaries, or to rank all the participating players.

In existing literatures, evaluating the tournaments has not been easy, and there are various criteria used to evaluate them. A popular approach is to compare the rank of the tournament result with the starting rank order which is assumed to be determined before starting the tournament. Since we consider a tournament as a design of experiment, it may be natural to use criteria also used in evaluating ordinary experimental designs. But using ordinary criteria are not enough, for the following reasons.

The first reason is that the objective of the tournament is not only to estimate the effects of the items or the strengths of the players, but also to estimate the rank order of the effects. They are often weighted, that is, the ranking in the top part of the participants are more important.

The second reason is that the final ranking should fit better with the results performed by players than the starting ranks. Using terms in design of experiment, the result of the tournament should express not the estimated true value of the parameter but the realized value of the random variables. In ordinary experimental design, the realized value of the random variables and the estimated parameter correspond easily to each other. That will not happen in the tournament designs. This is the essential difference between ordinary paired comparison experiments and the tournaments.

The third reason is that the tournament must be fair to all the players. The expected rank in the tournament should be the same for the players whose performance is at the same level of strength. This is a difficult problem to solve because it means seeding and prior ranking should not affect the outcome of the tournament strength-wise even though they will be used in actual tournaments.

Considering the above reasons, we propose ***to evaluate tournaments by the consistency between the tournament result and the performance***. In other words, the tournament result rankings should coincide with the strengths performed by the players. In this paper, we used Spearman's rank correlation coefficient as a criterion to evaluate the tournaments.

4.2 Evaluation Procedure

It would be convenient if we could evaluate the tournament system by mathematical means. But since the nature of the tournament system, especially the Swiss system, is complicated, we need to evaluate the tournament systems by simulation methods. As described in the previous section, we need to consider various criteria for evaluating the tournament systems. Therefore, a reproducible simulation method is preferred. In this regard, we need a procedure that can record the performance of each game for every player. In other words, we need the win-loss model which includes a performance measure that can be expressed numerically.

4.3 Probabilistic Model for Determining Result of a Match

The first thing needed is to express the performance numerically. This is best done by the Thurstone model. The effect of the item in the Thurstone's model corresponds to the player's true strength. The performance for each game is defined to be the sum of the player's true strength and the error. The performance for a tournament is calculated by averaging the performance for each game.

Elo (1978) applied the Thurstone's model to chess. He named the true strength 'rating'. In his theory, the error is normally distributed with zero mean and 200 standard deviation. In this assumption, it is easy to express the performance numerically. Also, the winning probability between two players can be expressed easily using the normal distribution function.

Bradley-Terry model and the Jackson's model as described in Marchand (2002) are sufficient in terms of giving the winning percentages but they are not suitable for expressing the performance. In some literatures, the probability of 50% for all the matches or 'stronger player always wins' is applied. In these cases, we have to be careful what is evaluated through these assumptions.

In this regard, it is reasonable to assume Elo's model.

4.4 Distribution of Effects of Sample Population

We also need to set the population of the players to actually compare the tournament systems. Evaluating tournament systems cannot be done with the performance model only. We have to take into account the distribution of the player strengths. It is of course impossible to examine the all possible distributions, but examination is possible for typical patterns.

If we can assume the distribution of strengths for all the players in any sport, we have to start with normal distribution. When we consider a part of the whole population is taking part in a tournament, four cases can be considered as Nozaki et al. (2006).

1. All the players participate at a constant rate: In this case, the distribution of the participating players will be normally distributed.
2. Top players participate: In any competition, interest is in strong players, so this pattern may actually be seen often. In this case, the distribution will be skewed to the right. Simple triangle distribution is to be considered.
3. Bottom players participate: In this case the distribution is the opposite compared to 2.
4. Middle players participate where the range of the strengths is small: In this case, uniform distribution is to be considered.

In all of the four cases, we still have to decide the variance of the distributions. We investigated both cases where the variance of the participants is large and small.

4.5 An Evaluation Example

We have evaluated the tournaments. We evaluated two Swiss tournaments using two criteria. The two Swiss tournaments examined are as follows.

1. Normal Swiss tournament: We have implemented the pairing system according to FIDE rules. The basic rule is that players with the same score are ranked according to rating. Then the top half is paired with the bottom half according to the rating. For instance, if there are eight players in a score

group, number 1 is paired with number 5, number 2 is paired with number 6 and so on.

2. Proposed Swiss tournament: This system is proposed by Nozaki et al. (2007). It is intended to overcome the deficit in the normal Swiss system found by Nozaki et al. (2006). The grouping is done as the same as FIDE rules. The top half is paired with the bottom half but order of the bottom half is reversed. For instance, if there are eight players in a score group, number 1 is paired with number 8, number 2 is paired with number 7 and so on.

The two criteria used to evaluate tournaments are Spearman's rank correlation coefficient between the tournament ranking of a player at the end of a tournament and;

a) the starting rank of the player at the beginning of the tournament.
b) the performance of the player during the tournament.

The latter criterion is the criterion proposed in Section 4.1. We compared the two tournaments. We assumed a uniform distribution for the population of the players. We examined cases where the number of the participants is 32 and 100, with large (rating difference between players is 10) and small (rating difference between players is 2) population variance of the participants. We simulated each tournament 1000 times. The result is shown in Table 1 when criterion a) is used and in Table 2 when criterion b) is used.

Table 1. Comparison of Swiss tournaments by criterion a)

Number of Players	Variance of players	Normal Swiss	Proposed Swiss
32	small	0.203	0.252
32	large	0.689	0.724
100	small	0.581	0.654
100	large	0.917	0.930

Table 2. Comparison of Swiss tournaments by criterion b)

Number of Players	Variance of players	Normal Swiss	Proposed Swiss
32	small	0.460	0.476
32	large	0.704	0.739
100	small	0.626	0.670
100	large	0.918	0.929

From these tables, we can see that the proposed Swiss system is superior in all the cases. The choice of the criterion will not make a big difference when evaluating tournaments with large population variance. This is because the performance of the players and the starting rank will have very high correlation. The choice of the criterion is notable when evaluating tournaments with small population variance. For example when the number of player is 32, the Spearman's rank correlation coefficient for normal Swiss system is 0.203 using criterion a) and 0.460 using criterion b). This happened because the criterion b) better explains who played well in the tournament. It can evaluate the tournament according to the performance of the players. Items for future work include increasing accuracy of the simulation results as well as investigation on other simulation conditions.

5 Summary

In this paper, we proposed a framework to evaluate the tournament systems by considering them as experimental design to rank players. The main point is to evaluate the tournament, that is, the experimental design, by the performance of the participants. We also presented an evaluation example.

In any competitions, many argue who is the best and so on. We hope our work will contribute to objective ranking of the tournament results.

References

1. Bradley RA, Terry ME (1952) Rank Analysis of Incomplete Block Designs. Biometrika 39:324-345
2. Elo AE (1978) The Rating of Chessplayers Past and Present (second edition). ARCO
3. FIDE Handbook http://www.fide.com/official/handbook.asp
4. Fujino Y, Takeuchi K (1988) Mathematical Science in Sports (in Japanese), Kyoritsu Publishing
5. Glickman ME (2005) Adaptive paired comparison design. Journal of statistical planning and inference 127:279-293
6. Hashimoto T, Nagashima J, Iida H (2002) A proposal of tournament system verification by the simulation –a case study using the World Computer Shogi Championship– (in Japanese). Proc. The 7th Game Programming Workshop, pp 101-108
7. ISO 5492:1992 (1992) Sensory Analysis –Vocabulary. ISO
8. Japan Chess Association. http://www.jca-chess.com/
9. Kawai T, Kawashima S, Iida H (2005) A new ranking measurement based on tournament records (in Japanese). IPSJ SIG Technical Report (2005-GI-13), pp 27-34
10. Kayama K (2003) Evaluation of Pairing Systems on Swiss-system Tournament by Simulation (in Japanese). Proc. The 8th Game Programming Workshop
11. Marchand E (2002) On the Comparison between Standard and Random Knockout Tournament. The Statistician 51:169-178
12. McGarry T, Schutz RW (1997) Efficacy of traditional sport tournament structures. Journal of the Operational Research Society 48:65-74
13. Nozaki Y, Minakawa H, Suzuki T (2006) A Study on Ranking Participants in Competitions (in Japanese). Proc. The 36th JSQC Annual Congress, pp 187-190
14. Nozaki Y, Minakawa H, Suzuki T (2007) Experimental Design for Adaptive Paired Comparisons. Proc. The 5th ANQ Congress, 10 pages (CD-ROM)
15. Sakoda Y, Suzuki T (2005) A Study on Sports Tournament Systems. Proc. The 19th Asia Quality Symposium

16. Scheffé H (1952) An Analysis of Variance for Paired Comparisons. Journal of American Statistical Association 147:381-400
17. Swiss-Manager Home Page. http://swiss-manager.at/, http://schach.wienerzeitung.at/
18. Swiss Perfect Home Page. http://www.swissperfect.com/
19. Takizawa T, Kakinoki Y (2002) A Pairing System and Its Effectiveness in the World Computer Shogi Championships (in Japanese). Proc. The 7th Game Programming Workshop, pp 93-100
20. Thurstone LL (1927) A Law of Comparative Judgment. Psychological Review 34:278-286

Approximated Interval Estimation in the Staggered Nested Designs for Precision Experiments

Motohiro Yamasaki[1], Michiaki Okuda[2], Yoshikazu Ojima[3], Seiichi Yasui[4], and Tomomichi Suzuki[5]

Department of Industrial Administration, Tokyo University of Science, 2641 Yamazaki, Noda, Chiba, 278-8510, Japan,
Yamasaki.Motohiro@md.mt-pharma.co.jp

Summary. Staggered nested experimental designs are the most popular class of unbalanced nested designs in practical fields. Reproducibility is one of the important precision measures. In our study, interval estimation of reproducibility is proposed and evaluated by precision experiments, which we call the staggered nested experimental design. In this design, the reproducibility estimator is expressed as linear combination of the variance components from ANOVA (analysis of variance). By using a gamma approximation, the shape parameter that is needed for the approximation is introduced in our study. Additionally, general formulae for the shape parameter are proposed. Applying the formulae, we constructed the confidence interval for the reproducibility of three-factor and four-factor staggered nested designs. The performance of the proposed gamma approximations is evaluated with the goodness of fit and compared with each other. The interval estimation of reproducibility is evaluated with the coverage probability through a Monte-Carlo simulation experiment. We also compared it to the method ignoring the covariance terms. As a result, the proposed approximations were better than the method without the covariance terms. The performance of the proposed interval estimation was also superior to that without covariance terms. Additionally, some practical recommendations were obtained for designing precision experiments, including the number of participating laboratories.

1 Introduction

In general, precision is one of the most important performance measures to evaluate measurement methods and measurement results. Measurement methods and measurement results are important basis for trading,

H.-J. Lenz et al. (eds.), *Frontiers in Statistical Quality Control 9*,
DOI 10.1007/978-3-7908-2380-6_23,

manufacturing and quality control. There are several precision measures such as repeatability, intermediate precision, and reproducibility. Their importance, the ways of determination and point estimations are described in ISO 5725-3 (1994).

Repeatability, intermediate precision measure, and reproducibility are defined in ISO 5725-1 (1994) as follows. Repeatability is precision under repeatability conditions where independent test results are obtained with the same method on identical test items in the same laboratory by the same operator using the same equipment within short intervals of time. Reproducibility is precision under reproducibility conditions where test results are obtained with the same method on identical test items in different laboratories by different operators using different equipment. Intermediate precision measures are precision under intermediate precision conditions that are between the two extreme conditions of precision: the repeatability and reproducibility. These precisions are statistically defined and obtained from measurement results that are reported by an inter-laboratory experiment called the precision experiment.

From the statistical view point, a measurement result y can be expressed as:

$$y = \mu + \alpha + \beta + \gamma + \epsilon, \tag{1}$$

where μ is a general mean or the true value, α is a random effect due to a laboratory, β is a random effect due to a day, γ is a random effect due to operator, ϵ is a random effect due to replication under repeatability conditions. We usually assume that μ is an unknown constant while α, β, γ and ϵ are normal random variables with expectation 0 and variances σ_4^2, σ_3^2, σ_2^2 and σ_1^2, respectively. Hence, repeatability, intermediate precision measures, and reproducibility are expressed as linear combinations of variance components corresponding to each definition, e.g., $\sigma_4^2 + \sigma_3^2 + \sigma_2^2 + \sigma_1^2$ is the statistical expression of reproducibility.

Nested experiments are commonly used to estimate precisions. Estimates are often obtained through ANOVA (analysis of variance) estimators of variance components. The estimators are expressed as linear combinations of the mean squares from the ANOVA. They are also unbiased estimators and obtained without any assumption of the distribution. Interval estimation is important to evaluate the precision as well as point estimators. However, it is difficult to construct the exact interval estimation for the precision except for repeatability. In the case of balanced nested experiments, the mean squares are mutually independent chi-squared random variables; the distribution of the linear combination is often approximated by chi-square distribution multiplied by a constant through the method of moments. This method is sometimes designated as Satterthwaite's approximation. It can be refered here to Satterthwaite, F. E. (1946). For the case of staggered nested experiments, the approximation is complicated because the mean squares are chi-squared random variables with correlation. Kaniwa and Ojima (2000) derived the approximate distribution of the precision estimators in staggered nested experiments. However, the estimators are approximated by chi-square distribution multiplied by a constant without the covariance terms.

In this paper, using the method of moments with covariance terms, the distribution of the estimator of the reproducibility is approximated by a gamma distributions multiplied by a constant for the case of m-factor staggered nested experiments. That is, general formulae for the shape parameter are proposed by using Satterthwaite (1946)'s approximation and Ojima (1998)'s canonical form. Applying the formulae, we constructed the confidence interval for the reproducibility of three-factor and four-factor staggered nested designs. The performance of the proposed gamma approximations is evaluated with the goodness of fit and compared with each other. The interval estimation of reproducibility is evaluated with the coverage probability through a Monte-Carlo simulation experiment. We also compare it to the method ignoring the covariance terms.

2 The staggered nested design

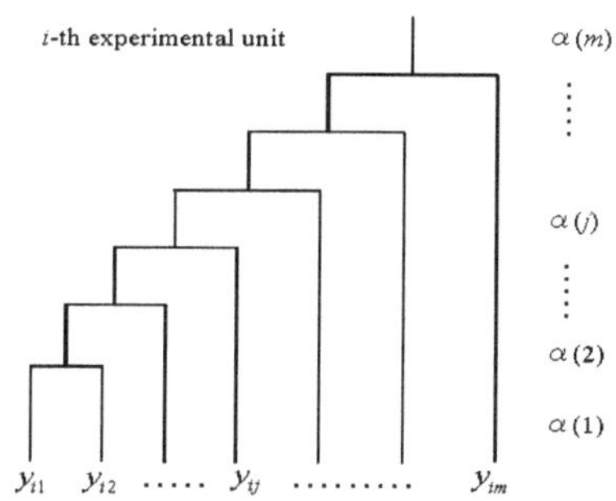

Fig. 1. Experimental unit of general m-factor staggered nested design

Let A_1, A_2,..., A_m be the random effect factors, with A_k nested within A_{k+1}, $k = 1,\ldots, m-1$; the corresponding random effects are denoted by $\alpha(1)$, $\alpha(2)$,..., $\alpha(m)$. The factor A_k is the kth factor from the bottom of the hierarchy, as mentioned before. All effects $\alpha(k)$ ($k = 1, \ldots, m$) are assumed to be completely independent normal random variables with zero means and variances σ_k^2.

The model for the m-factor staggered nested design as shown in Figure. 1 can be written as

$$y_{ij} = \mu + \sum_{l=1}^{m} \alpha(l)_{ij} = \mu + \sum_{l=1}^{j-1} \alpha(l)_{ij} + \sum_{l=j}^{m} \alpha(l)_{i1} \tag{2}$$

where $\alpha(l)_{ij} = \alpha(l)_{i1}$, for all $l \geq j$, $j = 1, \ldots, m$; $i = 1, \ldots, n$. This model is expressed by Ojima (1998). Let $\mathbf{y}_i$ be a data vector of the ith experimental unit, with $\mathbf{y}_i^{'} = (y_{i1}\ y_{i2}\ \cdots\ y_{im})$, where $\mathbf{A}^{'}$ is the transposed matrix of $\mathbf{A}$. The expectation and the dispersion matrix of $\mathbf{y}_i$ are

$$\begin{aligned} \mathrm{E}(\mathbf{y}_i) &= \mu \mathbf{j}_m \\ \mathrm{Var}(\mathbf{y}_i) &= \sum_{l=1}^{m} \sigma_l^2 (\mathbf{G}_i \oplus \mathbf{I}_{m-l}) \equiv \Sigma \\ \mathrm{Cov}(\mathbf{y}_i, \mathbf{y}_{i'}) &= \mathbf{O}, \text{ for } i \neq i^{'} \end{aligned} \tag{3}$$

where $\mathbf{j}_m$ is the $m \times 1$ vector of units, $\mathbf{G}_l$ is the $l \times l$ matrix of units, $\mathbf{I}_l$ is the identity matrix of order $l \times l$, $\mathbf{O}$ is the zero matrix, and $\mathbf{A} \oplus \mathbf{B}$ is the direct sum of matrices $\mathbf{A}$ and $\mathbf{B}$. Furthermore, let $\mathbf{y}$ be a vector of all data with $\mathbf{y}^{'} = (\mathbf{y}_1^{'}\ \mathbf{y}_2^{'}\ \cdots\ \mathbf{y}_n^{'})$. The expectation and the dispersion matrix of $\mathbf{y}$ are

$$\begin{aligned} \mathrm{E}(\mathbf{y}) &= \mu \mathbf{j}_{mn} \\ \mathrm{Var}(\mathbf{y}) &= \mathbf{I}_n \otimes \Sigma = \sum_{l=1}^{m} \sigma_l^2 \mathbf{I}_n \otimes (\mathbf{G}_i \oplus \mathbf{I}_{m-l}) \end{aligned}$$

where $\mathbf{A} \otimes \mathbf{B}$ is the direct product of matrices $\mathbf{A}$ and $\mathbf{B}$. For example, for the three-factor staggered nested design, the dispersion matrix of $\mathbf{y}_i$ is as follows.

$$\mathrm{Var}(\mathbf{y}_i) = \sigma_3^2 \begin{pmatrix} 1 & 1 & 1 \\ 1 & 1 & 1 \\ 1 & 1 & 1 \end{pmatrix} + \sigma_2^2 \begin{pmatrix} 1 & 1 & \\ 1 & 1 & \\ & & 1 \end{pmatrix} + \sigma_1^2 \begin{pmatrix} 1 & & \\ & 1 & \\ & & 1 \end{pmatrix}$$

3 Gamma approximation of the reproducibility estimator

3.1 Canonical form

ANOVA is usually used for the variance components estimation. Concerning the staggered nested designs, general formulae can be referred to Ojima (1998). We review the results briefly for the convenience of constructing general formulae for the shape parameter by using a gamma approximation.

To construct an orthogonal transformation in an experimental unit of $\mathbf{y}_i$, let $\mathbf{p}_k$ be a standardized vector of order m that represents a contrast that corresponds to A_k, i.e.

$$\mathbf{p}_k^{'} = \frac{1}{[k(k+1)]^{1/2}} \left(\mathbf{j}_k^{'} \quad -k \quad \mathbf{0}_{m-k-1}^{'}\right) \quad \text{for } k = 1, \ldots, m-1$$
$$\mathbf{p}_k = \frac{1}{m^{1/2}} \mathbf{j}_m \tag{4}$$

The matrix $\mathbf{P}_U = (\mathbf{P}_1 \ \mathbf{P}_2 \ \ldots \ \mathbf{P}_m)$ is an orthogonal matrix of order $m \times m$. Using $\mathbf{P}_U$, $\mathbf{y}_i$ is orthogonally transformed into $\mathbf{u}_i = \mathbf{P}_U^{'} \mathbf{y}_i$. Based on the equation (3), the expectations, variances, and covariances of the elements of $\mathbf{u}_i$ are obtained as follows.

$$\mathrm{E}(u_{ik}) = 0, \qquad \text{for } k = 1, \ldots, m-1$$
$$\mathrm{E}(u_{im}) = m^{1/2}\mu, \tag{5}$$

$$\mathrm{Var}(u_{ik}) = \sum_{l=1}^{k} \frac{l(l-1) + k(k+1)}{k(k+1)} \sigma_l^2, \qquad \text{for } k = 1, \ldots, m-1$$
$$\mathrm{Var}(u_{im}) = \sum_{l=1}^{m} \frac{l(l-1) + m}{m} \sigma_l^2, \tag{6}$$

$$\mathrm{Cov}(u_{ik}, u_{ik'}) = \sum_{l=1}^{k} \frac{l(l-1)}{[k(k+1)k'(k'+1)]^{1/2}} \sigma_l^2, \qquad k < k' < m$$
$$\mathrm{Cov}(u_{ik}, u_{im}) = \sum_{l=1}^{k} \frac{l(l-1)}{[k(k+1)m]^{1/2}} \sigma_l^2, \qquad k < m \tag{7}$$

For each value of the subscript k, $u_{ik} (i = 1, \ldots, n)$ is normally, independently and identically distributed, because $\mathbf{y}_i$ is independent of $\mathbf{y}_{i'}$ for all $i = i'$.

3.2 Gamma approximation of the estimator of reproducibility variance

The shape parameter needed for the estimate of reproducibility variance is derived from approximation by a gamma distribution. The approximation method is the method of moments. By this method, the distribution of the estimator of reproducibility variance is approximated by a gamma distribution with the shape parameter ϕ^* multiplied by a constant c. The approximation method of the shape parameter is sometimes called Satterthwaite (1946)'s method.

We constructed general formulae for the shape parameter as follows, by equations (4), (5), (6) and (7) and the method of moments from the characteristic of the gamma distribution. The following formulae are derived from the expectation and the variance:

$$\sum_{l=1}^{m} \sigma_l^2 = c\phi^* \tag{8}$$

$$\frac{2}{m^2 n}\left[\sum_{k=1}^{m} \mathrm{Var}(u_{ik})^2 + \frac{n}{n-1}\mathrm{Var}(u_{im})^2 + 2\sum_{k=1}^{m-2}\sum_{k'=k+1}^{m-1} \mathrm{Cov}(u_{ik}, u_{ik'})^2 + 2\sum_{k=1}^{m-1} \mathrm{Cov}(u_{ik}, u_{im}^2)\right] = 2c^2\phi^* \tag{9}$$

From equations (8) and (9), we have ϕ^* and c as follows:

$$\phi^* = \frac{m^2\left(\sum_{l=1}^{m}\sigma_l^2\right)^2}{\frac{1}{n}\left[\sum_{k=1}^{m} \mathrm{Var}(u_{ik})^2 + \frac{n}{n-1}\mathrm{Var}(u_{im})^2 + 2\sum_{k=1}^{m-2}\sum_{k'=k+1}^{m-1} \mathrm{Cov}(u_{ik}, u_{ik'})^2 + 2\sum_{k=1}^{m-1} \mathrm{Cov}(u_{ik}, u_{im}^2)\right]} \tag{10}$$

$$c = \frac{\sum_{l=1}^{m}\sigma_l^2}{\phi^*} \tag{11}$$

3.3 Three-factor staggered nested design

Figure. 2 shows an example of the experimental units of the three-factor staggered nested design. To construct an orthogonal transformation in the experimental unit $\mathbf{y}_i$, let $\mathbf{p}_k$ be a standardized vector of order $m = 3$ that represent contrasts corresponding to factors defined by the equation (4), respectively, as follows:

$$\mathbf{p}_1' = \frac{1}{2^{1/2}}(1\ -1\ 0)\,,\ \mathbf{p}_2' = \frac{1}{6^{1/2}}(1\ 1\ -2)\,,\ \mathbf{p}_3' = \frac{1}{3^{1/2}}(1\ 1\ 1)$$

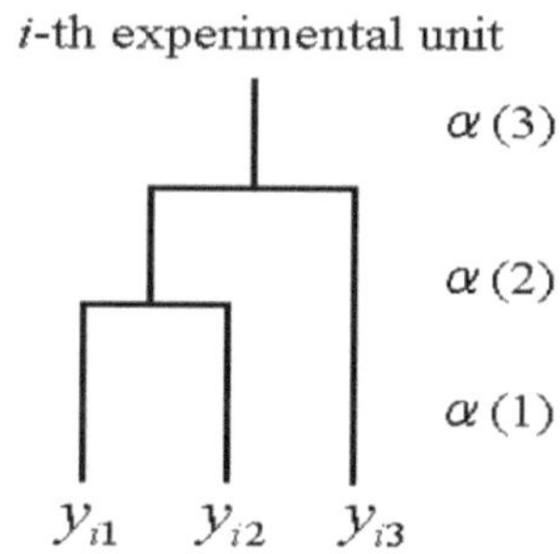

Fig. 2. Experimental unit of the three-factor staggered nested design

It should be noted that they are uniquely determined without signs. The matrix $\mathbf{P}$ composed by $\mathbf{p}_k$ is an orthogonal matrix of order 3×3 with $\mathbf{P} = (\mathbf{p}_1\ \mathbf{p}_2\ \mathbf{p}_3)$. And using $\mathbf{P}$, $\mathbf{y}_i$ is orthogonally transformed into $\mathbf{u}_i = (u_{i1}\ u_{i2}\ u_{i3})'$ as $\mathbf{u}_i = \mathbf{P}'\mathbf{y}_i$. Then the dispersion matrix of $\mathbf{u}_i$ can be expressed by equations (6) and (7) as follows:

$$\mathrm{Var}(\mathbf{u}_i) = \mathbf{P}'\Sigma\mathbf{P} = \begin{pmatrix} \sigma_1^2 & 0 & 0 \\ 0 & \sigma_1^2 + (4/3)\sigma_2^2 & (2^{1/2}/3)\sigma_2^2 \\ 0 & (2^{1/2}/3)\sigma_2^2 & \sigma_1^2 + (5/3)\sigma_2^2 + 3\sigma_3^2 \end{pmatrix}$$

Table. 1 shows the ANOVA table for a three-factor staggered nested design. MS(=Mean Square) V is expressed as $SS/d.f.$, hence MS is $V_1 = SS_1/n, V_2 = SS_2/n$, and $V_3 = SS_3/(n-1)$, respectively. From Table 1, the ANOVA estimators are derived as:

$$\hat{\sigma}_3^2 = (1/3)V_3 - (5/12)V_2 + (1/12)V_1,\ \hat{\sigma}_2^2 = (3/4)(V_2 - V_1),\ \hat{\sigma}_1^2 = V_1. \tag{12}$$

Table 1. The ANOVA table for a three-factor staggered nested design

factor	SS	$d.f.$	MS	$E(MS)$
$m = 3$	SS_3	$n-1$	V_3	$\sigma_1^2 + (5/3)\sigma_2^2 + 3\sigma_3^2$
$k = 2$	SS_2	n	V_2	$\sigma_1^2 + (4/3)\sigma_2^2$
$k = 1$	SS_1	n	V_1	σ_1^2

On the other hand, the shape parameter ϕ_1^* can be expressed by the equation (10) as follows.

$$\phi_1^* = \frac{9\left(\sigma_l^2 + \sigma_2^2 + \sigma_3^2\right)^2}{\frac{1}{n}\left[\begin{array}{l}\sigma_1^4 + \{\sigma_1^2 + (4/3)\sigma_2^2\}^2 + \frac{n}{n-1}\{\sigma_1^2 + (5/3)\sigma_2^2 + 3\sigma_3^2\}^2 \\ + \frac{4}{9}\sigma_2^4\end{array}\right]}$$

Since variance components are unknown, the above equation is calculated as the estimates, by using the equation (12).

$$\hat{\phi}_1^* = \frac{(V_l + V_2 + V_3)^2}{\frac{1}{n}\left[V_1^2 + V_2^2 + \frac{n}{n-1}V_3^2 + \frac{1}{4}(V_2 - V_1)^2\right]} \tag{13}$$

3.4 Four-factor staggered nested design

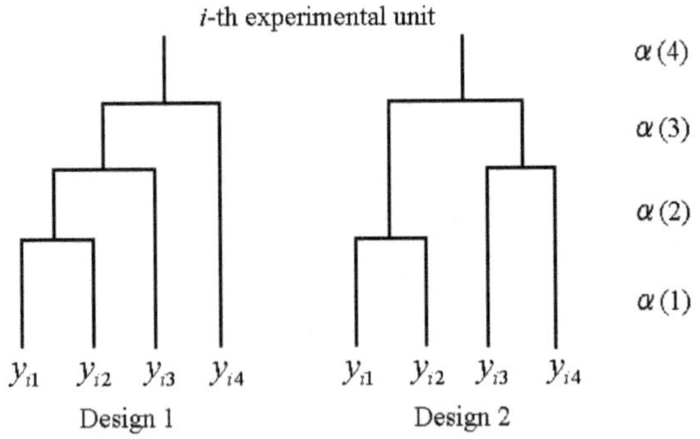

Fig. 3. Experimental unit of the four-factor staggered nested design

Figure. 3 shows an example of the experimental units of the four-factor staggered nested design, including the generalized staggered nested design. These designs can be referred to Ojima (2000). We constructed an orthogonal transformation, the dispersion matrix of $\mathbf{u}_i$, and the shape parameter of the staggered nested design (Design 1), and the generalized staggered nested design (Design 2), respectively.

Firstly, to construct an orthogonal transformation in the experimental unit $\mathbf{y}_i$, let $\mathbf{p}_k$ be standardized vectors of order $m = 4$ that represent contrasts corresponding to factors defined by the equation (4) as follows:

(Design 1)

$$\mathbf{p}_1' = \frac{1}{2^{1/2}}\left(1\ -1\ 0\ 0\right),\ \mathbf{p}_2' = \frac{1}{6^{1/2}}\left(1\ 1\ -2\ 0\right),$$
$$\mathbf{p}_3' = \frac{1}{12^{1/2}}\left(1\ 1\ 1\ -3\right),\ \mathbf{p}_4' = \frac{1}{2}\left(1\ 1\ 1\ 1\right)$$

(Design 2)

$$\mathbf{p}_1' = \frac{1}{2^{1/2}}\left(1\ -1\ 0\ 0\right),\ \mathbf{p}_2' = \frac{1}{2^{1/2}}\left(0\ 0\ 1\ -1\right),$$
$$\mathbf{p}_3' = \frac{1}{2}\left(1\ 1\ -1\ -1\right),\ \mathbf{p}_4' = \frac{1}{2}\left(1\ 1\ 1\ 1\right)$$

Secondly, dispersion matrixes of $\mathbf{u}_i$ can be expressed by equations (6) and (7) as follows. However, in Design 2, the matrix is based on the equation (3) and the standardized vector.

(Design 1)

$$\begin{aligned}
\mathrm{Var}(\mathbf{u}_i) &= \mathbf{P}'\Sigma\mathbf{P} \\
&= \begin{pmatrix}
\sigma_1^2 & 0 & 0 & 0 \\
0 & \sigma_1^2 + \frac{4}{3}\sigma_2^2 & \frac{1}{3\sqrt{2}}\sigma_2^2 & \frac{1}{\sqrt{6}}\sigma_2^2 \\
0 & \frac{1}{3\sqrt{2}}\sigma_2^2 & \sigma_1^2 + \frac{7}{6}\sigma_2^2 + \frac{3}{2}\sigma_3^2 & \frac{1}{2\sqrt{3}}\sigma_2^2 + \frac{3}{2\sqrt{3}}\sigma_3^2 \\
0 & \frac{1}{\sqrt{6}}\sigma_2^2 & \frac{1}{2\sqrt{3}}\sigma_2^2 + \frac{3}{2\sqrt{3}}\sigma_3^2 & \sigma_1^2 + \frac{3}{2}\sigma_2^2 + \frac{5}{2}\sigma_3^2 + 4\sigma_4^2
\end{pmatrix}
\end{aligned}$$

(Design 2)

$$\mathrm{Var}(\mathbf{u}_i) = \mathbf{P}'\Sigma\mathbf{P}$$
$$= \begin{pmatrix} \sigma_1^2 & 0 & 0 & 0 \\ 0 & \sigma_1^2+\sigma_2^2 & 0 & 0 \\ 0 & 0 & \sigma_1^2+\frac{3}{2}\sigma_2^2+2\sigma_3^2 & \frac{1}{2}\sigma_2^2 \\ 0 & 0 & \frac{1}{2}\sigma_2^2 & \sigma_1^2+\frac{3}{2}\sigma_2^2+2\sigma_3^2+4\sigma_4^2 \end{pmatrix}$$

Table. 2 shows the ANOVA table for the four-factor generalized staggered nested designs. MS (=Mean Square) V is expressed as $SS/d.f.$, hence MS is $V_1 = SS_1/n$, $V_2 = SS_2/n$, and $V_3 = SS_3/n$, $V_4 = SS_4/(n-1)$, respectively. From Table 2, the ANOVA estimators are derived as:

Table 2. The ANOVA tables for a four-factor staggered nested designs

	factor	SS	$d.f.$	MS	$\mathrm{E}(MS)$
Design 1	$m=4$	SS_4	$n-1$	V_4	$\sigma_1^2+(3/2)\sigma_2^2+(5/2)\sigma_3^2+4\sigma_3^2$
	$k=3$	SS_3	n	V_3	$\sigma_1^2+(7/6)\sigma_2^2+(3/2)\sigma_3^2$
	$k=2$	SS_2	n	V_2	$\sigma_1^2+(4/3)\sigma_2^2$
	$k=1$	SS_1	n	V_1	σ_1^2
Design 2	$m=4$	SS_4	$n-1$	V_4	$\sigma_1^2+(3/2)\sigma_2^2+2\sigma_3^2+4\sigma_3^2$
	$k=3$	SS_3	n	V_3	$\sigma_1^2+(3/2)\sigma_2^2+2\sigma_3^2$
	$k=2$	SS_2	n	V_2	$\sigma_1^2+\sigma_2^2$
	$k=1$	SS_1	n	V_1	σ_1^2

(Design 1)

$$\begin{aligned} \hat{\sigma}_4^2 &= (1/4)V_4 - (5/12)V_3 + (1/12)V_2 + (1/12)V_1, \\ \hat{\sigma}_3^2 &= (2/3)V_3 - (7/12)V_2 - (1/12)V_1, \\ \hat{\sigma}_2^2 &= (3/4)(V_2 - V_1),\ \hat{\sigma}_1^2 = V_1. \end{aligned} \tag{14}$$

(Design 2)

$$\begin{aligned} \hat{\sigma}_4^2 &= (1/4)(V_4 - V_3),\ \hat{\sigma}_3^2 = (1/2)V_3 - (3/4)V_2 + (1/4)V_1, \\ \hat{\sigma}_2^2 &= V_2 - V_1,\ \hat{\sigma}_1^2 = V_1. \end{aligned} \tag{15}$$

Lastly, shape parameters ϕ_2^* (Design 1) and ϕ_3^* (Design 2) can be expressed by the equation (10) as follows.

(Design 1)

$$\phi_2^* = \frac{16\left(\sigma_1^2+\sigma_2^2+\sigma_3^2+\sigma_4^2\right)^2}{\frac{1}{n}\left[\begin{array}{l}\sigma_1^4+\left\{\sigma_1^2+(4/3)\sigma_2^2\right\}^2+\left\{\sigma_1^2+(7/6)\sigma_2^2+(3/2)\sigma_3^2\right\}^2+\\ \frac{n}{n-1}\left\{\sigma_1^2+(3/2)\sigma_2^2+(5/2)\sigma_3^2+4\sigma_4^2\right\}^2+\\ (1/9)\sigma_2^4+(1/3)\sigma_3^4+2\left\{(1/2\sqrt{3})\sigma_2^2+(3/2\sqrt{3})\sigma_3^2\right\}^2\end{array}\right]}$$

(Design 2)

$$\phi_3^* = \frac{16\left(\sigma_1^2+\sigma_2^2+\sigma_3^2+\sigma_4^2\right)^2}{\frac{1}{n}\left[\begin{array}{l}\sigma_1^4+\left\{\sigma_1^2+\sigma_2^2\right\}^2+\left\{\sigma_1^2+(3/2)\sigma_2^2+2\sigma_3^2\right\}^2+\\ \frac{n}{n-1}\left\{\sigma_1^2+(3/2)\sigma_2^2+2\sigma_3^2+4\sigma_4^2\right\}^2+(1/2)\sigma_2^4\end{array}\right]}$$

Since the variance components are unknown, the above parameters are calculated as the estimates through equations (14) and (15):

(Design 1)

$$\hat{\phi}_2^* = \frac{(V_1+V_2+V_3+V_4)^2}{\frac{1}{n}\left[\begin{array}{l}V_1^2+V_2^2+V_3^2+\frac{n}{n-1}V_4^2+\\ 2\left\{(1/8)(V_2-V_1)^2+(1/12)(2V_3-V_2-V_1)^2\right\}\end{array}\right]} \tag{16}$$

(Design 2)

$$\hat{\phi}_3^* = \frac{(V_1+V_2+V_3+V_4)^2}{\frac{1}{n}\left[V_1^2+V_2^2+V_3^2+\frac{n}{n-1}V_4^2+2\left\{(1/4)(V_2-V_1)^2\right\}\right]} \tag{17}$$

4 Interval estimation of reproducibility variance

In this section, interval estimation of the reproducibility variance is introduced. The reproducibility variance is estimated as the sum of the ANOVA estimators of the variance components as follows.

(Three-factor staggered nested design: $R = 3$)

$$\hat{\sigma}^2_{R=3} = (1/3)(V_1+V_2+V_3) \tag{18}$$

(Four-factor staggered nested design (including the generalized staggered nested design): $R = 4$)

$$\hat{\sigma}^2_{R=4} = (1/4)(V_1+V_2+V_3+V_4) \tag{19}$$

By equations (18) and (19), the reproducibility variance is exactly equal to the average of MS (=Mean Square) V. This is a general property of the staggered nested design including the generalized staggered nested design.

Here, the gamma distribution has scale parameter θ and shape parameter k. Let $\theta = 2$ and $k = \phi^*/2$, where the gamma distribution is identical to the chi-square distribution with ϕ^* degrees of freedom. Therefore, the approximate 95% confidence interval of the reproducibility variance σ_R^2 is expressed as follows.

$$\frac{\phi^* \times \hat{\sigma}_R^2}{G(k = \phi^*/2, \theta = 2, 0.025)} \leq \sigma_R^2 \leq \frac{\phi^* \times \hat{\sigma}_R^2}{G(k = \phi^*/2, \theta = 2, 0.975)} \qquad (20)$$

The approximate shape parameters are equations (12), (16) and (17), respectively. And estimators of the reproducibility variance are equations (18) and (19). Here, $G(k = \phi^*/2, \theta = 2, 0.025)$ means a quantile when the upper probability of the gamma distribution is 0.025 under $\theta = 2$ and $k = \phi^*/2$. $G(k = \phi^*/2, \theta = 2, 0.975)$ means a quantile when the upper probability of the gamma distribution is 0.975 under $\theta = 2$ and $k = \phi^*/2$.

5 Evaluation of the methods

5.1 Evaluation of the gamma approximation

To evaluate the accuracy of the approximation by the method of moments, the Kolmogorov-Smirnov Test (KS test) is often used. The KS test statistic D_s is defined by

$$D_s = \sup_x |F(x) - F_s(x)|,$$

where s is the sample size, $F(x)$ is a hypothesized CDF with fixed parameters, and $F_s(x)$, also called the empirical distribution function, is a step-function that increases by $1/n$ at each data value.

The evaluation based on the KS test is investigated by Monte-Carlo simulation experiments. Hence, estimates of the reproducibility variance are obtained by generating random effects $\alpha(1), \alpha(2), \ldots, \alpha(m)$ using the normal random variates. $F_s(x)$ is obtained by $s = 1000$ times iteration and their order statistics. The approximate distribution function $F(x)$ is obtained by the gamma function, the shape parameter (e.g. equations (14), (17), and (18)), and the constant c. We show the case of $m = 4$, that is, the four-factor staggered nested design, under experimental conditions $\sigma_4^2/\sigma_1^2 = 0.5, 1.0$, and 2.0; $\sigma_3^2/\sigma_1^2 = 0.5, 1.0$, and 2.0; $\sigma_2^2/\sigma_1^2 = 0.5, 1.0$, and 2.0; number of participating laboratories $n = 4, 5, 6, 7, 8, 9, 10, 20$, and 30, and we assume $\sigma_1^2 = 1.0$ without loss of generality.

The performance of the KS test is evaluated with the probability of the goodness of fit by 10,000 times iteration. Additionally, the critical value C_0 is

$$C_0 \cong \sqrt{\frac{1}{2s}\ln\frac{2}{\alpha}} = 0.042947;$$

the significance level of the test, α, is 0.05.

Table. 3 shows the probability of the goodness of fit in conditions of $\sigma_4^2/\sigma_1^2 = 1.0$. It indicates that the performance of the approximation is good when n is greater than 5. A similar tendency appears in the four-factor staggered nested design without covariance terms, the three-factor staggered nested design with covariance terms, and without covariance terms. Also in the four-factor generalized designs with covariance terms and without them, we are confirming similar results. Here, the design without covariance terms is that the covariance terms are ignored in expressions for shape parameters and their hat.

Table 3. Probability(%) of the goodness of fit by the KS test

Condition			n								
σ_4^2	σ_3^2	σ_2^2	4	5	6	7	8	9	10	20	30
1.0	0.5	0.5	66.75	78.88	82.90	85.35	86.66	88.43	88.98	93.69	95.43
		1.0	73.96	82.52	85.92	88.48	89.12	89.91	90.48	93.81	93.10
		2.0	81.65	86.68	89.54	90.10	91.43	91.27	91.75	94.41	93.33
1.0	1.0	0.5	69.72	80.04	84.75	87.41	88.28	89.04	90.37	91.91	94.12
		1.0	74.35	82.22	85.54	87.89	89.54	90.71	90.32	93.20	94.43
		2.0	81.00	85.87	88.48	89.97	90.77	91.71	92.05	93.78	93.67
1.0	2.0	0.5	73.53	82.34	86.45	88.17	89.29	89.88	90.79	92.96	94.97
		1.0	75.43	82.99	86.61	87.90	89.69	90.23	90.40	93.27	93.99
		2.0	79.80	85.30	88.11	89.93	90.00	91.29	91.45	93.94	94.02

5.2 Evaluation of the interval estimation

The interval estimation of the reproducibility variance is evaluated with the coverage probability by the equation (20) through a Monte-Carlo simulation experiment. We show the case of $m = 4$, that is, the four-factor staggered nested design with experimental conditions $\sigma_4^2/\sigma_1^2 = 0.5, 1.0$, and 2.0; $\sigma_3^2/\sigma_1^2 = 0.5, 1.0$, and 2.0; $\sigma_2^2/\sigma_1^2 = 0.5, 1.0$, and 2.0; number of participating laboratories $n = 4, 5, 6, 7, 8, 9, 10, 15, 20$, and 30, and $\sigma_1^2 = 1.0$ (without loss of generality). The generation of simulation data is repeated 10,000 times.

Table. 4 shows the coverage probabilities including the true values for the 95% confidence interval under conditions of $n = 5, 10, 15$, and 20. Case 1 is the four-factor staggered nested design without covariance terms, and Case 2 is the one with covariance terms. It indicates that Case 2 has better performance than Case 1 in the real situation; i.e. $\sigma_4^2/\sigma_1^2 > \sigma_3^2/\sigma_1^2 > \sigma_2^2/\sigma_1^2$. The performance of Case 2 is good regardless of the condition unless σ_4^2/σ_1^2 is small (e.g. 0.5). And the number of participating laboratories will be sufficient with more than $n = 4$ when $\sigma_4^2/\sigma_1^2 > \sigma_3^2/\sigma_1^2 > \sigma_2^2/\sigma_1^2$.

On the other hand, Table. 5 shows the result for the four-factor generalized staggered nested design same as conditions of the four-factor staggered nested design. It indicates that Case 1 has a good performance as well as Case 2. From this, we think that there is little effect of covariance terms in this design. We can confirm this by the equation (17). But if the value of covariance terms is large, Case 2 will probably have better performance than Case 1.

Table 4. Coverage probability(%) including the true value for 95% confidence interval in the four-factor staggered nested design

Condition			$n = 5$		$n = 10$		$n = 15$		$n = 20$	
σ_4^2	σ_3^2	σ_2^2	Case 1	Case 2	Case 1	Case 2	Case 1	Case 2	Case 1	Case 2
0.5	0.5	0.5	96.02	96.56	95.09	95.65	95.02	95.31	95.00	95.35
		1.0	96.51	97.03	95.64	96.22	95.22	95.80	94.83	95.36
		2.0	96.37	96.78	95.63	96.28	95.13	95.77	94.85	95.48
	1.0	0.5	95.74	96.38	94.85	95.37	94.86	95.44	94.59	95.12
		1.0	95.63	96.42	95.51	96.11	95.10	95.63	94.92	95.44
		2.0	96.15	96.87	95.02	95.68	94.71	95.57	94.60	95.44
	2.0	0.5	94.70	95.62	94.42	95.32	94.61	95.57	94.51	95.23
		1.0	95.19	96.17	94.44	95.46	94.63	95.60	94.23	94.96
		2.0	95.29	96.14	94.81	95.78	94.72	95.53	94.65	95.50
1.0	0.5	0.5	95.12	96.00	94.86	95.25	95.01	95.36	94.98	95.25
		1.0	95.40	96.04	95.00	95.50	95.12	95.53	94.61	95.05
		2.0	95.95	96.51	95.17	95.88	94.93	95.49	94.71	95.29
	1.0	0.5	94.85	95.63	94.51	95.21	94.32	94.94	94.40	94.82
		1.0	95.20	96.02	94.73	95.42	94.75	95.20	94.30	94.89
		2.0	95.57	96.32	94.83	95.85	95.31	95.89	94.76	95.35
	2.0	0.5	94.46	95.33	93.87	94.82	94.47	95.24	94.24	95.02
		1.0	95.09	95.96	94.46	95.36	94.55	95.33	94.42	95.13
		2.0	95.65	96.43	94.82	95.68	94.52	95.34	94.59	95.57
2.0	0.5	0.5	93.61	94.38	93.68	94.13	93.85	94.05	94.52	94.68
		1.0	94.38	95.27	94.11	94.56	94.18	94.53	94.42	94.57
		2.0	95.26	96.08	94.98	95.41	94.68	95.05	94.58	95.00
	1.0	0.5	93.41	94.16	94.26	94.62	94.15	94.54	94.10	94.36
		1.0	94.46	95.22	94.14	94.54	94.54	94.88	94.61	94.97
		2.0	95.26	96.25	94.63	95.28	94.72	95.12	94.95	95.36
	2.0	0.5	93.79	94.72	93.93	94.70	94.20	94.91	94.18	94.75
		1.0	93.89	94.96	94.36	95.02	94.58	95.15	94.21	94.75
		2.0	94.94	95.79	94.08	94.89	94.48	95.16	94.17	94.84

6 Conclusions

In this paper, using the method of moments with covariance terms, the distribution of the estimator of the reproducibility was approximated by a gamma distribution for the case of m-factor staggered nested experiments. That is, general formulae for the shape parameter were proposed. Applying the formulae, we constructed the confidence interval for the reproducibility in the three-factor and four-factor staggered nested designs including the generalized staggered nested design.

The design with covariance terms exceeds the performance of the one without covariance terms when $\sigma_4^2/\sigma_1^2 > \sigma_3^2/\sigma_1^2 > \sigma_2^2/\sigma_1^2$. However, in the generalized staggered nested design, the design without covariance terms has the performance as well as the one with covariance terms.

Table 5. Coverage probability(%) including the true value for 95% confidence interval in the four-factor generalized staggered nested design

Condition			$n=5$		$n=10$		$n=15$		$n=20$	
σ_4^2	σ_3^2	σ_2^2	Case 1	Case 2	Case 1	Case 2	Case 1	Case 2	Case 1	Case 2
0.5	0.5	0.5	96.38	96.62	95.54	95.71	95.26	95.44	95.35	95.50
		1.0	96.18	96.57	95.93	96.08	95.46	95.68	95.23	95.41
		2.0	96.11	96.52	95.60	95.96	95.15	95.51	95.12	95.45
	1.0	0.5	96.56	96.84	95.47	95.59	95.57	95.64	95.16	95.22
		1.0	95.95	96.35	95.58	95.82	95.62	95.79	95.45	95.62
		2.0	96.02	96.31	95.41	95.64	95.02	95.31	95.04	95.31
	2.0	0.5	95.77	95.97	95.34	95.42	95.71	95.74	95.50	95.54
		1.0	96.10	96.31	95.13	95.29	95.64	95.76	95.26	95.36
		2.0	95.90	96.29	95.55	95.75	95.49	95.66	95.33	95.53
1.0	0.5	0.5	95.55	95.89	95.47	95.62	95.16	95.26	95.00	95.11
		1.0	95.49	95.87	95.24	95.36	95.21	95.42	94.95	95.07
		2.0	96.02	96.40	95.44	95.70	95.02	95.28	94.90	95.28
	1.0	0.5	95.51	95.85	95.34	95.48	95.01	95.11	94.97	95.04
		1.0	95.81	96.03	95.18	95.31	95.24	95.36	94.99	95.14
		2.0	95.77	96.21	95.53	95.83	95.52	95.72	95.21	95.49
	2.0	0.5	95.21	95.43	95.67	95.74	95.35	95.40	95.17	95.20
		1.0	95.91	96.03	95.27	95.36	95.15	95.20	95.11	95.18
		2.0	95.86	96.08	95.59	95.77	95.07	95.24	95.15	95.33
2.0	0.5	0.5	94.03	94.43	94.01	94.13	94.05	94.12	94.55	94.60
		1.0	94.64	94.96	94.53	94.68	94.37	94.45	94.29	94.42
		2.0	95.49	96.03	94.95	95.24	94.93	95.21	94.62	94.81
	1.0	0.5	93.82	94.22	94.64	94.69	94.65	94.70	94.78	94.80
		1.0	95.13	95.44	94.11	94.30	95.05	95.12	94.86	94.93
		2.0	95.61	95.98	94.87	95.14	95.20	95.36	95.34	95.47
	2.0	0.5	94.63	94.77	94.67	94.74	94.78	94.83	94.99	95.00
		1.0	94.76	95.09	94.77	94.88	95.43	95.47	95.13	95.15
		2.0	95.25	95.59	94.86	95.10	95.00	95.18	94.95	95.09

Additionally, we recommended the necessary number of participating laboratories in the three-factor and four-factor staggered nested design. Our recommendation will contribute to the planning of precision experiments.

References

1. ISO 5725-1: 1994, Accuracy (trueness and precision) of measurement methods and results - Part1: General principles and definitions, International Organization for Standardization, Geneva, Switzerland.
2. ISO 5725-3: 1994, Accuracy (trueness and precision) of measurement methods and results - Part3: Intermediate measures of the precision of a standard measurement method, International Organization for Standardization, Geneva, Switzerland.

3. Satterthwaite, F. E. 1946, An approximate distribution of estimates of variance components, Biometrics Bulletin 2, 110-114.
4. Nahoko KANIWA, Yoshikazu OJIMA, 2000, Experimental design for evaluating accuracy / trueness and precision of analytical for drugs and drug products (1) and (2), PHARM TECH JAPAN, Vol. 16, No. 2, pp. 171-179 and No. 4, 541-565.
5. Yoshikazu OJIMA, 1998, General formulae for expectations, variances and covariances of the mean squares for staggered nested designs, Journal of Applied Statistics, Vol. 25, No. 6, 785-799.
6. Yoshikazu OJIMA, 2000, Generalized staggered nested designs for variance components estimation, Journal of Applied Statistics, Vol. 27, No. 5, 541-553.